Vibration Monitoring of Induction Motors

Master the art of vibration monitoring of induction motors with this unique guide to on-line condition assessment and fault diagnosis, building on the author's 50 years of investigative expertise.

It includes:

- Robust techniques for diagnosis of a wide range of common faults, including shaft misalignment and/or soft foot, rolling element bearing faults, sleeve bearing faults, magnetic and vibrational issues, resonance in vertical motor drives and vibration and acoustic noise from inverters.
- Detailed technical coverage of 30 real-world industrial case studies, from initial vibration spectrum analysis through to fault diagnosis and final strip-down.
- An introduction to real-world vibration spectrum analysis for fault diagnosis, and practical guidelines to reduce bearing failure through effective grease management.

This definitive book is essential reading for industrial end users, engineers and technicians working in motor design, manufacturing and condition monitoring. It will also be of interest to researchers and graduate students working on condition monitoring.

William T. Thomson is the Director of EM Diagnostics Ltd, and a former Professor of Electrical Machines at Robert Gordon University. He has over 50 years of experience in condition monitoring of induction motor drives, and is a Senior Member of the IEEE, a Chartered Engineer and a Fellow of the IET.

Vibration Monitoring of Induction Motors

Practical Diagnosis of Faults via Industrial Case Studies

WILLIAM T. THOMSON
EM Diagnostics Ltd, Scotland

CAMBRIDGE
UNIVERSITY PRESS

University Printing House, Cambridge CB2 8BS, United Kingdom

One Liberty Plaza, 20th Floor, New York, NY 10006, USA

477 Williamstown Road, Port Melbourne, VIC 3207, Australia

314–321, 3rd Floor, Plot 3, Splendor Forum, Jasola District Centre, New Delhi – 110025, India

79 Anson Road, #06–04/06, Singapore 079906

Cambridge University Press is part of the University of Cambridge.

It furthers the University's mission by disseminating knowledge in the pursuit of education, learning, and research at the highest international levels of excellence.

www.cambridge.org
Information on this title: www.cambridge.org/9781108489973
DOI: 10.1017/9781108784887

© Cambridge University Press 2020

This publication is in copyright. Subject to statutory exception and to the provisions of relevant collective licensing agreements, no reproduction of any part may take place without the written permission of Cambridge University Press.

First published 2020

Printed in the United Kingdom by TJ Books Limited, Padstow Cornwall

A catalogue record for this publication is available from the British Library.

Library of Congress Cataloging-in-Publication Data
Names: Thomson, William T., 1946- author.
Title: Vibration monitoring of induction motors : practical diagnosis of faults via industrial case studies / William T. Thomson, EM Diagnostics Ltd, Scotland.
Description: First edition. | Cambridge ; New York, NY : Cambridge University Press, 2020. | Includes index.
Identifiers: LCCN 2020007147 (print) | LCCN 2020007148 (ebook) | ISBN 9781108489973 (hardback) | ISBN 9781108784887 (epub)
Subjects: LCSH: Electric motors, Induction–Vibration. | Electric machinery–Monitoring.
Classification: LCC TK2785 .T463 2020 (print) | LCC TK2785 (ebook) | DDC 621.46–dc23
LC record available at https://lccn.loc.gov/2020007147
LC ebook record available at https://lccn.loc.gov/2020007148

ISBN 978-1-108-48997-3 Hardback

Cambridge University Press has no responsibility for the persistence or accuracy of URLs for external or third-party internet websites referred to in this publication and does not guarantee that any content on such websites is, or will remain, accurate or appropriate.

Contents

Preface	page xv
About the Author	xviii
Acknowledgements	xix
Biographies of Personnel in the Acknowledgements	xx
Nomenclature	xxii
Acronyms and Abbreviations	xxiv
Relevant Units of Equivalence Useful for this Book	xxvi

1 **Vibration Monitoring of Induction Motors and Case Histories on Shaft Misalignment and Soft Foot**   1

   1.1 Introduction   1
      1.1.1 Overview of Causes of Vibration Problems in Induction Motors   2
      1.1.2 Overview of Shaft Misalignment   7
      1.1.3 Vibration due to Shaft Misalignment and/or Soft Foot   10
      1.1.4 Introductory Industrial Case History – Normal Shaft Misalignment and no Soft Foot in a 230 kW/308 H.P. SCIM Pump Drive   12
      1.1.5 Conclusions   19
   1.2 Industrial Case History – Diagnosis of Misalignment in a 554 kW/743 H.P. SCIM Driving a Gas Recirculating Fan   19
      1.2.1 Introduction   19
      1.2.2 Overall Vibration Measurements   21
      1.2.3 Vibration Spectrum Analysis Detected Abnormal Misalignment   23
      1.2.4 Main Conclusions   23
   1.3 Industrial Case History – Diagnosis of Misalignment on a 7.5 kW/10 H.P. SCIM – Pump Drive used to Lubricate Sleeve Bearings in a 35 MVA Generator on an Offshore Oil Production Platform   27
      1.3.1 Introduction   27
      1.3.2 Overall Vibration Results   27
      1.3.3 Vibration Spectrum Analysis   29
      1.3.4 Conclusions   31

1.4		Industrial Case History – Vibration Analysis Identified Soft Foot – Shaft Misalignment in a 110 kW/147 H.P. SCIM Pump Drive	31
	1.4.1	Introduction	31
	1.4.2	Vibration Analysis on Repaired Motor in Repair Shop	32
	1.4.3	On-Site Vibration Analysis during a Coupled Run of Repaired Motor	32
	1.4.4	On-Site Vibration Analysis during an On-Site Uncoupled Run of Repaired Motor	36
	1.4.5	On-Site Vibration Analysis during a Coupled Run after Re-alignment of the Drive Train	36
	1.4.6	Conclusions	37
1.5		Industrial Case History – Vibration Spectrum Analysis (VSA) Diagnosed Soft Foot and Abnormal Shaft Misalignment in a 180 kW/240 H.P. SCIM Pump Drive	37
	1.5.1	Summary	37
	1.5.2	Overall Vibration Measurements and Vibration Spectrum Analysis (VSA) – Motor Run Uncoupled and Coupled	37
	1.5.3	Overall R.M.S. Velocities on the Motor as a Function of Slackening and Tightening the Motor's Fixing Bolts During an Uncoupled Run – Soft Foot was Diagnosed	39
	1.5.4	Overall Vibration Measurements and VSA – Comparison between before and after Removal of Misalignment and Soft Foot	41
	1.5.5	Conclusions	42
References and Further Reading			42

2 Rolling Element Bearings for Induction Motors — 47

2.1		Basic Construction and Types of Rolling Element Bearings for Induction Motors	47
	2.1.1	Basic Construction of a Deep Groove Ball Bearing	47
2.2		Cylindrical Roller Element Bearings	53
	2.2.1	Main Operational Features of a Cylindrical Roller Element Bearing	54
2.3		Angular Contact and 4-Point Contact Ball Bearings	55
	2.3.1	Angular Contact Ball Bearings	55
	2.3.2	Main Operational Features of an Angular Contact Ball Bearing	57
	2.3.3	4-Point Angular Contact Ball Bearings – The QJ Series	57
2.4		Miscellaneous Rolling Element Bearings	58
	2.4.1	Tapered Roller Bearings	58
	2.4.2	Spherical Roller Bearings	59
2.5		Bearing Arrangements	61
References			61

3	**Types of Defects in Rolling Element Bearings**		63
	3.1 Bearing Life and Fatigue		63
		3.1.1 Prediction of the L_{10} Life of a Rolling Element Bearing	63
		3.1.2 Fatigue Failure	65
		3.1.3 Excessive Loads	66
		3.1.4 False Brinelling	66
		3.1.5 True Brinelling	67
		3.1.6 Skidding or Slipping Tracks	67
		3.1.7 Contamination – Corrosion – Fretting	68
		3.1.8 Shaft Currents	68
	References		73
4	**Introduction to Vibration Spectrum Analysis to Diagnose Faults in Rolling Element Bearings in Induction Motors**		75
	4.1 Summary		75
		4.1.1 Liquid Natural Gas (LNG) Processing Plant	75
	4.2 Vibration Analysis to Diagnose Bearing Faults		78
		4.2.1 Idealized Stages of Bearing Degradation and Vibration Analysis Techniques	78
		4.2.2 Vibration Spectrum Analysis (VSA) as Applied in the Industrial Case Histories	79
	4.3 Flow Chart for Vibration Measurements and VSA to Diagnose Bearing Defect Frequencies from Faulty Rolling Element Bearings in SCIMS		81
	4.4 Introductory Industrial Case History (1988) – Illustration of a VSA Procedure to Diagnose Bearing Defect Frequencies in SCIMs		83
		4.4.1 Stage One – Drive Train and Nameplate Data	83
		4.4.2 Stage Two – Photographs of Motor and Positions of Accelerometers	83
		4.4.3 Stage Three – Select an Accelerometer	83
		4.4.4 Stage Four – Predict the Bearing Defect Frequencies at the Full-Load Rated Speed	83
		4.4.5 Overall R.M.S. Velocities and VSA to Detect the $1X$ Frequency Component and the Bearing Frequencies using Velocity versus Frequency Spectra	84
	4.5 Conclusions		87
	References		88
5	**Industrial Case Histories on VSA to Diagnose Cage Faults in Rolling Element Bearings of SCIMS**		90
	5.1 Introduction		90
		5.1.1 Motor Data and Positions for Vibration Measurements	90
		5.1.2 Overall Velocity Levels and VSA	92
		5.1.3 Inspection of the NDE Bearing	95
		5.1.4 Conclusions	95

Appendix 5A	– Photos of the Faulty Bearing Parts	96
5.2	Industrial Case History – VSA Detected a Broken Cage in a Polyamide Cylindrical Roller Bearing in a 75 kW/100 H.P. SCIM	98
	5.2.1 Introduction	98
	5.2.2 Overall R.M.S. Velocities	99
	5.2.3 VSA Predicted a Broken Cage in the DE Bearing	100
	5.2.4 Inspection of the Faulty Bearing	101
	5.2.5 Conclusions	101
Reference		102

6 Industrial Case Histories – VSA Detected Inner and Outer Race Faults in Rolling Element Bearings in SCIMS — 103

6.1	Introduction	103
	6.1.1 Overall Vibration Measurements	103
	6.1.2 Vibration Spectrum Analysis	106
	6.1.3 Visual On-Site Inspection of the NDE of Motor B	107
	6.1.4 Interpretation of Logarithmic Spectrum and Predictions	107
	6.1.5 Conclusions	108
6.2	Industrial Case History – VSA Diagnosed Outer Race and Ball Defects in a 7324 B Single-Row Angular Contact Ball Bearing in the NDE of a Vertical 1193 kW/1600 H.P. SCIM Driving a Thruster Propeller	109
	6.2.1 Introduction	109
	6.2.2 Vibration Measurements and Overall Velocity Levels	110
	6.2.3 Vibration Spectrum Analysis	113
	6.2.4 Inspection – Photos – Conclusions	114
Appendix 6A	Prediction of Bearing Defect Frequencies	115
6.3	Industrial Case History – VSA Diagnosed Outer Race Defects in a Rolling Element Bearing via Vibration Measurements on the Drive End Frame of a 160 kW/215 H.P. SCIM Driving a Boiler Forced Draft Fan	115
	6.3.1 Summary	115
	6.3.2 On-Site Vibration Measurements and Spectrum Analysis before New Bearings were Fitted	117
	6.3.3 Inspection of DE Bearing	119
	6.3.4 Conclusions	120
6.4	Industrial Case History – VSA of Vibration Measured on the Outer Frame Predicted an Outer Race Bearing Defect in a Vertically Mounted 75 kW/100 H.P. SCIM	120
	6.4.1 Summary	120
	6.4.2 Overall Vibration Velocity Measurements	121
	6.4.3 Vibration Spectrum Analysis – Diagnosis of *BPFO*	123
	6.4.4 Conclusion and Recommendations	124
	6.4.5 Vibration Analysis Predicted Misalignment in Drive Train B	124

7	**Industrial Case Histories – VSA Diagnosed False Brinelling and Problems in Cylindrical Roller Bearings in SCIMs**		127
	7.1	Introduction	127
		7.1.1 Motor Data and Overall Vibration Velocity Measurements	127
		7.1.2 Diagnosis of Bearing Defect Frequencies	129
		7.1.3 Conclusions	131
	7.2	Industrial Case History – VSA Diagnosed Skidding from an Extra Capacity Cylindrical Roller Bearing (N234E M C3) in a 225 kW/300 H.P. SCIM	133
		7.2.1 Historical Perspective and Summary	133
		7.2.2 Vibration Measurements before and after New Bearings were Fitted	133
	7.3	Conclusions	138
	References		139
8	**Industrial Case Histories on VSA to Diagnose Miscellaneous Faults in Rolling Element Bearings in SCIMS**		140
	8.1	VSA Detected Corroded Deep Groove Ball Bearing in Vertically Mounted 1.5 kW/2 H.P. SCIMs	140
		8.1.1 Motor Data and Description of Installation Layout	141
		8.1.2 Vibration Measurements	141
		8.1.3 Inspection of the Motor and DE Bearing	147
		8.1.4 Proposed New Motor Design	148
		8.1.5 Conclusions	150
	8.2	Industrial Case History – Envelope Analysis used by a Vibration Sub-Contractor Produced a False Diagnosis of a Cage Fault in a Cylindrical Roller Element Bearing in a 225 kW/300 H.P. SCIM	152
		8.2.1 Introduction	152
		8.2.2 Nameplate Data and Construction of the Motor	153
		8.2.3 Vibration Measurements and VSA Applied to the SCIM before Strip-Down and Inspection of the DE Bearing	156
		8.2.4 Conclusions	159
	8.3	Bearing Failures in 800 kW/1072 H.P. SCIMs Driving Sulphate Removal Pumps (SRP) and a FAT on Repaired Motor	159
		8.3.1 Catastrophic Bearing Failure and Broken Shaft	159
		8.3.2 RCFA of a Faulty 6316 C3 Bearing and FAT of the Repaired Motor (B) with New Bearings	162
		8.3.3 Conclusions	164
	8.4	Industrial Case History – False Brinelling and FAT of an 800 kW/1072 H.P. SCIM and Attenuation of Vibration between the Bearing Housing and the Outer Periphery of the End Frame	166
		8.4.1 Summary	166
		8.4.2 Vibration Factory Acceptance Test (FAT) of Repaired Motor	169

		8.4.3	Attenuation of Bearing Defect Frequencies	171
		8.4.4	Conclusions and Final Outcome	173
	References			174

9 Industrial Case Histories on Vibration Measurements and Analysis Applied to Large Induction Motors with Sleeve Bearings — 175

9.1 Introduction and Basic Operation of a Sleeve Bearing — 175
    9.1.1 Illustrations of the Practical Installation of an Oil-Fed Pressure Sleeve Bearing — 177
    9.1.2 Brief Overview of Operational Problems with Sleeve Bearings — 180

9.2 Introductory Case History – Vibration Factory Acceptance Test of a New 6800 kW/9115 H.P. SCIM — 182
    9.2.1 Measurement of Shaft Displacement — 183
    9.2.2 Motor Nameplate Data and Vibration Test Specification — 185
    9.2.3 No-Load FAT Results from Vibration Measurements on the Bearing Housing — 186
    9.2.4 No-Load FAT Results from Shaft Displacement Probe Measurements — 189
    9.2.5 Full-Load Shaft Displacement Results — 190

9.3 Industrial Case History – Analysis of Shaft Displacement and Subsequent Strip-Down Inspections Diagnosed Faults in a Journal Bearing of a 6250 kW/8380 H.P. Slip Ring Induction Motor — 192
    9.3.1 Background — 192
    9.3.2 On-Site Operation and Motor Nameplate Data — 193
    9.3.3 Deliverables and Vibration Test Results — 194
    9.3.4 Ingress Protection (IP) Rope Seal Malfunction – Source of High Shaft Displacement — 198
    9.3.5 Shaft Displacement with the IP Rope Seals Removed from the DE and NDE — 200
    9.3.6 Shaft Displacements with New IP Rope Seals at the DE and NDE – Full-Load Heat Run — 200
    9.3.7 Conclusions — 200

9.4 Industrial Case History – Excessive Shaft Displacement during First 100 Minutes of a Five-Hour Heat Run of a Re-Furbished 2-Pole 6800 kW/9115 H.P. SCIM – Function of Temperature Change and Rotor Design — 202
    9.4.1 Summary — 202
    9.4.2 Shaft Displacements at the NDE during the Full-Load Heat Run — 202
    9.4.3 Conclusions — 206

References — 206

10	**Industrial Case Histories on Magnetic Forces and Vibration from Induction Motors**	**208**
	10.1 Electromagnetic Forces and Vibration in Induction Motors	208
	10.1.1 Twice Supply Frequency Vibration	208
	10.1.2 Vibration Components Caused by Rotor Slotting	210
	10.2 Industrial Case History – Magnetic Forces and Vibration on the Stator Core and Outer Frame of a New 4000 kW/5362 H.P. SCIM	211
	10.2.1 Vibration Measurement and VSA during Factory Acceptance Tests	212
	10.2.2 Time Domain and VSA of Vibration on the Stator Core Back	213
	10.2.3 Bearing Housing Vibration – Comparison with Stator Core Vibration	216
	10.2.4 Outer Frame Vibration and Analysis at Full Load	218
	10.3 Industrial Case History – False Positive of Cage Winding Breaks by a Vibration Condition Monitoring (CM) Sub-Contractor – Identified the True Cause as Normal and Inherent Stator Core Vibration at a Rotor Slot Passing Frequency	221
	10.3.1 History and Summary	221
	10.3.2 Overall Vibration Measurements	222
	10.3.3 Vibration Spectrum Analysis	223
	10.3.4 Conclusions	226
	10.4 Case Study – Measurement of the Stator Core Vibration Proved that Broken Rotor Bars in a SCIM Modulate the Vibration Rotor Slot Passing Frequencies at Twice the Slip Frequency	226
	10.4.1 Background	226
	10.4.2 Measurement of Vibration Rotor Slot Passing Frequencies on the Stator Frame	227
	10.4.3 VSA of the Stator Frame Vibration with Broken Rotor Bars	230
	10.4.4 Conclusions on Motor Current Signature Analysis (MCSA) versus Vibration Analysis to Diagnose Broken Rotor Bars in SCIMs	231
	10.5 Case Study – Vibration Monitoring to Identify Changes in the Stator Frame Vibration of a SCIM due to Supply Voltage Unbalance	232
	10.5.1 Introduction	232
	10.5.2 Measurement of the Velocity and Acceleration of the Twice Supply Frequency Component on the Stator Frame as a Function of Position during No-Load and Full-Load Operation	233
	10.5.3 Measurement of the Velocity and Acceleration of the Twice Supply Frequency Component on the Stator Frame at Full Load as a Function of Unbalanced Voltage Supplies	235
	10.5.4 Conclusions	236
	Appendix 10A Derivation of Twice Supply Frequency Vibration	237
	References	238

11	**Miscellaneous Industrial Case Histories on Vibration Analysis Applied to Induction Motor Drives**	**241**
	11.1 Industrial Case History – Structural Resonance in a Vertically Mounted 265 kW/355 H.P. SCIM Driving a Fire-Water Pump	241
	11.1.1 Background	241
	11.1.2 Phase One – Vibration Measurements During On-Site Coupled Run	244
	11.1.3 Phase Two – Vibration Measurements – On-Site Uncoupled Run	245
	11.1.4 Phase Three – Motor FAT Vibration Measurements – Rotor Re-balanced to ISO G0.4 Grade – New Bearings Fitted	247
	11.1.5 Phase Four – On-Site Vibration Measurements on the Refurbished Motor during a Coupled Run – Operating Current of 300 Amperes	248
	11.1.6 Phase Five – On-Site Vibration Measurements on the Refurbished Motor during a Coupled Run – Operating Current of 400 Amperes	249
	11.1.7 Conclusions	250
	Appendix 11A Rotor Balance Certificate	252
	11.2 Industrial Case History – Investigation into the Cause of Loud Acoustic Noise from an Inverter-Fed 3.3 kV, 4500 kW/6032 H.P. Vertically Mounted SCIM Driving a Multi-Phase Pump	252
	11.2.1 Background	252
	11.2.2 On-Site Vibration Measurements and Analysis	253
	11.2.3 Current Spectrum – Magnetic Flux and Electromagnetic Forces	256
	11.2.4 Vibration Spectrum Analysis	258
	Appendix 11B Human Perception of Acoustic Noise	260
	11.3 Industrial Case History – VSA of a Repaired 1000 kW/1340 H.P. SRIM Diagnosed a High Twice Supply Frequency Vibration and Cracks in the Concrete Mounting Plinth	260
	11.3.1 Background and Objectives	260
	11.3.2 Vibration Measurements and Analysis – Uncoupled Run at Repair Shop	262
	11.3.3 Vibration Measurements and Analysis – Uncoupled and Coupled Runs at the Cement Factory	265
	References	269
12	**Overview of Key Features of Vibration Monitoring of SCIMS**	**270**
	12.1 Appraisal on VSA to Diagnose Faults in Rolling Element Bearings Used in SCIMS	270
	12.2 Predictions and Prognosis of Remaining Run Life	271
	12.2.1 Variables that Affect the Remaining Run Life of Faulty Rolling Element Bearings in SCIMS	272

12.3 Difficulties of Access to Measure Vibration Directly on the Bearing
    Housings of Rolling Element Bearings in Induction Motors   272
12.4 Guidelines for Successful Grease Management of Rolling Element
    Bearings in Induction Motors   277
12.5 Incorrect Bearings Fitted by Motor Repair Shop   278
References   278

Index   280

Preface

Vibration condition monitoring has been used for many years to diagnose mechanical faults in bearings, gearboxes, compressors and pumps. Numerous text books cover these topics but none focus on vibration monitoring to diagnose faults in induction motors.

This book is unique because it is solely dedicated to vibration monitoring and analysis to diagnose faults in induction motors.

There are 30 industrial case histories which include both theoretical and practical knowledge for on-line condition assessment of induction motors. A key feature of these case histories is closure of the loop between the diagnosis of faults using vibration spectrum analysis and subsequent strip-down of the motors and accompanying photographic evidence of, for example, faulty bearings. The industrial case histories include:

Five on the diagnosis of shaft misalignment and/or soft foot.
Fifteen on the diagnosis of rolling element bearing faults.
Three on vibration problems in sleeve bearings.
Four on problems due to magnetic forces and vibration.
Three on miscellaneous problems including resonance in a vertical induction motor drive, vibration and acoustic noise from induction motors supplied from inverters.

The case histories are presented in detail because a broad-brush, superficial presentation that lacks clarity and evidence as to how the fault was diagnosed is meaningless to the reader. Each case history is stand-alone and does not require the reader to scroll backwards or forwards within the book to understand the vibration measurements, spectrum analysis and predictions.

Chapter 1 provides an overview of publications on sources of vibration in electrical machines. Problems in rolling element bearings account for the largest number of failures of induction motors, therefore a review is presented on the use of vibration measurements and spectrum analysis to diagnose faults in rolling element bearings. The practical difficulty of accessing the housings of rolling element bearings is strongly emphasised, because if bearing housings cannot be accessed then it is more difficult to diagnose bearing faults.

A brief overview of shaft misalignment and soft foot in induction motor drives is included and five industrial case histories are presented on vibration analysis to detect shaft misalignment and/or soft foot.

Chapter 2 is a preparatory chapter for Chapters 3 and 4 and presents the main types and basic features of rolling element bearings that are used in induction motors.

Chapter 3 presents an overview of the types and causes of defects that occur in rolling element bearings to support the case histories in Chapters 4 to 8.

Chapter 4 presents an introduction to vibration spectrum analysis to diagnose faults in rolling element bearings in induction motors. This chapter is a precursor to the presentation of industrial case histories in Chapters 5 to 8 using conventional vibration spectrum analysis to diagnose the onset of faults in rolling element bearings before actual bearing failures occur.

It is emphasised that previously published books and papers that cover the theory and application of vibration monitoring to diagnose faults in rolling element bearings assume that vibration transducers such as accelerometers can always be mounted on the bearing housings. In many cases access to bearing housings is not practically possible on induction motors and this is a key fact that is demonstrated in this book.

Chapter 5 presents industrial case histories using VSA to diagnose cage faults in bearings of SCIMs.

Chapter 6 presents industrial case histories using VSA to diagnose inner and outer race faults in bearings of SCIMs.

Chapter 7 presents industrial case histories using VSA to diagnose false brinelling and skidding problems in cylindrical roller bearings in SCIMs.

Chapter 8 presents industrial case histories using VSA to diagnose miscellaneous faults in rolling element bearings in SCIMs.

Chapter 9 presents fundamental knowledge on the construction and operation of sleeve bearings. The practical measurement of shaft displacement is also described, to support the case histories in this chapter on vibration monitoring to diagnose problems in sleeve bearings.

Chapter 10 presents the fundamental causes of electromagnetic forces and consequential vibration in induction motors, including the twice supply frequency component and its harmonics, and the classical rotor slot passing frequency components.

Chapter 11 presents industrial case histories on miscellaneous problems such as mechanical resonance in a vertical induction motor drive and vibration and acoustic noise from induction motors supplied by inverters.

Chapter 12 presents an appraisal on VSA to diagnose faults in rolling element bearings. A discussion is provided on the key outcome that end users hope to achieve from vibration monitoring, which is the prognosis of remaining operational life of a SCIM after a fault is diagnosed.

It is also reiterated, via more photographic evidence, that access to mount temporary accelerometers directly on the bearing housings of rolling element bearings used in induction motors can be difficult.

Guidelines are given for grease management of rolling element bearings because of the predominance of failures being caused by incorrect greasing practice.

Industrial end users, motor manufacturers, condition monitoring companies and repairers of induction motors are the target market for this book. It is not a classical academic text for undergraduate university or college courses. However, it may well be of interest to post-graduate research students and academic staff with a specific interest in condition monitoring of induction motor drives because the case histories demonstrate what the real challenges are in industry.

This book is not suited to having questions at the end of each chapter and readers can contact the author directly if they have questions on its content.

William T. Thomson (Bill)
Telephone: +44 19755 62446
e-mail: bill@emdiagnostics.com

About the Author

Bill was born in Scotland in 1946 and started his career in 1961 as a maintenance electrician. He has worked with induction motors at all levels from craft apprentice through to appointment as a professor in electrical machines in 1990. For the past 20 years he has been, and still is, the managing director of his own company providing consulting services in condition monitoring of electrical machines and drives.

Evening class study provided the vocational qualifications to enter the University of Strathclyde (Glasgow, Scotland) in 1970. In 1973 he graduated with an Honours degree in Electrical & Electronic Engineering specialising in electrical machines. From 1973 to 1977 he was a noise and vibration engineer with Hoover Ltd. In 1977 he was awarded a master's degree from the University of Strathclyde for a research thesis entitled 'Reduction of Acoustic Noise and Vibration from Small-Power Electric Motors'. From 1977 to 1979, Bill was a lecturer in electrical power at the Hong Kong Polytechnic and from 1979 until 2001 he was a lecturer (1979–83), senior lecturer (1983–90) and professor (1990–2001) at Robert Gordon University in Aberdeen, Scotland. In 1980, Bill initiated his research on condition monitoring of induction motors and received research funding from power utilities and major oil companies. The focus of the research was on the industrial application of vibration analysis and Motor Current Signature Analysis (MCSA) to diagnose faults in induction motors and drives before failures occur. Bill successfully supervised ten PhD and eight MPhil students.

He left academia in 2001 to start his own company and is the Director and principal consultant of EM Diagnostics Ltd providing consultancy services on the operation and condition monitoring of induction motors to power stations, petrochemical refineries, natural gas refineries and offshore oil and gas production platforms. He has published 72 papers on condition monitoring of induction motors in engineering journals such as *IEEE Transactions* (USA), *IEE Proceedings* (UK), and at International IEEE and IEE (IET, UK) conferences.

He is the co-author of an IEEE Press Wiley publication in 2017: William T. Thomson and Ian Culbert, *Current Signature Analysis for Condition Monitoring of Cage Induction Motors: Industrial Application and Case Histories.* Bill is a SMIEEE, FIEE (IET) in the UK and a chartered professional engineer registered in the UK. He was awarded the Queen's award for technological achievement in 1992 for knowledge input to 'Motor monitor' marketed by Entek, USA. In 1999, Bill provided access to his knowledge on Motor Current Signature Analysis, via a licence from Robert Gordon University, Scotland, to Iris Power, Canada, for the development of an MCSA instrument.

Acknowledgements

Sincere thanks are expressed to Mr Archie Low and Mr Donald Sutherland, formerly of Robert Gordon University (RGU), Scotland, for their indispensable contributions to the design and construction of fault producing, 3-phase induction motor test rigs.

The author greatly acknowledges Mr Ellis Hood (see biography), former senior lecturer at RGU for his invaluable input on checking the manuscript and for the many technical suggestions he made to improve its content and style of presentation.

Very sincere thanks are expressed to Dr Muni Gunawardene (see biography), consultant, EM Diagnostics Ltd, for his excellent drawings and presentation of vibration spectra and for his technical contributions.

Mr Alistair Carr is acknowledged for his excellent advice and contributions that he provided to the author to produce Sections 9.1.1 through to 9.1.2.4, which cover an introduction to sleeve bearings in large HV induction motors. Alistair is a senior electrical machines consultant with EM Diagnostics Ltd (author is the Director) and was appointed in November 2014. He was formerly Head of Test with Parsons Peebles, Scotland, and was employed by this company for 42 years.

Mr Bill Lockley, Life Fellow of IEEE, Calgary, Canada, is duly acknowledged for his invaluable and private review of the manuscript, which was by personal invitation from the author.

Bill dedicates this book to his wife, Mary Thomson, for her irreplaceable support, patience and encouragement throughout his career which made the writing of this book become a reality; without Mary, it would not have been possible.

William T. Thomson

Biographies of Personnel in the Acknowledgements

Ellis Hood BSc (Eng) Hons 2.1 MIEE CEng FTC (coms) CGLI

1953–55: Telecommunications Engineering training.
1955–57: National Service in RAF, wireless training, second line maintenance on V Bomber gear.
1957–60: UK Post Office Technical Officer in Training – POED.
1960–66: Post Office Technical Officer, Line Transmission. Passed Post Office Executive Engineers Board (1970).
1970–73: Lecturer in Telecomms and Electrical Engineering, Aberdeen Technical College.
1973–77: Napier University, Edinburgh, Lecturer in Electrical Engineering.
1977–79: Secondary School Teacher of Mathematics – Hilton Academy Aberdeen.
1979–90: Robert Gordon University, Aberdeen. Lecturer (1979–84), Senior Lecturer (1984–90) in Telecomms andElectrical Engineering, Course leader on HND course in Electrical Engineering, 1979–88 and course leader on combined BSc/HND course – the first of its kind in Scotland, 1988–90.

Education

1953: Scottish Higher Leaving Certificate.
1953–55/64–66: Full Technological Certificate in Telecommunications CGLI.
1966–70: BSc Electrical Engineering (Hons 2.1), University of Aberdeen, Scotland.
1971–72: Further Education Teaching Certificate in Engineering and Mathematics 1971–72.
1977: PGSCE Secondary Mathematics Teaching Certificate.
1984: MIERE CEng, MIEE 1987.

Dr G. W. D. M. Gunawardene BSc, MSc, PhD

1973–74: Gresham Lion Electronics, Twickenham, Middlesex, Test Technician.
1974–75: INSPEC, IEE, Hitchin, Hertfordshire, Technical Publications Inspector.
1976–79: Sheffield City Polytechnic, now Sheffield Hallam University, Research Assistant.
1979–2010: Robert Gordon University, Lecturer.

Education

1955–67: Moratu Maha Vidyalaya, Moratuwa, Sri Lanka, Primary and Secondary Education.
1967–69: Katubedda Technical College, Sri Lanka, Technician Course.
1969–73: Kingston Polytechnic, Middlesex, BSc Electrical Engineering.
1975–76: University of Bradford, MSc Control Engineering.
1976–79: Sheffield City Polytechnic, now Sheffield Hallam University, PhD.

Publications

(1) G. W. D. M. Gunawardene and M. J. Grimble, Development of a Static Model for a Sendzimir Cold Rolling Mill, *IMACS Symposium on Control Systems*, Technical University, Vienna, Sept. 1978.
(2) G. W. D. M. Gunawardene, M. J. Grimble and A. Thomson, Static Model for Sendzimir Cold Rolling Mill, *Metals Technology*, July 1981.
(3) Static Model of a Sendzimir Mill for Use in Shape Control, PhD thesis, 1982.
(4) G. W. D. M. Gunawardene and E. Forest, Controllability of Linear Control Systems using Gilbert and Kalman Criteria, *International Journal of Mathematical Education*, Sept. 1996.

Achievements

A programme to calculate the conditions of Sendzimir Cold Rolling Mill when loaded was developed during the research period which was bought by two steel companies in USA and Sweden.

Nomenclature

Quantity	Quantity symbol	Unit	Unit symbol
Acceleration	a	Metres per second squared	m/s^2
Angular frequency	ω	Radians per second	rad/s
Ball diameter in rolling element bearing	BD	Millimetres	mm
Bearing dynamic capacity	C	Newtons	N
Bearing rating life	L_{10}	Hours	hr
Centrifugal force	$C.F.$	Newtons	N
Coil distribution factor	k_d	Number	–
Coil span factor	k_s	Number	–
Contact angle of bearing surface to outer race	β	Degrees	e.g. 40°
Current (r.m.s.)	I	Ampere	A
Decibel	–	–	dB
Displacement	d	Metres or millimetres	m or mm
Electromagnetic force	F_{em}	Newtons	N
Equivalent bearing load	P	Newtons	N
Frequency of mains supply	f_1	Hertz	Hz
Frequency of rotational speed of rotor	f_r	Hertz	Hz
Fundamental rotational frequency of vibration	$1X$	Hertz	Hz
Flux density	B	Tesla	T
Full-load current	$I_{f.l.}$	Amperes	A
Induced voltage in stator winding per phase	E	Volts	V
Line current	I_L	Amperes	A
Line voltage	V_L	Voltage	V
Lubricating viscosity	v_c	Centistokes	cSt
Magnetic flux	Φ	Weber	Wb
Mass	m	Kilograms	kg
Number of turns	N_T	Integer	–
Number of rolling elements	n_e	Integer	–
Pole pairs	p	Integer	–
Power output	P	Kilowatts or horsepower	kW or H.P.
Power input to the induction motor	P_{in}	Kilowatts	kW

(cont.)

Quantity	Quantity symbol	Unit	Unit symbol
Power factor	p.f.	Number	–
Pitch diameter of rolling element bearing	PD	Millimetres	mm
Rotor slotting vibration frequency component	f_{rv}	Hertz	Hz
Rotor speed	N_r	Revolutions per minute	r/min
Relative permeability of free space	μ_0	Henry per metre	H/m
Rotor bars or slots	R	Integer	–
Synchronous speed	N_s	Revolutions per minute	r/min
Slip	s	–	Number or %
Slip at full-load speed	$s_{f.l.}$	–	Number or %
Speed	n	Revolutions per minute	r/min
Time	t	Seconds	s
Torque	T	Newton metre	N·m
Torque – full-load	$T_{f.l.}$	Newton metre	N·m
Torque – starting	T_s	Newton metre	N·m
Turns per phase	N_{ph}	Integer	–
Voltage	V	Volts or kilovolts	V or kV
Volt-amps	VA	Volt-amperes	VA
Velocity	v	Metres per second or millimetres per second	m/s or mm/s

Acronyms and Abbreviations

API	American Petroleum Institute, USA
ASWLP	Auxiliary Sea Water Lift Pump
BD	Ball Diameter
BPFI	Ball Pass Frequency Inner Race
BPFO	Ball Pass Frequency Outer Race
BS	British Standard
BSF	Ball Spin Frequency
CCGT	Combined Cycle Gas Turbine
C.F.	Centrifugal Force
CM	Condition Monitoring
DE	Drive End
DEA	Drive End Axial
DEH	Drive End Horizontal
DEV	Drive End Vertical
FAT	Factory Acceptance Test
FFT	Fast Fourier Transform
FLC	Full Load Current
FLT	Full Load Torque
FPSO	Floating Production and Oil Off-loading ship
FTF	Fundamental Train Frequency
GRF	Gas Recirculating Fan
HV	High Voltage
H.P.	Horse Power
ISO	International Standards Organization
LNG	Liquid Natural Gas
LV	Low Voltage
MCSA	Motor Current Signature Analysis
NDE	Non Drive End
NDEA	Non Drive End Axial
NDEH	Non Drive End Horizontal
NDEV	Non Drive End Vertical
NEMA	National Electrical Machines Association, USA
OEM	Original Equipment Manufacturer
OIM	Offshore Installation Manager

PD	Pitch Diameter
PM	Planned Maintenance
QA	Quality Assurance
QC	Quality Control
r.m.s. (or R.M.S.)	Root Mean Square
R&D	Research and Development
RCFA	Root Cause Failure Analysis
RTD	Resistance Temperature Detector
SCIM	Squirrel Cage Induction Motor
SPM	Shock Pulse Measurement or Method
SRIM	Slip Ring Induction Motor
SRP	Sulphate Removal Pump
SWIP	Sea Water Injection Pump
TEFC	Totally Enclosed Fan Cooled
VSA	Vibration Spectrum Analysis

Relevant Units of Equivalence Useful for this Book

Metric/SI units	Imperial
1.0 m	39.4 inches
25.4 mm	1.0 inch
1.0 mm	0.0394 inches/39.4 mils ($\cong$40 thou/mils)
50 μm	2.0 thou/mils
25 mm/s	1.0 inch/s
1.0 mm/s	$\cong$0.04 inches/s
1.0 kg	2.2046 lbs ($\cong$2.2 lbs)
1.0 N	0.2248 lbsf ($\cong$0.225 lbsf)
1.0 Nm	0.73756 lbsf-ft ($\cong$0.738 lbsf-ft)
1.0 kg-m^2	0.042 lbs-ft^2
1.0 N/m^2	145×10^{-6} lbsf/inch^2
745.7 W ($\cong$746 W)	1.0 H.P.

1 Vibration Monitoring of Induction Motors and Case Histories on Shaft Misalignment and Soft Foot

1.1 Introduction

Induction motors dwarf all other types of electric motors because they are used in their hundreds of millions throughout industry around the world. Without the induction motor, which was invented and successfully patented in the USA by Nikola Tesla in 1888 (see references [1.1] and [1.2]), modern society as we know it today would probably not exist. It is without doubt one of the greatest inventions of all time but one which is sadly not common knowledge worldwide. It is well known that induction motors typically consume between 40% and 50% of the generated electricity [1.3] in an industrialised nation and are often referred to as the workhorses of industry. A condition assessment of their operational integrity via condition monitoring is normal practice by industrial users to prevent the following:

(i) catastrophic failures
(ii) unscheduled downtime and lost (delayed) production and income
(iii) hazardous conditions that may lead to major accidents.

Maintenance strategies can be streamlined to be more reliable and cost effective.

The primary objective of this book is as follows:

To present industrial case histories on vibration measurements and analysis to diagnose faults in induction motors.

This book avoids duplicating material already published on the theory of induction motors (see references [1.23] to [1.29]), and on basic concepts and theory of vibration measurements, signal processing and vibration analysis, since there is an abundance of published literature on these topics and typical samples are given in references [1.4] to [1.20]. The World Wide Web has many sites that can provide information on vibration spectrum analysis [1.21]. A glossary of terms is unnecessary due to easy access to definitions of vibration terms via the Web; for example, an excellent vibration analysis dictionary can be accessed via reference [1.22].

1.1.1 Overview of Causes of Vibration Problems in Induction Motors

A seminal textbook [1.30] on the nature of induction motors for information on the theoretical principles of electromagnetic forces and consequential vibration and acoustic noise from induction motors, and a paper [1.33] by the same author, titled Magnetic Noise in Poly-phase Induction Motors, are both recommended. A paper [1.34] titled Understanding the Vibration Forces in Induction Motors, which focuses on the causes of electromagnetic forces and associated electromechanical aspects, is recommended for further reading. This paper [1.34] presents five industrial case histories, and these are:

(i) stator core 120 Hz vibration transmitted to shaft and bearing housings
(ii) rotor thermal bow due to smeared laminations
(iii) rotor bar breakage on a compressor drive
(iv) demonstration of non-linear damping of oil film
(v) demonstration of vibration modulation at one times slip speed.

Several papers have also been published — see references [1.35] to [1.44] — on motor vibration problems that also include vibration monitoring and analysis used to diagnose a range of faults. The reader is encouraged to read these papers, particularly reference [1.40], titled An Analytical Approach to Solving Motor Vibration Problems. The author of this new book has also published papers on electromagnetic forces and vibration produced by cage induction motors as a function of airgap eccentricity, broken rotor bars and voltage unbalance – samples of which are given in references [1.46] to [1.54]. A book titled *Condition Monitoring of Rotating Electrical Machines* [1.55] presents a chapter (No. 8, 33 pages) on *vibration monitoring* that covers the principal sources of vibration from electrical machines, namely:

- the dynamic behaviour of the rotating rotor
- the response of the shaft bearings to forces and vibration from the rotor
- the response of the stator core due to electromagnetic forces and mechanical forces from the rotor
- the response of the stator end windings in large induction motors due to electromagnetic forces on the conductors.

However, no actual industrial case histories were presented in this reference [1.55] on vibration monitoring and analysis, which diagnosed faults in induction motors. The author has reviewed the contents of numerous books on vibration monitoring and there is no evidence of any book which has, as a central theme, the presentation of industrial case histories on vibration monitoring and analysis used to diagnose faults in induction motors. The reader can find a very extensive list of references on motor vibration and acoustic noise in reference [1.45].

1.1.1.1 Brief Review of Vibration Monitoring to Diagnose Faults in Rolling Element Bearings in Induction Motors

The references cited in Section 1.1.1 do not, nor were they intended to, present industrial case histories on vibration monitoring and analysis to *diagnose faults in rolling element bearings* used in induction motors. The published literature indicates that bearing failures account for approximately 50% of failures in induction motors, as shown in the pie chart in Figure 1.1 [1.57].

An excellent presentation (70 slides) by ABB via the World Wide Web [1.58] also states that approximately 50% of motor failures are caused by rolling element bearing failures. Owing to this statistic, a review is presented on the application of vibration analysis to diagnose faults in rolling element bearings. References [1.59] and [1.60] state that approximately 50% of motor failures are due to bearing failures but they do not subdivide that 50% to show the relative percentages of failures of rolling element bearings compared to failures of sleeve bearings.

This needs to be given some consideration. It is very well known and accepted that the number of rolling element bearings used in induction motors far exceeds those with sleeve bearings and Chapter 4 expands on this with examples from industrial sites. Note that failure statistics of electric motors can vary widely between industries, their applications and maintenance strategies of the end users. Consequently, statistics on motor failures need to be treated with some caution and cannot be generally applied to all industries and different end users because there are numerous complex variables that influence the statistics and types of motor failures; see references [1.56] to [1.62]. Reference [1.62] can be easily accessed via the Web and gives a very worthwhile

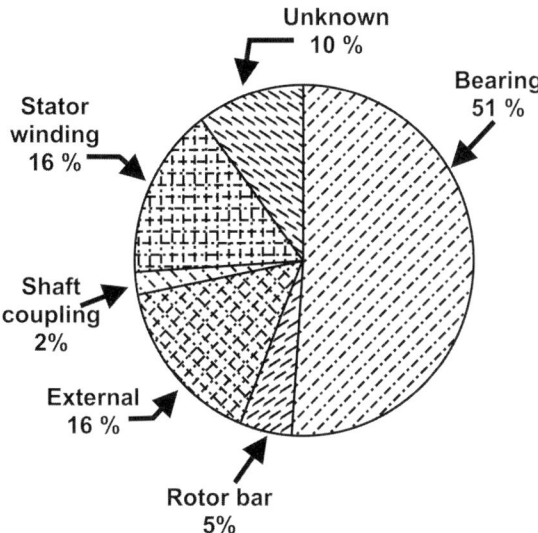

Figure 1.1 Typical distribution of failures in induction motors [1.57]. (A. H. Bonnett, Root Cause Methodology for Induction Motors; A Step-by-Step Guide to Examining Failure, *IEEE Industry Applications Magazine*, 18 (6), 2012, pp. 50–62.)

review titled *Large Electric Motor Reliability: What Did the Studies Really Say?*, and the reader is encouraged to read this reference.

Sleeve or journal bearings are mainly, but not exclusively, used in induction motors with ratings from, typically, 1000 kW/1340 H.P. upwards, and this will be further discussed and explained in Chapter 9. Because of the predominance of induction motors with rolling element bearings it is obvious that there will be more failures in these than in sleeve bearings. However, that is certainly not the sole statistical reason.

Displacement probes are normally fitted in the housings of sleeve bearings as shown in Figure 1.2 to monitor the peak–peak shaft displacement. An alarm level is set to alert staff to possible problems and prevents a catastrophic bearing failure because the motor is tripped out at a pre-set peak–peak displacement level, which is a function of the bearing clearances and rotor speed.

Many strategic induction motors in industrial plants with rolling element bearings do not have *permanent* vibration sensors fitted on the bearing housings as illustrated in the photographs of Figures 1.3 (a and b), 1.4 and 1.5. For example, many end users rely on monthly vibration surveys, using temporarily mounted accelerometers to measure and trend the vibration from rolling element bearings.

It is naïve to suggest that end users should simply retrofit permanent accelerometers after the induction motors are commissioned, because this would require a significant capital cost investment.

For example, in the petrochemical industry the vibration sensors would often have to be certified as being *intrinsically safe* and approved for the hazardous operating zone

Figure 1.2 Photograph of the NDE of a sleeve bearing assembly on a 3-phase, 6.6 kV, 2800 kW, 290 A, 60 Hz, 3575 r/min, p.f. = 0.85, eff = 96%, SCIM fitted with displacement probes.

1.1 Introduction 5

Figure 1.3a Photograph of the NDE of a 3-phase, 6.6 kV, 750 kW/1000 H.P., SCIM.

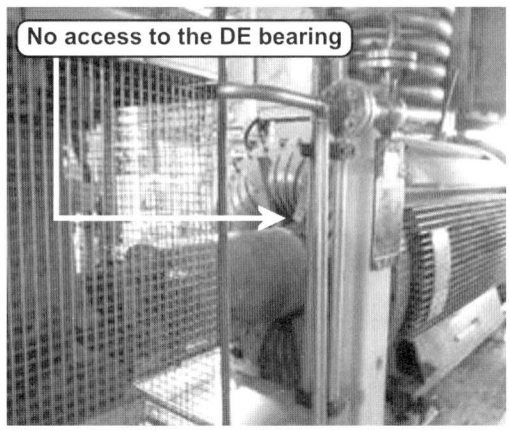

Figure 1.3b Photograph of the DE of a 3-phase, 6.6 kV, 750 kW/1000 H.P., SCIM.

Figure 1.4 Photograph of a 3-phase, 415 V, 75 kW/100 H.P., SCIM.

Figure 1.5 Photograph of a 3-phase, 415 V,185 kW/250 H.P., SCIM – coupling guard prevents access to the DE bearing housing.

which further increases the capital costs. Flame-proof motors and certified hazardous zone SCIMs must be re-certified by the OEM to permit mounting permanent vibration sensors on the bearing housings. The reality is that, in many industrial plants, where permanent vibration sensors were not installed on the bearing housings of induction motors by the OEM before shipment to industry, they are rarely retrofitted.

A *crucial practical factor* that is very rarely mentioned in published papers or in condition monitoring books is as follows:

It is often impossible to mount temporary accelerometers directly on both the DE and NDE rolling element bearing housings of induction motors.

This is illustrated in Figures 1.3 (a and b) to 1.5 and will also be discussed and further verified in case histories presented in Chapters 4 to 8. When the vibration cannot be measured directly on the bearing housings, the vibration and spectra at the positions where it is measured will obviously be different from that on the bearing housing. Therefore, the analysis is more complex because the mechanical impedance (stiffness) between the bearing housing and the positions of vibration measurements can attenuate or amplify the vibration, the latter being the case from a fan cowl.

There is an additional complexity in the content of vibration from an induction motor compared to the vibration from a pump, fan or compressor because the former produces inherent electromagnetic forces and consequential vibration [1.30] to [1.44] – see also Chapter 10 in this book. The closer the accelerometer is to the stator frame (and core), for example on the outer periphery of an end frame, the greater the effects that electromagnetic forces can have on the measured vibration.

One of the most informative books on vibration analysis to diagnose faults in rolling element bearings is by Taylor and Kirkland, titled *The Bearing Analysis Handbook* [1.63]. Case histories are presented in three chapters with definitive predictions. However, there are only four of 21 case histories on diagnosing faults in rolling element bearings in electric motors.

1.1 Introduction

Taylor and Kirkland's book [1.63] is not dedicated to diagnosing rolling element bearing faults in induction motors. It was never intended to, since it covers a wide range of rotating mechanical plant with rolling element bearings. The largest electric motor, which was covered in the four case histories on detecting bearing faults, in Taylor and Kirkland's [1.63] book was 224 kW/300 H.P., whereas there is a case history presented in Section 6.2 in this book on the diagnosis of rolling element bearing defects in an 1139 kW/1600 H.P. SCIM.

Condition monitoring companies and OEMs of vibration instruments and software often present end users with *vibration diagnostic charts*, which give vibration characteristics caused by faults in electric motors. A comprehensive vibration diagnostic chart was published in reference [1.40], which makes the following points, with which the author of this book agrees.

(i) Many different problems either electrical or mechanical in nature can cause vibration at the same or similar frequencies.
(ii) One must look closely to differentiate between the true sources of vibration.

During the past 40 years the author has reviewed numerous reports on behalf of end users that were written by vibration monitoring companies who conclude by stating the problem *could be, may be or might be this, that or the other* and the end user was left to make the final diagnosis and decide what action to take. The end users do not want uncertainties. The case histories in this book give clear predictions and actions to be taken.

1.1.2 Overview of Shaft Misalignment

Theoretical coverage of three-dimensional misalignment forces is not required to support the case histories presented in this chapter because the focus is on the application of vibration spectrum analysis to detect shaft misalignment. For technical information on shaft alignment and its practical application in industry the reader is referred to the following books, references [1.64] and [1.65].

There is also an excellent white paper (62 pages) on a practical guide to shaft alignment; see reference [1.66]. With respect to the industrial case histories that follow, it is adequate to simply illustrate the different types of misalignment which are shown in Figures 1.6 to 1.10.

Parallel Misalignment

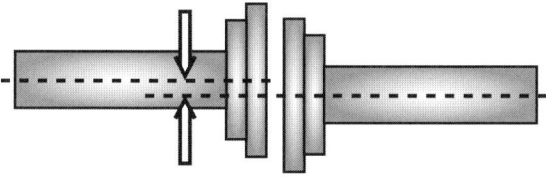

Figure 1.6 Illustration of parallel misalignment.

Axial Misalignment

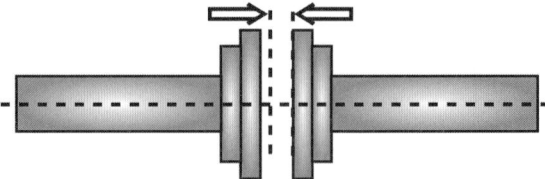

Figure 1.7 Illustration of axial misalignment.

Angular Misalignment

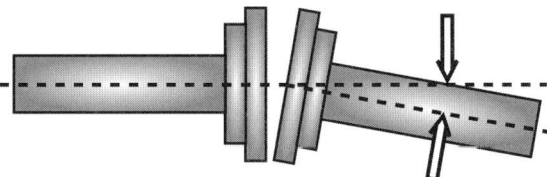

Figure 1.8 Illustration of angular misalignment.

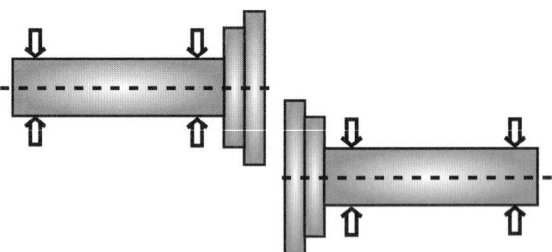

Figure 1.9 Misaligned shafts prior to coupling being made up.

Horizontal parallel misalignment is often the most critical and gives rise to shaft reaction forces in the direction of displacement as shown in Figures 1.9 and 1.10.

One of the major effects of shaft misalignment in an induction motor drive is the production in a specific radial direction of *rotor pre-load*, which produces a *radial force* that can push the rotor to the side. The rotor can in fact become displaced from its original position resulting in a higher eccentricity level inside the seals and bearings. In extreme cases it is also possible for the rotor to become bowed and rotate in a bow configuration. This means that misalignment can change the radial airgap length due to a change in dynamic airgap eccentricity [1.32]. It is well known that misalignment can cause coupling failure and premature wear and consequential failure of bearings in induction motors.

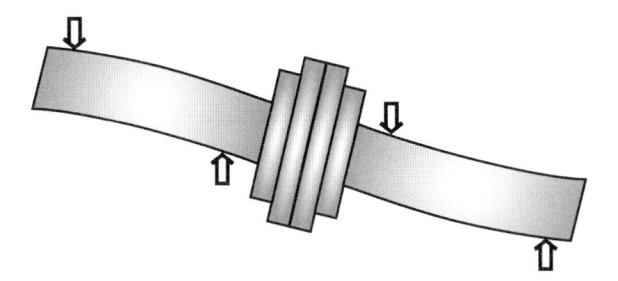

Figure 1.10 Deflected shape after coupling is made up.

1.1.2.1 Motor Soft Foot

Soft foot is the term used to describe the following conditions:

(i) The mounting feet of the motor are uneven as illustrated in Figure 1.11a.
(ii) The surface of the mounting base for the motor is uneven as shown in Figure 1.11b.

When soft foot exists, and the motor is coupled to the driven load this fault can cause consequential shaft misalignment. Note that soft foot can develop over time due to corrosion and rusting of shims and/or the mounting base. First-line overall vibration measurements and VSA applied to an induction motor drive train can detect the existence of abnormal misalignment.[1]

This will be shown in the case histories in this chapter, but identification of the type of shaft misalignment and its fundamental cause cannot be reliably identified by solely using VSA. The drive train needs to be shut down and alignment measurements carried out, using optical or laser alignment or by traditional manual alignment; see references [1.64] and [1.65].

In the first instance, it is wise to carry out a visual inspection of the motor's base-plate for any evidence of rust and loose fixing bolts. An initial indication of soft foot can be achieved by using a dial gauge on each foot mount and each bolt sequentially loosened and tightened.

Typically, more than 0.05 mm (i.e. 2 mils/thou) movement normally indicates soft foot. The vibration measurements should be repeated during an uncoupled run followed by tightening all the bolts to the same applied torque. Each bolt should be sequentially slackened and retightened in sequence and vibration measurements repeated. A soft foot problem will normally be obvious because the overall vibration levels on the bearing housings and foot mounts will normally change as this check proceeds.

---

[1] **Abnormal misalignment**: end users often set a misalignment limit of 0.05 mm/0.002 inches in each of the parallel, axial and radial planes.

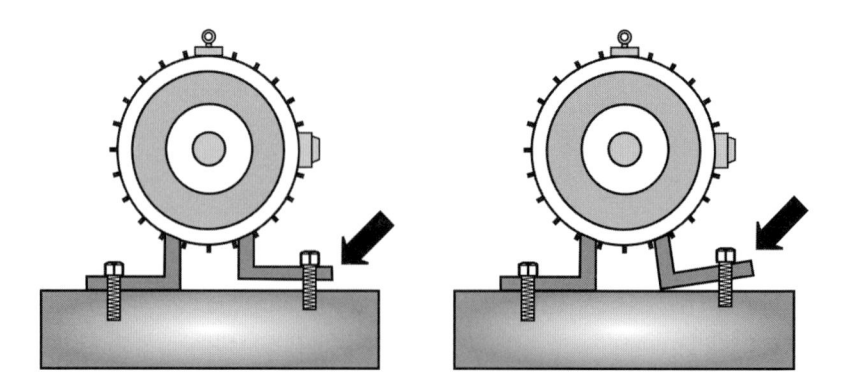

Figure 1.11a Illustrations of soft foot produced by the motor.

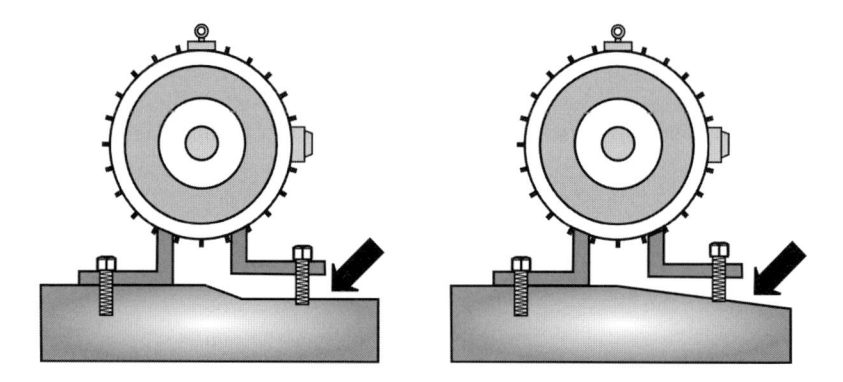

Figure 1.11b Illustrations of soft foot produced by the mounting base, note that soft foot can also occur in the axial direction.

1.1.3 Vibration due to Shaft Misalignment and/or Soft Foot

Industrial case histories will verify that VSA can diagnose shaft misalignment problems in induction motor drives. Vibration spectrum analysis, with no phase measurements, was used in the case histories which follow and only the key vibration characteristics are considered, since VSA alone cannot reliably determine the actual type of shaft misalignment.

There is still some debate on the vibration components which are truly affected or produced by each specific type of shaft misalignment, but it is incontestable that misalignment produces changes in the vibration spectra, which are identifiable. This is demonstrated in each case history in this chapter, because subsequent re-alignment produced a normal vibration spectrum at the bearings.

The centrifugal force (C.F.) produced by inherent rotor imbalance in an induction motor is constant provided the rotor's mechanical imbalance and speed are constant. The C.F. rotates at the speed of the rotor and produces the fundamental *1X* vibration component. For the avoidance of doubt the use of *1X* in this book is defined as:

The fundamental rotational speed frequency in Hz produced by the rotor.

The *1X* component is not given in units of r.p.m. or c.p.m. because it applies to vibration frequency spectra.

If the motor does not have soft foot and the shaft misalignment between the motor and mechanical load in all planes is a minimum, then the velocity of the *1X* component should not normally vary significantly around the periphery of the bearing housings at the DE and NDE.

In contrast, the forces produced by shaft misalignment are unlikely to be similar around the radius of the bearing housing or outer periphery of the end frames and therefore the magnitude of the velocity in one radial direction can be different from the velocity in another radial direction. The generic condition of shaft misalignment can cause the following vibration characteristics:

- An increase in the magnitude of the *1X* velocity component on the motor's feet and/or the bearing housing in the radial direction.
- A high magnitude of the *2X* velocity component, for example two or three times greater than the velocity of the *1X* component, for example, in the axial direction at the DE.
- Harmonics of the *1X* component may occur due to shaft misalignment, which may also be a by-product of soft foot in the motor mounting.
- Shaft misalignment may cause high horizontal velocity at one end of the motor whereas it can be high in the vertical direction at the other end. An increase in rotor imbalance does not normally produce this vibration characteristic.

1.1.3.1 Preparatory Guidelines for Vibration Monitoring of Induction Motors

Before carrying out vibration condition monitoring of induction motors there are two preparatory stages, namely, before and during the first on-site visit and these are as follows.

Prior to visiting the site, the end user should provide the following:

A single source on-site contact.
The drive designations and tag numbers of the induction motors.
The full nameplate data that is held in the end user's data base.
Information on records of greasing the bearings in induction motors.
History of planned maintenance and faults with dates and repairs.
Have new bearings been installed in any of the motors? If yes, the end user should provide the numbers of the new bearings installed.
Date when each induction motor drive was last aligned.
Historical records of vibration measurements from induction motors.

During the first on-site visit carry out the following:

Record the nameplate data directly from all induction motors to be monitored and compare with the end user's data base.

A *look and listen tour* (this excludes touch because surfaces can be hot) of the motors to be vibration monitored noting any unusual noises and also evidence of rust on the base-plates and at the edges of mounting shims.

If permitted take photos of the motors and focus on accessibility of the DE and NDE bearing housings and the size of accelerometers required to be mounted on them via magnets.

1.1.4 Introductory Industrial Case History – Normal Shaft Misalignment and no Soft Foot in a 230 kW/308 H.P. SCIM Pump Drive

Abstract – This introductory case history illustrates that meticulous attention to detail is essential for reliable vibration measurements and VSA to diagnose faults in induction motors.[2] This case history is deliberately prescriptive and *a detailed flow chart* (Section 1.1.4.1) for vibration measurements and VSA presents the method used by the author.

1.1.4.1 Procedures for Vibration Measurements and Vibration Spectrum Analysis (VSA) of SCIMs

For This Case History
Three SCIMs Driving Condensate Extraction Pumps
Tag numbers: DM 2501A, B, C; serial numbers: 8110004336.01/1/2/3.
   Nameplate data: 3-phase, SCIM, 11 kV, 230 kW/300 H.P., 14 A, 50 Hz, 2957 r/min, 0.9 p.f., eff = 95.8%,
   DE and NDE bearings, 6317 M C3 deep groove ball bearings.[*]

⬇

Preparations for On-Site Vibration Measurements
The accelerometer shown in Figures 1.14 to 1.16 with its magnetic attachment can be mounted at the positions shown in Figures 1.12 and 1.13.
   Its main features are: 100 mV/g ± 10%; linear frequency response: 2–15 kHz ± 5%; mounted base resonance nominally: 22 kHz ± 10%; 25 grams/ 1.0 ounce.
   This accelerometer can be mounted directly on the bearing housings via a magnet and owing to its small dimensions and light weight will not alter the response of the vibrating surface.

⬇

On-Site Vibration Measurements – Basic Preparations
Operating currents: D2501A = 10.8 A, D2501B = 10.8 A, D2501C = 10.6 A.
   Full-load speed from nameplate = N_r = 2957 r/min.

---

[2] **Normal shaft misalignment**: typically less than 0.05 mm/0.002 inches in each of the parallel, axial and radial planes.
*See Section 2.1.1: M = brass cage, C3 = radial clearance; bore diameter = 5 × 17 = 85 mm.

(*cont.*)

Full-load slip, $s_{f.l.} = \frac{N_s - N_r}{N_s} = \frac{3000 - 2957}{3000} = 0.0143$ pu (1.43%).

The full-load *1X* rotational speed frequency $f_r = 2957/60 = 49.28$ Hz.

The operational current of each motor < full-load current; therefore, the rotational speed N_r will be greater than the rated full-load speed at 2957 r/min and likewise the *1X* component will be greater than 49.28 Hz.

⬇

Vibration Instrument Specification and Spectrum Analysis Settings

The vibration instrument should be able to measure overall r.m.s., peak and peak–peak velocity.

The instrument should have the facility to produce a linear and a dB velocity spectrum (mm/s or dB vs frequency) with 12,800 spectral lines over the frequency span selected. A vibration time domain display is also required.

Select a frequency span from 10–1000 Hz, and 12,800 spectral lines and measure the overall r.m.s. or peak velocities at the positions shown in Figures 1.12 and 1.13.

For this motor select the velocity spectrum mode with a span of 10–1000 Hz, 12,800 spectral lines and therefore a line frequency resolution of 0.078 Hz/line and record the spectra at all positions on each motor.

Select the velocity spectrum with a span of 10–120 Hz, 12,800 spectral lines and a line frequency resolution of 0.0086 Hz/line and record the spectra at all positions on each motor. The reason for this span and frequency resolution is to ensure reliable separation of frequency components, *2X* and *2f₁*.

*See Section 2.1.1: M = brass cage, C3 = radial clearance; bore diameter = 5 × 17 = 85 mm.

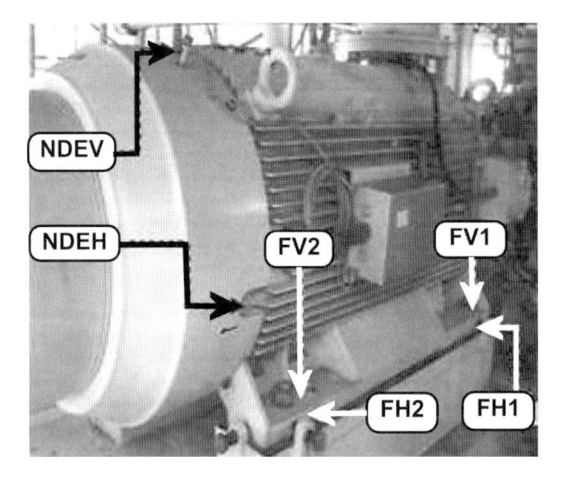

Figure 1.12 Positions of accelerometers on the NDE end frame and the motor's mounting base.

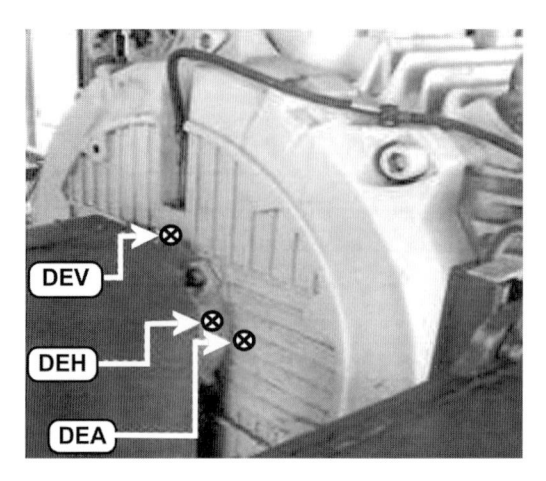

Figure 1.13 Positions of accelerometers on the DE bearing housing and end frame.

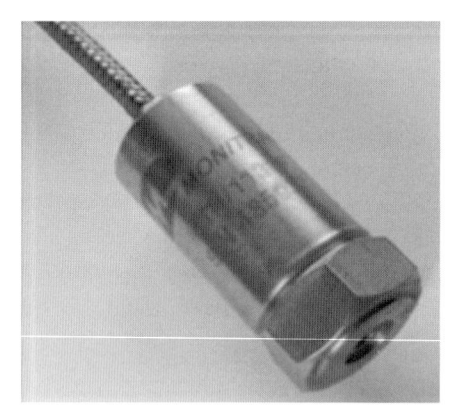

Figure 1.14 Accelerometer.

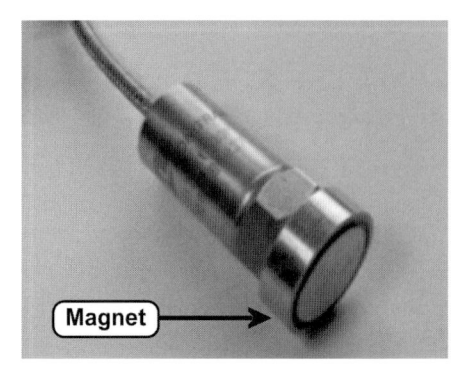

Figure 1.15 Accelerometer with magnet (flux density 0.32 tesla/3200 gauss).

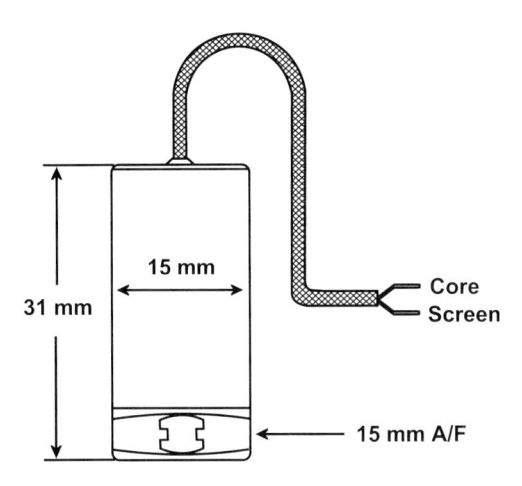

Figure 1.16 Dimensions of accelerometer.

1.1.4.2 Overall R.M.S. Velocities

The results are presented in Table 1.1 and the highest velocity measured from the three SCIMs was only 1.0 mm/s r.m.s. ±10% at the DEH position on DM-2501B. The British Standard (BS) 60034-14 -2004 [1.68] allows up to 2.3 mm/s r.m.s. with this motor running uncoupled on a large, solid steel base plate (rigid mounting) – see Table 1.2.

Note that the NEMA MG 1-2006 [1.69] vibration standard allows 3.0 mm/s peak (see sections 7.8.1, 7.8.2 and table 7-1 in the standard) on a rigid mounting – see Table 1.3. Therefore, these high-speed 2-pole SCIMs were running very smoothly indeed.

1.1.4.3 Vibration Spectrum Analysis

The velocity spectra were recorded using two different baseband frequency spans of 10–1000 Hz and 10–120 Hz to provide base-line spectra for future comparisons, and the Fast Fourier Transform (FFT) [1.19, 1.20], had 12,800 spectral lines to give line resolutions of 0.078 Hz/line and 0.0086 Hz/line respectively for the frequency spans used.

As an illustration, only one set of velocity spectra is presented, which is for the SCIM in slot DM2501A, since the spectra for the other two motors are very similar.

(i) Figure 1.17 shows the *1X* component at 49.5 Hz at a velocity of 0.4 mm/s r.m.s., which is due to inherent mechanical imbalance in the rotor. The OEM confirmed that the rotor was balanced to G1.0 grade, ISO 1940-1:2003 [1.67], hence the low velocity of this component.

(ii) The analysis of the 10–500 Hz spectrum displays only one component at 100 Hz of 0.46 mm/s r.m.s. (the highest) because the spectral line resolution was 0.078 Hz/line in Figure 1.17. The *2f₁* and *2X* components cannot be reliably separated in this spectrum.

Table 1.1 Overall r.m.s. velocity levels in mm/s r.m.s., to within ±10%. Frequency span: 10–1000 Hz

Condensate export circulation pump DM-2501A		8110004336.01/1	
	Vertical	Horizontal	Axial
DE bearing housing	0.7	0.6	0.6
NDE end frame	0.6	0.66	Not accessible

Condensate export circulation pump DM-2501B		8110004336.01/2	
	Vertical	Horizontal	Axial
DE bearing housing	0.6	1.0	0.86
NDE end frame	0.75	0.7	Not accessible

Condensate export circulation pump DM-2501C		8110004336.01/3	
	Vertical	Horizontal	Axial
DE bearing housing	0.86	0.7	0.8
NDE end frame	0.7	0.5	Not accessible

Table 1.2 Limits of maximum velocity in mm/s r.m.s. – shaft height H

BS 60034–14–2004 [1.67], rigid mounting

Shaft height in mm $56 \leq H \leq 132$ 1.3 mm/s r.m.s.	Shaft height in mm $132 \leq H \leq 2802$ 1.8 mm/s r.m.s.	Shaft height in mm $H > 280$ 2.3 mm/s r.m.s.

Table 1.3 Unfiltered vibration limits – resiliently mounted motors USA NEMA standards. MG 1, section 7.8 [1.68]

r/min @ 60 Hz	Velocity inches/s peak	Velocity mm/s peak	r/min @ 50 Hz	Velocity inches/s peak	Velocity mm/s peak
3600	0.15/0.12*	3.8/3.0*	3000	0.15/0.12*	3.8/3.0*
1800	0.15/0.12*	3.8/3.0*	1500	0.15/0.12*	3.8/3.0*
1200	0.15/0.12*	3.8/3.0*	1000	0.13/0.1*	3.3/2.6*
900	0.12/0.096*	3.0/0.24*	750	0.1/0.08*	2.5/2.0*
720	0.09/0.072*	2.3/0.18*	600	0.08/0.064*	2.0/1.6*
600	0.08/0.064*	2.0/0.16*	500	0.07/0.056*	1.7/1.4*

NOTE: For motors with a rigid mounting, multiply values in Table 1.3 by 0.8: the corrected values for rigid mounting have an asterisk *.

1.1 Introduction

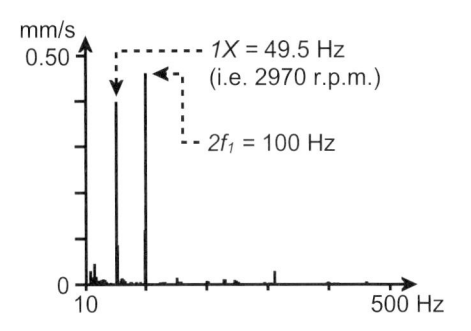

Figure 1.17 DE vertical bearing housing velocity spectrum.

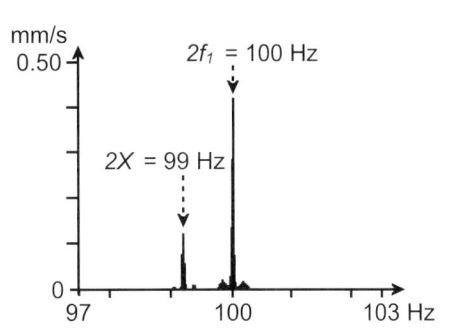

Figure 1.18 DE vertical bearing housing velocity zoom spectrum.

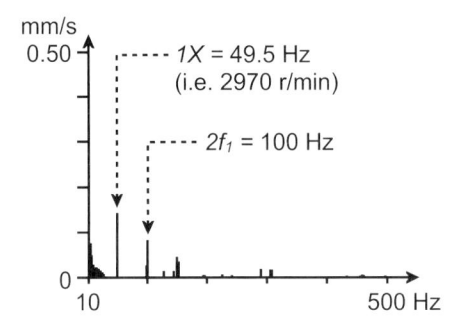

Figure 1.19 DE horizontal bearing housing velocity spectrum.

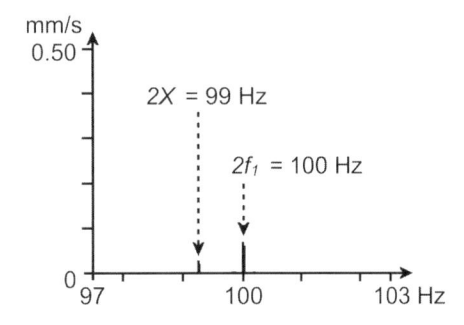

Figure 1.20 DE horizontal bearing housing velocity zoom spectrum.

(iii) Note that a measured component at 100 Hz suggests it is the twice supply frequency component at $2f_1$, which is caused by the fundamental electromagnetic force in a SCIM, see Section 10.1.1 and Appendix 10A.

(iv) Separation of the $2f_1$ and 2X was achieved by analysis of a 10–120 Hz spectrum, with a line resolution of 0.0086 Hz/line and the zoom analysis is presented in Figure 1.18.

(v) The 2X component is only 0.12 mm/s r.m.s., whereas the $2f_1$ component at 100 Hz is 0.42 mm/s r.m.s., which is 3.3 times greater than the velocity at 2X.

The reason for focusing on separating the frequency components at $2f_1$ and 2X (for a 2-pole SCIM) is that the component at 100 Hz in Figure 1.17 is often referred to as the 2X component which in this case, it is not, and its frequency is not given.

When diagnosing a shaft misalignment problem, the 2X component can, but certainly not always, be an indicator of shaft misalignment if it has a substantially higher velocity, for example, two to three times higher than the velocity of the 1X component. This is certainly not the case in this case history. The velocity spectra for the other positions on motor DM-2501A are shown in Figures 1.19 to 1.26.

The overall r.m.s. velocities measured next to the fixing bolts on each motor are presented in Table 1.4.

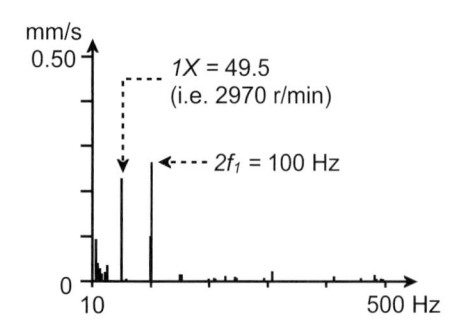

Figure 1.21 DE axial bearing housing velocity spectrum.

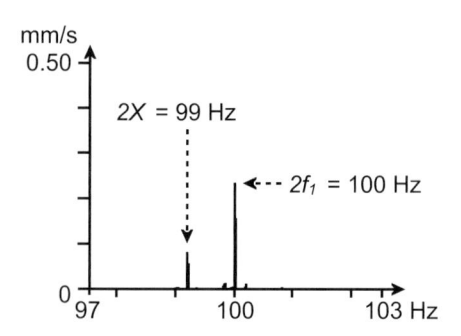

Figure 1.22 DE axial bearing housing velocity zoom spectrum.

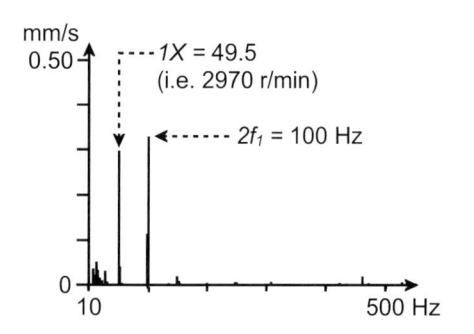

Figure 1.23 NDE vertical bearing housing velocity spectrum.

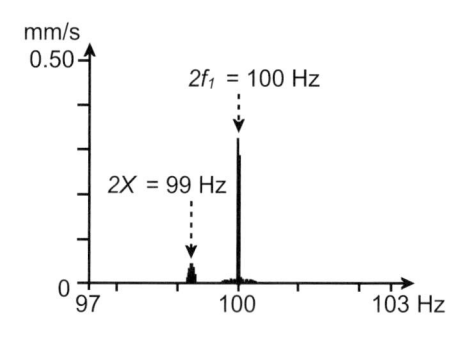

Figure 1.24 NDE vertical bearing housing velocity zoom spectrum.

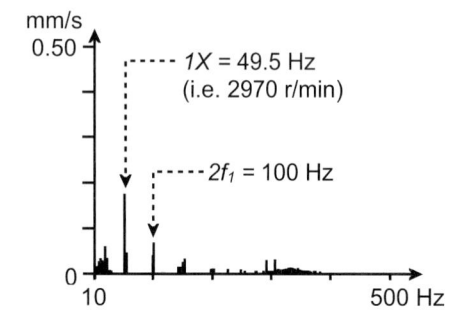

Figure 1.25 NDE horizontal bearing housing velocity spectrum.

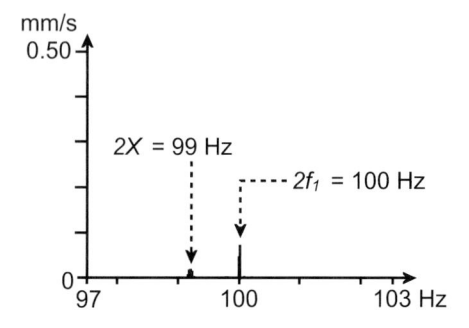

Figure 1.26 NDE horizontal bearing housing velocity zoom spectrum.

Table 1.4 Overall r.m.s. velocities ±10%, next to the motor's fixing bolts; 10–1000 Hz

Motor Base NDE DE V,H⊗	Motor Base NDE DE ⊗V,H	⊗V,H NDE DE Motor Base	V,H⊗ NDE DE Motor Base
DM-2501A FV1: 0.7 mm/s FH1: 0.24 mm/s DM-2501B FV1: 0.7 mm/s FH1: 0.5 mm/s	DM-2501A FV2: 0.6 mm/s FH2: 0.3 mm/s DM-2501B FV2: 0.67 mm/s FH2: 0.6 mm/s	DM-2501A FV3: 0.7 mm/s FH3: 0.26 mm/s DM-2501B FV3: 0.8 mm/s FH3: 0.4 mm/s	DM-2501A FV4: 0.6 mm/s FH4: 0.3 mm/s DM-2501B FV4: 0.74 mm/s FH4: 0.3 mm/s

1.1.5 Conclusions

All the vibration spectra were acceptable on the DE bearing housing and on the outer periphery of the NDE's end frame of the motor in Tag slot DM-2501A. The highest velocity of the *1X* component was only 0.4 mm/s r.m.s. in the vertical direction on the bearing housing at the DE. The velocities of the *2X* components at all positions are very low with the highest being only 0.12 mm/s r.m.s. in the vertical direction on the DE bearing housing of DM-2501A.

There is no evidence whatsoever of an above normal shaft misalignment (e.g. > 0.05 mm/2 mils in of the alignment planes). The velocity of the *$2f_1$* component in each of the spectra due to the fundamental electromagnetic force was normal for a 2-pole SCIM, with the highest being only 0.42 mm/s r.m.s. in the vertical direction on the DE bearing housing. The other two motors had very similar velocity spectra.

There are no abnormal velocity levels at any of the positions next to the fixing down bolts on the three motors and an inspection of Table 1.4 shows that the velocities in the vertical positions do not differ by more than 0.1 mm/s r.m.s. and likewise in the horizontal directions. The velocity levels in the vertical and horizontal directions next to the fixing bolts are virtually the same and there is therefore no evidence of soft foot in the motor mounting.

The case histories that follow in this chapter will confirm that abnormal levels of shaft misalignment and soft foot can be diagnosed via VSA.

1.2 Industrial Case History – Diagnosis of Misalignment in a 554 kW/743 H.P. SCIM Driving a Gas Recirculating Fan

1.2.1 Introduction

Abstract – Four nominally identical SCIMs were used to drive gas recirculating (GRF) fans in a Combined Cycle Gas Turbine (CCGT) fired power station. This station was originally designed and built in the 1970s to use oil as the fuel source but has had numerous changes during the past 40 years and is presently a CCGT station.

The nameplate data of the motors was as follows:

- 3-phase, induction motor
- special Trislot [1.32] rotor design
- 3.3 kV, 554 kW/743 H.P., 110 A, 50 Hz, 740 r/min, star connected.

There were no permanently fitted displacement probes (see Section 5.2.3) in the housings of the sleeve bearings of the fans or accelerometers on the bearing housings of the rolling element bearings in the SCIMs.

A photograph of one of the GR fan motors is shown in Figure 1.27 and it is obvious that access to measure vibration at or close to the NDE bearing is not possible. This is the practical reality of applying vibration monitoring to SCIMs in industry and note that there are internal motor cooling fans at the DE and NDE of the motor, which means that the actual bearing housings are not accessible. There was very limited access to the end frame of the DE bearing assembly due to the coupling guard. The only accessible position was next to the grease pipe nipple as shown in Figure 1.28 at approximately 13.00 hours looking from the DE of the motor through to the NDE.

The vertical, horizontal and axial vibration on the DE sleeve bearing housings of the fans could be measured via accelerometers mounted via magnets as shown in Figures 1.29 to 1.31 but note that the top half of the bearing assembly is a non-magnetic alloy.

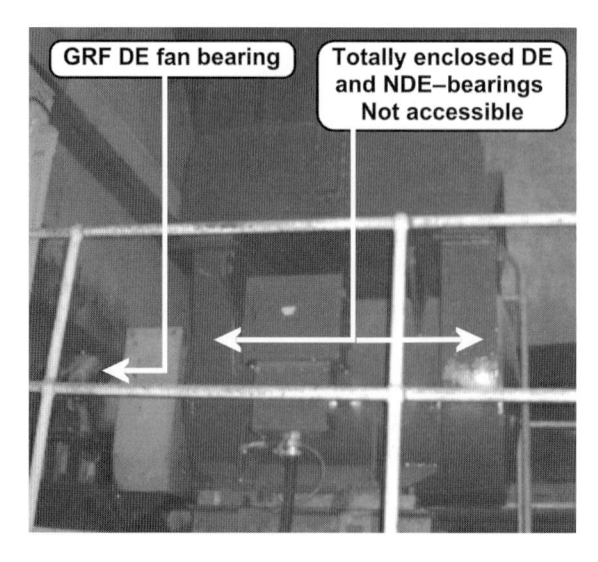

Figure 1.27 Photograph of gas recirculating fan (GRF) motor.

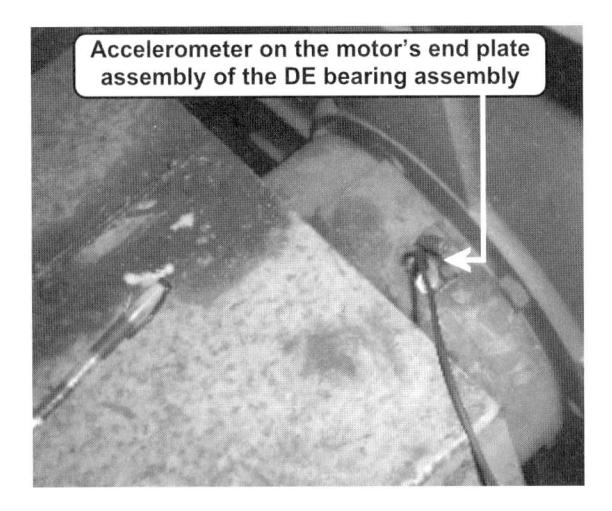

1.2 Industrial Case History – Diagnosis of Misalignment in a 554 kW/743 H.P. SCIM

Figure 1.28 Photograph of the motor's end plate at the DE bearing assembly of a GR fan motors.

Figure 1.29 Photograph of accelerometer in vertical direction on the DE bearing assembly of GRF2A.

1.2.2 Overall Vibration Measurements

The overall r.m.s. velocities on the motors' DE and on the DE and NDE sleeve bearings of the GR fans are presented in Table 1.5. Recall that accelerometers initially measure acceleration, which is electronically integrated to provide the required velocity data.

(i) The velocity in the horizontal direction on the fan's DE bearing of GRF2A was 6.4 mm/s r.m.s., which was five times higher than the velocity (1.3 mm/s) on the DE bearing housing of the motor.

Table 1.5 Overall velocities in mm/s r.m.s., frequency span 10–1000 Hz

Drive train unit tag slot number: GRF1A

Motor NDE bearing	**Vertical**	**Horizontal**	**Axial**
	Not accessible	Not accessible	Not accessible
Motor DE bearing	Radial position as shown in Figure 1.28: 1.8 mm/s		
Fan DE bearing	0.5 mm/s	0.73 mm/s	0.74 mm/s
Fan NDE bearing	0.3 mm/s	0.5 mm/s	0.3 mm/s

Drive train unit tag slot number: GRF1B

Motor NDE bearing	**Vertical**	**Horizontal**	**Axial**
	Not accessible	Not accessible	Not accessible
Motor DE bearing	Radial position as shown in Figure 1.28: 1.4 mm/s		
Fan DE bearing	0.9 mm/s	2 mm/s	1.0 mm/s
Fan NDE bearing	0.6 mm/s	1.0 mm/s	0.4 mm/s

Drive train unit tag slot number: GRF2A

Motor NDE bearing	**Vertical**	**Horizontal**	**Axial**
	Not accessible	Not accessible	Not accessible
Motor DE bearing	Radial position as shown in Figure 1.28: 1.3 mm/s		
Fan DE bearing	2.0 mm/s	**6.4 mm/s**	**4.0 mm/s**
Fan NDE bearing	0.5 mm/s	1.2 mm/s	0.5 mm/s

Drive train unit tag slot number: GRF2B

Motor NDE bearing	**Vertical**	**Horizontal**	**Axial**
	Not accessible	Not accessible	Not accessible
Motor DE bearing	Radial position as shown in Figure 1.28: 1.3 mm/s		
Fan DE bearing	1.0 mm/s	3 mm/s	2.0 mm/s
Fan NDE bearing	1.3 mm/s	1.1 mm/s	0.7 mm/s

Figure 1.30 Photograph of accelerometer in horizontal direction on the DE bearing assembly of GRF2A fan.

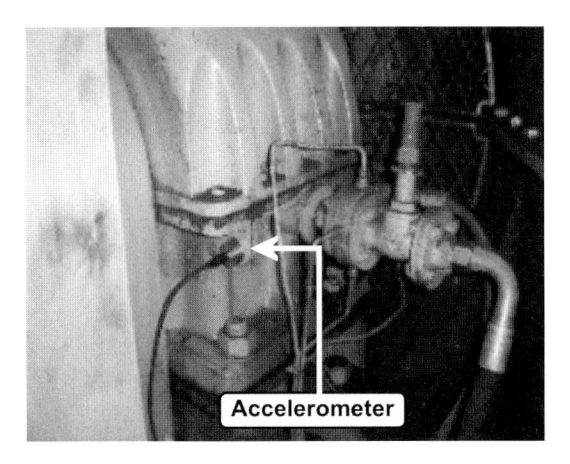

1.2 Industrial Case History – Diagnosis of Misalignment in a 554 kW/743 H.P. SCIM

Figure 1.31 Photograph of accelerometer in axial direction on the DE bearing assembly of GRF2A fan.

(ii) At the NDE of the GRF2A fan the overall r.m.s. velocity in the horizontal direction on the bearing housing was 1.2 mm/s compared to 6.4 mm/s r.m.s. in the same direction on the fan's DE bearing.
(iii) The overall r.m.s. velocity in the axial direction on the DE bearing of the GRF2A fan was 4 mm/s r.m.s., which was eight times higher than the velocity (0.5 mm/s r.m.s.) in the axial direction on the NDE bearing housing of the fan.

1.2.3 Vibration Spectrum Analysis Detected Abnormal Misalignment

Owing to the relatively high overall r.m.s. velocity levels of 6.4 mm/s and 4 mm/s on the GRF2A DE fan bearing compared to the other drives and the much lower levels on the NDE of GRF2A, there was clearly an indication of a vibration abnormality in the GRF2A drive train. Spectrum analysis was required to determine the cause of these abnormal levels.

For comparisons with future vibration measurements, which should be taken at the same load, the operational current of the motor was recorded at 85 amperes ($I_{f.l.}$ = 110 A), and a corresponding speed of 744 r/min (full-load speed of 740 r/min). The vibration velocity (r.m.s.) spectra in the vertical, horizontal and axial positions on the DE plain bearings of the GRF2A fan are presented in Figures 1.32 to 1.37.

1.2.4 Main Conclusions

(i) The *1X* frequency component at 12.4 Hz (744 r/min) dominates the vibration spectra in the horizontal and axial directions of the DE fan bearing of GRF2A at 6.3 mm/s and 3.7 mm/s respectively.
(ii) The horizontal and axial *1X* components at the NDE of the fan were only 0.5 and 1.0 mm/s.

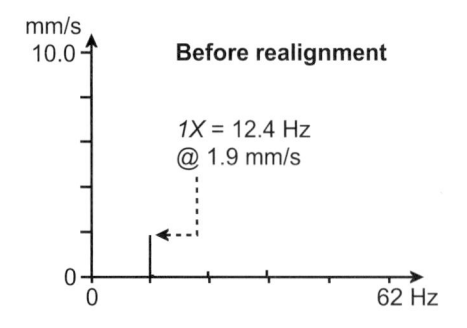

Figure 1.32 GRF2A vertical velocity spectrum, housing of DE fan bearing. (Before realignment, 1X = 12.4 Hz @ 1.9 mm/s)

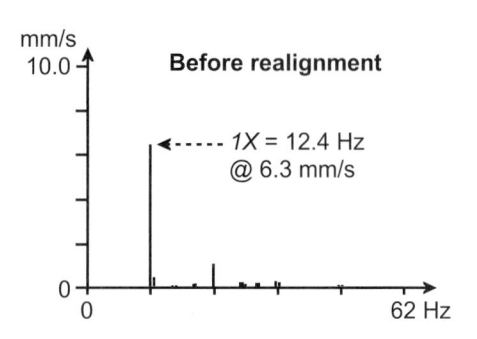

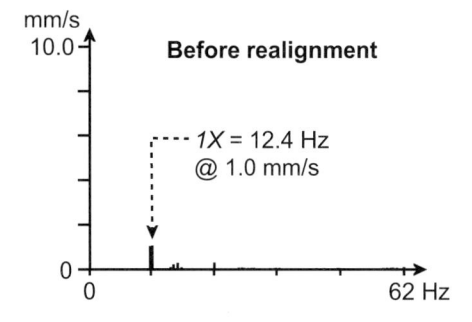

Figure 1.33 GRF2A horizontal velocity spectrum, housing of DE fan bearing. (Before realignment, 1X = 12.4 Hz @ 6.3 mm/s)

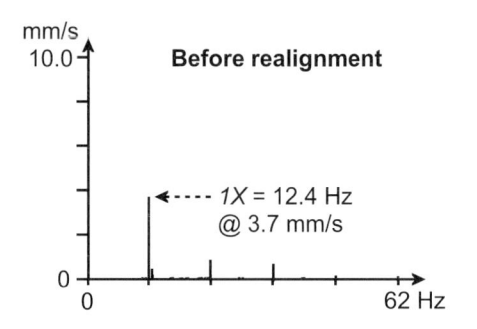

Figure 1.34 GRF2A axial velocity spectrum, housing of DE fan bearing. (Before realignment, 1X = 12.4 Hz @ 3.7 mm/s)

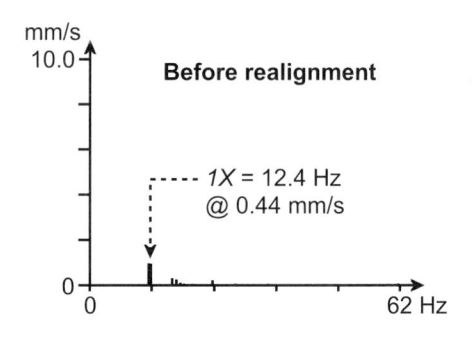

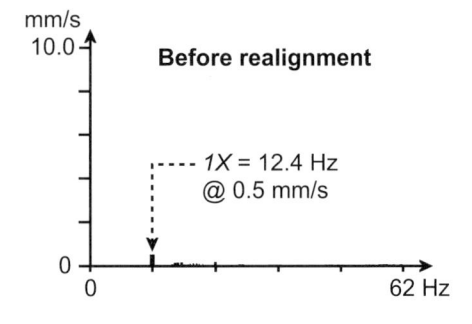

Figure 1.35 GRF2A vertical velocity spectrum, housing of NDE fan bearing. (Before realignment, 1X = 12.4 Hz @ 0.44 mm/s)

Figure 1.36 GRF2A horizontal velocity spectrum, housing of NDE fan bearing. (Before realignment, 1X = 12.4 Hz @ 1.0 mm/s)

Figure 1.37 GRF2A axial velocity spectrum, housing of NDE fan bearing. (Before realignment, 1X = 12.4 Hz @ 0.5 mm/s)

(iii) If the relatively high *1X* component at 6.3 mm/s r.m.s. in the horizontal direction on the DE fan bearing housing was due to an increase in the mechanical imbalance of the fan, then this should normally be also reflected in the horizontal direction on the NDE fan bearing but it was only 0.5 mm/s r.m.s. Thus, a mechanical imbalance fan problem was discounted.

(iv) As explained in Section 1.1.3 this vibration characteristic indicates that there is misalignment between the motor and the DE bearing of the fan. Recall only one vibration measurement was practically feasible on the motor. The alignment was checked by power station personnel and the following misalignments were reported to the author of this book.

From: Ian.Johnston@XXXXXXX
To: Bill Thomson mcsa@consultant.com
Subject: Re VIBRATION TESTING
Date: Thu, 24th August 2006 14:42:55 +0100
Bill,
  At long last we have finished the alignment checks on GR Fan 2A.
  The results are as follows, the motor was 10 thou high and the motor was 22 thou out towards the turbine hall. As discussed the motor plinth holding down bolts were slack, retightened approximately 2 turns.
  Hope this information is of use to you.
Regards
Ian

Note that towards the turbine hall means towards the DE fan bearing on GRF2A.
  Also 10 thou = 10 mils = 0.25 mm and 22 thou = 22 mils = 0.56 mm.
  The normal misalignment tolerance used by the power station is typically between 2 thou (0.05 mm) and a maximum of 4 thou (0.1 mm) on these drive trains.

1.2.4.1 Conclusions on VSA after Re-Alignment
With reference to Table 1.6:

(i) Before realignment the *1X* component was 6.3 mm/s r.m.s. in the horizontal direction on the DE bearing housing of the fan (Figure 1.33) and after re-alignment it dropped to 3 mm/s.

(ii) This reduction of 53%, confirmed that vibration analysis identified the misalignment problem. The levels were now acceptable, and the results are given in Table 1.6.

Vibration spectra on the fan's DE bearing before and after realignment are presented in Figures 1.38 to 1.43.

The two spectra to compare are presented in Figures 1.40 and 1.41.
The reason for this selection is to verify that the largest change in the *1X* component was in the horizontal direction on the fan's DE bearing housing.
There were no changes in the *1X* component at the vertical and axial positions on the DE bearing.

Table 1.6 Drive Train Unit Tag Slot Number: GRF2A

Velocity r.m.s. of the *1X* frequency component before and after re-alignment

Fan DE bearing	Vertical	Horizontal	Axial
	Before: 1.9 mm/s	**Before: 6.3 mm/s**	Before: 3.7 mm/s
	After: 1.9 mm/s	**After: 3 mm/s**	After: 3.6 mm/s
Fan NDE bearing	Vertical	Horizontal	Axial
	Before: 0.44 mm/s	Before: 1.0 mm/s	Before: 0.5 mm/s
	After: 0.23 mm/s	After: 0.76 mm/s	After: 0.26 mm/s

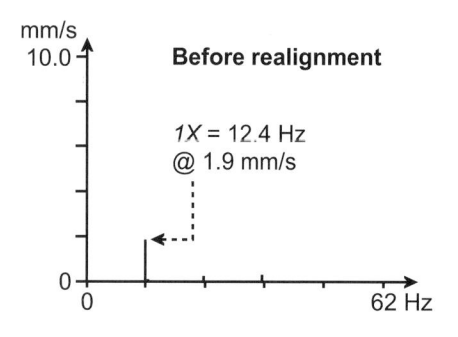

Figure 1.38 GRF2A vertical velocity spectrum, housing of DE fan bearing.

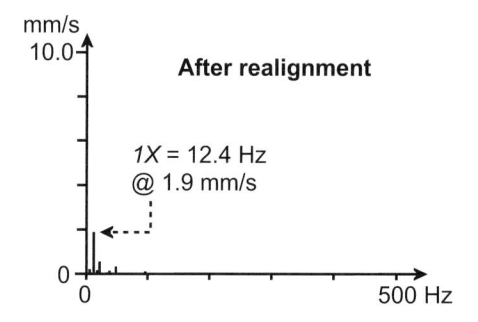

Figure 1.39 GRF2A vertical velocity spectrum, housing of DE fan bearing.

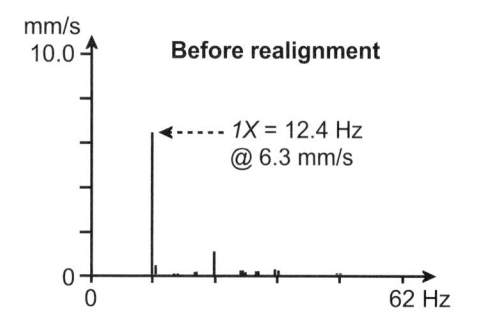

Figure 1.40 GRF2A horizontal velocity spectrum, housing of DE fan bearing.

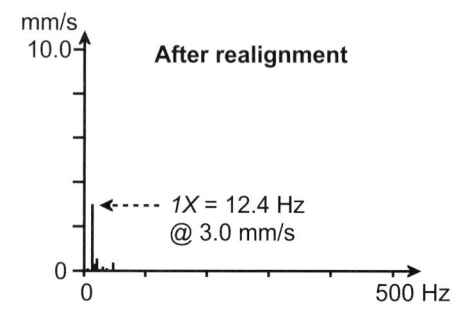

Figure 1.41 GRF2A horizontal velocity spectrum, housing of DE fan bearing.

Table 1.6 shows that the *1X* components on the fan's NDE bearing housing were normal before realignment and that they dropped to very low levels after realignment.

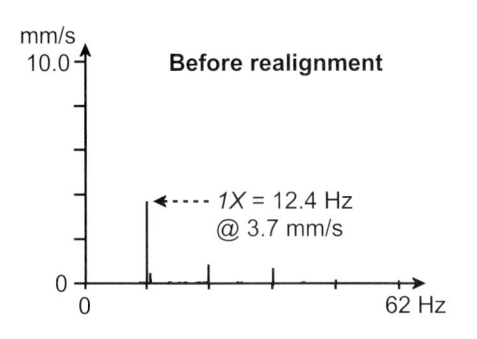

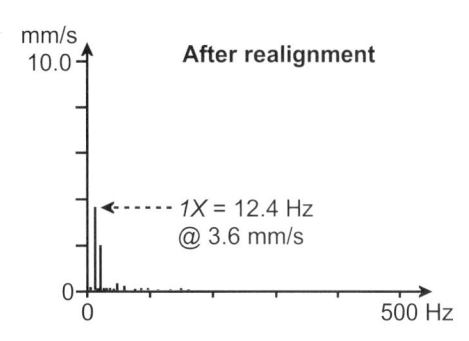

Figure 1.42 GRF2A axial velocity spectrum, housing of DE fan bearing.

Figure 1.43 GRF2A axial velocity spectrum, housing of DE fan bearing.

1.3 Industrial Case History – Diagnosis of Misalignment on a 7.5 kW/10 H.P. SCIM – Pump Drive used to Lubricate Sleeve Bearings in a 35 MVA Generator on an Offshore Oil Production Platform

1.3.1 Introduction

Abstract – Three SCIMs were used to drive pumps that circulate lubrication oil via a closed loop with an oil header tank to sleeve bearings in three 11 kV, 35 MVA, synchronous generators operating on an offshore oil production platform. These motors (see Figure 1.44) were highly critical because if one fails there is a loss of circulating oil to the plain bearings in the generator, which is then automatically disconnected.

The offshore installation manager (OIM) was informed by the incumbent vibration condition monitoring sub-contractor that one of the motors had a bearing fault. No detailed vibration analysis was presented to the OIM by the technician who took the vibration measurements, and the diagnosis was based solely on high overall r.m.s. velocity levels. The OIM contacted the author to provide an independent investigation, which included a one-off visit to the offshore installation and the main objective was as follows:

- To carry out vibration measurements and analysis to ascertain whether there was truly a faulty bearing in one of the small-power motors.

Neither the vibration measurement positions nor the velocity magnitudes on the frame of the motor were made available to the author so that no direct comparisons with earlier measurements could be made.

1.3.2 Overall Vibration Results

The three small-power SCIMs are referred to as motors A, B and C, and motor A was the subject of the investigation. Only A and C were available for testing as B was

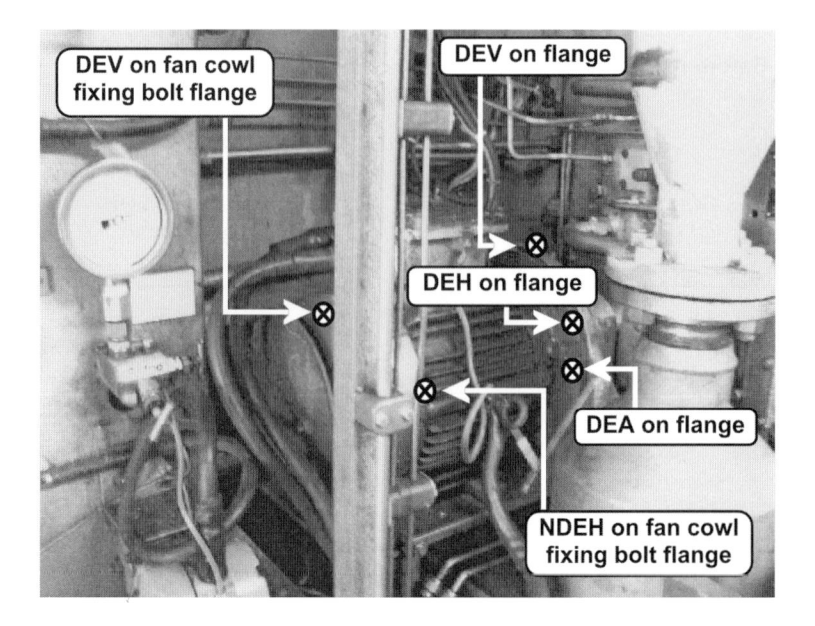

Figure 1.44 Photograph of one of the motors, showing the positions of the accelerometers – the bearing housings are inaccessible.

off-line because only two generators are required to deliver the load demand. The motor nameplate data for A and C are as follows:

- date of manufacture: 1984
- 415 V, 7.5 kW/10 H.P., 14.5 A, 50 Hz, 2880 r/min
- DE bearing: 6208-2Z-C3
- NDE bearing 6206-2Z-C3.

These are deep groove, sealed ball bearings.

Measurements were taken at the positions shown in Figure 1.44 because of the inaccessibility of the DE and NDE bearing housings. The accelerometers were mounted via strong magnets and at the NDE were positioned on the bolts securing the fan cowl directly to the NDE end frame, which provides a direct transmission path to the end frame and bearing. Axial vibration measurements could not be taken at the NDE because of the fan cowl.

The author had not previously analysed vibration from these motors. Note that an 'off-the-shelf' SCIM of 7.5 kW/10 H.P. is not normally supplied with actual Factory Acceptance Test (FAT) results from the OEM. The suppliers will merely state that the vibration levels meet the specifications set by, for example, NEMA MG 2 [1.68] or BS 60034-14-2004 [1.68]. The overall r.m.s. velocities from motors A and C are presented in Table 1.7.

From Table 1.7, the overall velocities at the DEH, NDEV and NDEH positions on motor A were 8, 9 and 17 mm/s r.m.s. respectively. Whereas at the same positions on the identical motor C the highest velocity was 3 mm/s r.m.s. at the NDEH position.

1.3 Industrial Case History – Diagnosis of Misalignment on a 7.5 kW/10 H.P. SCIM

Table 1.7 Overall r.m.s. velocities; ±10%, 10–1000 Hz

Motor A

Measurement position: DE mounting flange

| **DEV:** 1.5 mm/s | **DEH:** 8 mm/s | **DEA:** 4 mm/s |

Measurement position: NDE on bolts securing the fan cowl

| **NDEV:** 9 mm/s | **NDEH:** 17 mm/s | **NDEA:** Not accessible |

Motor C

| **DEV:** 1.5 mm/s | **DEH:** 2.4 mm/s | **DEA:** 1.4 mm/s |

Measurement position: NDE on bolts securing the fan cowl

| **NDEV:** 1.0 mm/s | **NDEH:** 3.0 mm/s | **NDEA:** Not accessible |

Motor A had unacceptably high velocities but before switching off the motor the velocity spectra were recorded for subsequent interpretation.

1.3.3 Vibration Spectrum Analysis

The velocity spectra from the problem motor A presented in Figures 1.45 to 1.49 were taken using a frequency span of 10–500 Hz, and a line resolution of 0.078 Hz/line.

Note in particular the comparison between motors A and C presented in Figures 1.49 and 1.50.

In motor A, the rotational speed frequency component ($1X$) dominates the velocity spectra at the DEH and NDEH positions with values of 7.3 (overall 8 mm/s) and 15 mm/s (overall 17 mm/s) r.m.s. respectively.

- Caution should be exercised with respect to the magnitude of the velocity measured on the holding down bolt for the fan cowl at the NDEH position as coming solely from the NDE end frame, because the cowl could be amplifying the vibration.
- At the same position (NDEH) on the identical motor C the overall r.m.s. velocity was only 3.0 mm/s and Figure 1.50 shows that the $1X$ component was only 2 mm/s, whereas at the same position on motor A it was 17 mm/s (see Figure 1.48). The bolts holding down the fan cowl on motors' A and B were all tightly secured.
- The overall r.m.s. velocities at the DEH and NDEH positions from motor A were unacceptably high (8 and 17 mm/s r.m.s. respectively), particularly at the NDE position and the high overall velocities were coming from the dominant $1X$ component in these velocity spectra. It was obvious from the velocity spectra that all other components were negligible in comparison to the $1X$ component

Motor A: DEV
Overall Velocity = 1.5 mm/s
1X = 49 Hz @ 0.9 mm/s

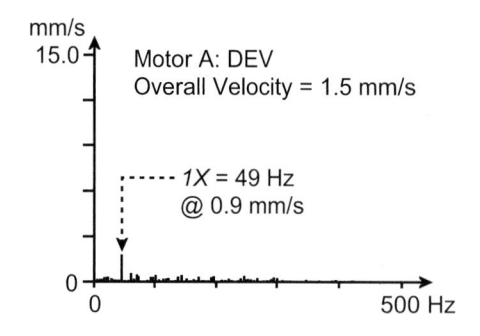

Figure 1.45 Motor A DEV velocity spectrum.

Motor A: DEH
Overall Velocity = 8.0 mm/s
1X = 49 Hz @ 7.3 mm/s

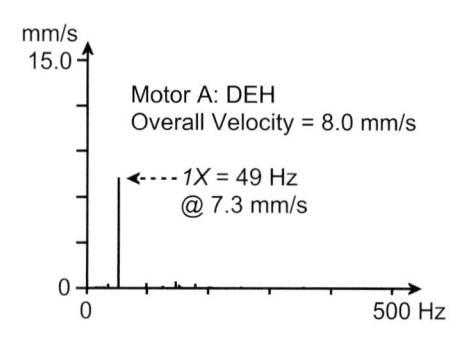

Figure 1.46 Motor A DEH velocity spectrum.

Motor A: DEA
Overall Velocity = 4.0 mm/s
1X = 49 Hz @ 3.5 mm/s

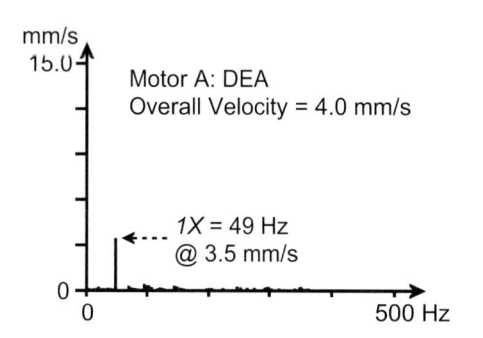

Figure 1.47 Motor A DEA velocity spectrum.

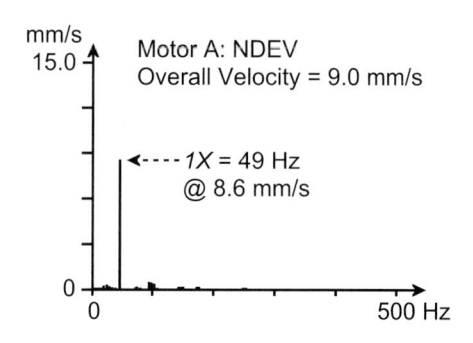

Motor A: NDEV
Overall Velocity = 9.0 mm/s
1X = 49 Hz @ 8.6 mm/s

Figure 1.48 Motor A NDEV velocity spectrum.

1X = 49 Hz @ 15.0 mm/s

Motor A: NDEH
Overall Velocity = 17.0 mm/s

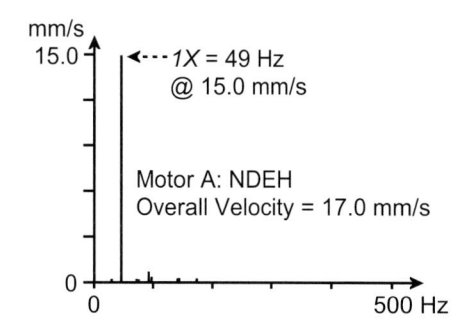

Figure 1.49 Motor A NDEH velocity spectrum.

Motor C: NDEH
Overall Velocity = 3.0 mm/s
1X = 49 Hz @ 2.0 mm/s

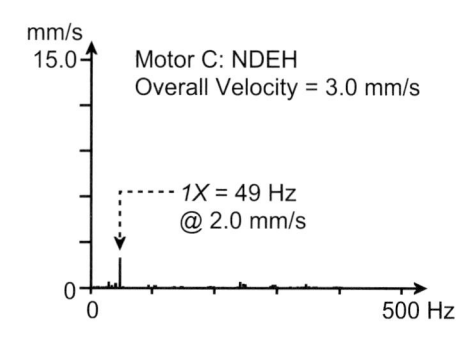

Figure 1.50 Motor C NDEH velocity spectrum.

1.3.4 Conclusions

(i) The high overall r.m.s. velocities at the DEH and NDEH positions were a result of the dominant *1X* component in each spectrum.

(ii) An increase in the mechanical imbalance in the rotor could possibly cause the high *1X* velocities at the DEH (7 mm/s r.m.s.) and NDEH (15 mm/s r.m.s.). However, this is highly unlikely in this 7.5 kW/10 H.P. SCIM because the squirrel cage winding is aluminium die cast. There are no internal cooling fins on the rotor that could be broken or lose material, since these are an integral part of the short-circuiting end ring [1.32].

(iii) There is therefore no likely reason for an increase in the mechanical imbalance of the rotor. The motor is cooled via an external fan and channels/ribs on the outer frame of the motor as shown in Figure 1.44. Therefore, an increase in the rotor's mechanical imbalance was ruled out as the cause of the high *1X* velocities. The high velocities in the horizontal direction at the DE and NDE suggest that there was an unacceptable level of misalignment between the motor and the pump.

(iv) It was subsequently confirmed by the OIM that the pump had been replaced nine months prior to a set of vibration readings being taken by the incumbent condition monitoring contractor followed by the investigation and analysis carried out by the author.

(v) The OIM reported that following the author's report the drive unit A was re-aligned, and the overall velocities were reduced to 3 mm/s and 5 mm/s at the DEH and NDEH positions thus confirming that the problem had been correctly diagnosed.

(vi) Note that optical or laser alignment instrumentation could not be used owing to the very restricted access to the drive train as shown in Figure 1.44, and re-alignment was achieved by trial and error and subsequent vibration measurements until the lowest velocities could be achieved at the DE and NDEH positions. The re-alignment and vibration measurements were carried out by on-site maintenance personnel on the offshore installation.

The reader is referred to references [1.64] to [1.66] for practical guidance on alignment procedures for drive trains.

1.4 Industrial Case History – Vibration Analysis Identified Soft Foot – Shaft Misalignment in a 110 kW/147 H.P. SCIM Pump Drive

1.4.1 Introduction

Abstract – This case history illustrates the industrial reality of an independent consultant (the author) having to deal with three different parties after a project was nominally completed but an ongoing vibration problem still existed in a SCIM drive train. The three parties were as follows:

(i) the end user
(ii) electric motor repair shop
(iii) the vibration monitoring contractor who surveyed the rotating plant once per month.

This case history will show that various problems existed such as:

- Shaft misalignment, due to *soft foot* in the motor mounting, when a repaired SCIM was coupled to the pump.
- The soft foot was caused by rusted shims under the motor's feet and a corroded mounting base.

Prior to the author's input the SCIM motor was overhauled and new bearings were fitted.

Motor nameplate information:

- 3-phase SCIM, 415 V
- 110 kW/147 H.P., 176 A, 0.88 p.f.
- 50 Hz. 2970 r/min
- DE and NDE bearings: 6314 C3 deep groove ball bearings.

1.4.2 Vibration Analysis on Repaired Motor in Repair Shop

The vibration was measured during an uncoupled run on a solid steel base-plate in the motor repair shop at the positions shown in Figure 1.51. At the NDE only one measurement was taken from a position on top of a bolt that secured the fan cowl to the NDE end frame. This ensured that there was a solid transmission path through the end frame to the bearing housing.

The overall r.m.s. velocities (highest 1.7 mm/s r.m.s.) are presented in Table 1.8 and they are perfectly normal for an overhauled motor with new bearings. These will be taken as the base-line readings during an uncoupled run on a solid base-plate. The SCIM was operating with a no-load current of 35 amperes.

The velocity spectra are shown in Figures 1.52 to 1.55.

1.4.3 On-Site Vibration Analysis during a Coupled Run of Repaired Motor

The repaired motor was returned to the LNG plant and the author emphasised that vibration measurements should be initially carried out during an uncoupled run, but the end user required the motor to be returned to continuous operation as soon as possible and therefore did not take the author's advice.

This was a constraint imposed by the end user.

The incumbent vibration condition monitoring vendor who carried out routine vibration measurements once per month subsequently reported to the end user that the overall vibration was higher than expected from a recently refurbished motor with new bearings.

1.4 Industrial Case History – Vibration Analysis Identified Soft Foot

Table 1.8 Overall r.m.s. velocities ±10% span 10–1000 Hz, 0.078 Hz/line, uncoupled run on a solid base-plate in the motor repair shop

DEV	DEH	DEA	NDE on the bolt holding down the fan cowl to the end frame
0.63 mm/s	0.4 mm/s	0.36 mm/s	1.7 mm/s

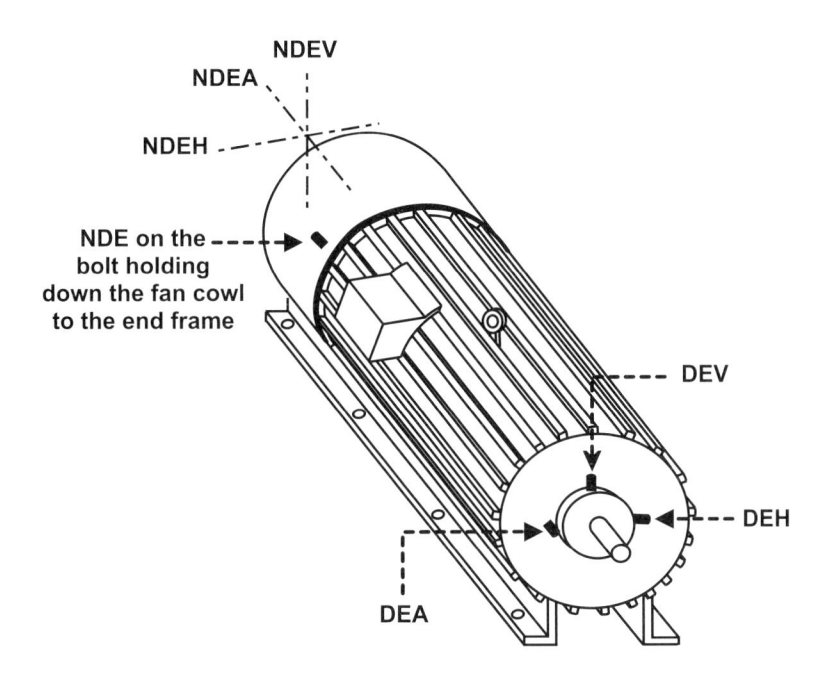

Figure 1.51 Positions of accelerometers.

DEV velocity (r.m.s.) spectrum:
- 1X = 49.98 Hz @ 0.58 mm/s
- 2X = 99.96 Hz @ 0.2 mm/s

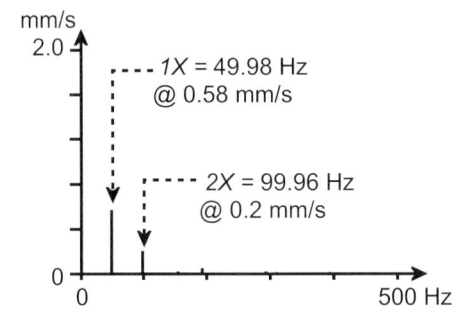

Figure 1.52 DEV velocity (r.m.s.) spectrum, uncoupled run in repair shop.

DEH velocity (r.m.s.) spectrum:
- 1X = 49.98 Hz @ 0.27 mm/s
- 2X = 99.96 Hz @ 0.1 mm/s

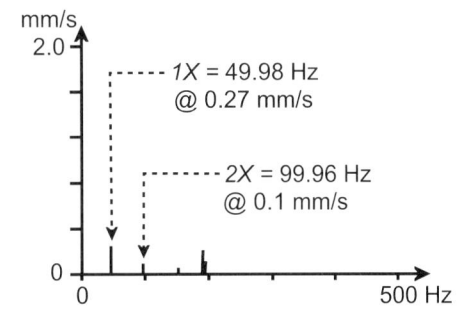

Figure 1.53 DEH velocity (r.m.s.) spectrum, uncoupled run in repair shop.

Table 1.9 Overall r.m.s. velocities ±10% span 10–1000 Hz, 12,800 lines, 0.078 Hz/line on-site coupled run

DEV	DEH	DEA	NDE on bolt holding down fan cowl
2.7 mm/s	5.4 mm/s	1.5 mm/s	2.8 mm/s

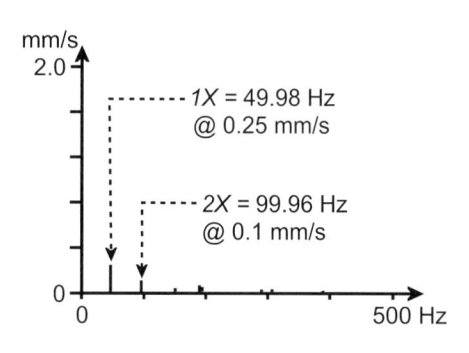

Figure 1.54 DEA velocity (r.m.s.) spectrum, uncoupled run in repair shop.

Figure 1.55 NDE vertical velocity (r.m.s.) spectrum, uncoupled run in repair shop.

The end user contacted the author to investigate the comments from the condition monitoring vendor, but no quantitative vibration readings were provided to the author. **This was a constraint imposed by the end user.**

The overall r.m.s. velocities measured by the author with the motor driving the pump with an operating current of 100 amperes, at the same positions as the results taken from the uncoupled run in the motor repair shop, are presented in Table 1.9.

1.4.3.1 Vibration Spectra – Coupled Run

Figures 1.56 to 1.59 give the velocity spectra using 12,800 lines and a frequency resolution of 0.078 Hz/line from the coupled run.

The rotor was balanced to the ISO grade of G1.0 during the overhaul. Although the overall velocities were acceptable on the SCIM, when driving a pump, it was of concern that the DEH overall velocity was 5.4 mm/s r.m.s. from a recently refurbished motor and note that this was double the magnitude of the overall r.m.s. velocity in the vertical direction.

The key features of the velocity spectra are as follows.

(i) Figure 1.56 shows that the velocity spectrum from the DEV position is dominated by the *2X* component at 2 mm/s r.m.s., which is 3.6 times the magnitude of the *1X* component. In contrast, Figure 1.57 shows that the velocity spectrum from the DEH position is dominated by the *1X* component at 4 mm/s r.m.s., which is seven times greater than the *1X* component in the DEV position.

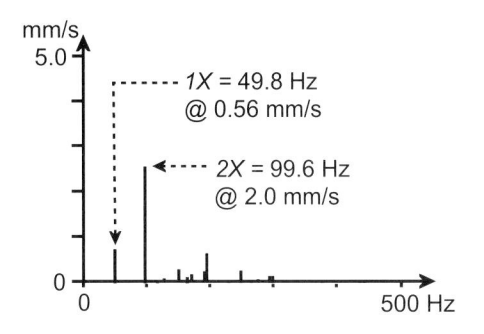

1.4 Industrial Case History – Vibration Analysis Identified Soft Foot

Figure 1.56 DEV velocity (r.m.s.) spectrum, on-site coupled run.

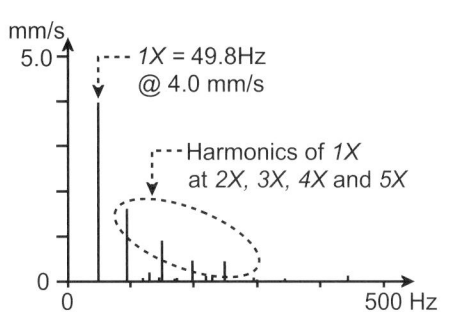

Figure 1.57 DEH velocity (r.m.s.) spectrum, on-site coupled run.

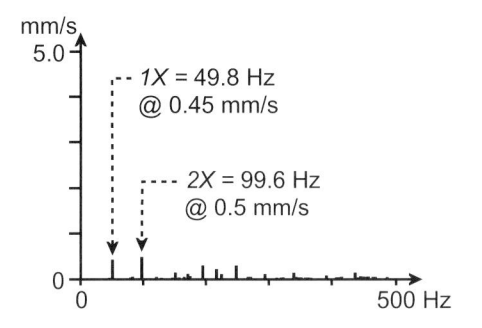

Figure 1.58 DEA velocity (r.m.s.) spectrum, on-site coupled run.

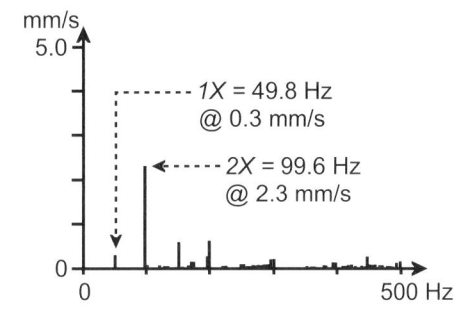

Figure 1.59 NDE vertical velocity (r.m.s.) spectrum, on-site coupled run.

(ii) There are also harmonics of $1X$ at $2X$, $3X$, $4X$ and $5X$ from the DEH position as shown in Figure 1.57.

These characteristics are not normal, particularly the $1X$ component at 4 mm/s r.m.s. in the DEH position in comparison to the much lower velocity of 0.56 mm/s r.m.s. of the $1X$ component in the DEV position because this was a very recently refurbished SCIM and the rotor was balanced to the ISO G1.0 grade.

Recall from Section 1.2, that rotor imbalance causes a centrifugal force, which is rotating at the rotor speed frequency of $1X$ Hz, and this causes the bearing housing to vibrate at that frequency, but such a large difference (by a factor of seven) between the velocities from the DEH and DEV positions is abnormal.

The author predicted that the shaft misalignment was unacceptable and was the cause of the vibration characteristics during the coupled run. He advised that the vibration measurements were repeated during an on-site uncoupled run of the SCIM for comparisons with uncoupled results from vibration measurements in the motor repair shop.

Table 1.10 Overall r.m.s. velocities ±10%, span 10–1000 Hz, 0.078 Hz/line

On-site uncoupled run and uncoupled run in the motor repair shop

Operating condition	DEV	DEH	DEA	NDE on bolt holding down the fan cowl
Uncoupled run in the motor repair shop	0.63 mm/s	**0.4 mm/s**	0.36 mm/s	1.7 mm/s
On-site uncoupled run	1.5 mm/s	**2.3 mm/s**	0.9 mm/s	2.6 mm/s

Maintenance staff at the LNG plant could or would not provide (perhaps through lack of records?) any alignment results when the repaired motor was initially re-installed, the author simply received a 'blank' response.

This was a constraint by the end user.

1.4.4 On-Site Vibration Analysis during an On-Site Uncoupled Run of Repaired Motor

Table 1.10 presents the overall r.m.s. velocities for the uncoupled runs in the repair shop and on-site.

There is a substantial difference between the velocities during uncoupled runs at the LNG plant and the motor repair shop. For example, the on-site DEH and DEV overall r.m.s. velocities were 5.75 and 2.5 times greater than those measured in the repair shop with the motor running uncoupled on a solid steel base plate. Owing to these large differences, it was predicted that there was a soft foot problem with the motor mountings on the on-site base-plate. The author observed that the base-plate was corroded, and the maintenance staff confirmed that they simply replaced the shims which were used before refurbishment of the motor. When challenged they did admit there was some rust on the shims.

1.4.5 On-Site Vibration Analysis during a Coupled Run after Re-alignment of the Drive Train

The author's recommendations were to clean up the base-plate, check the base of the motor for any signs of corrosion, fit new shims, re-align the drive train and carry out a coupled run. The motor was realigned so that the maximum shaft misalignment was 0.05 mm/2mils/0.002 inches in all planes and there was no indication of soft foot via the laser alignment measurements. The author repeated the vibration measurements at an operating current of 100 amperes for a direct comparison with velocity measurements presented from the initial coupled run. Table 1.11 presents the velocities before and after re-alignment.

Table 1.11 Overall r.m.s. velocities, span 10–1000 Hz, 0.078 Hz/line

Before and after correction of soft foot and shaft re-alignment

DEV	DEH	DEA	NDE on bolt securing fan cowl
Before 2.7 mm/s	**Before** 5.4 mm/s	Before 1.5 mm/s	**Before** 2.8 mm/s
After 1.3 mm/s	**After** 1.9 mm/s	After 0.67 mm/s	**After** 0.75 mm/s

1.4.6 Conclusions

The overall r.m.s. velocities have been reduced by factors of 2, 2.8, 2.2 and 3.7 at the DEV, DEH, DEA and NDE positions respectively after the drive train was realigned so that the misalignment was 0.05 mm/2 mils in all planes and the motor's soft foot had been removed.

Traditional vibration spectrum analysis without using phase measurements successfully diagnosed the problem.

1.5 Industrial Case History – Vibration Spectrum Analysis (VSA) Diagnosed Soft Foot and Abnormal Shaft Misalignment in a 180 kW/240 H.P. SCIM Pump Drive

1.5.1 Summary

Abstract – This case history verified that vibration analysis provided the evidence required to prove that a motor had to be removed because of soft foot due to rusted shims and a very rusty base-plate. The motor's nameplate data are as follows:

- 3-phase SCIM, 415 V, 180 kW/240 H.P., 290 A
- 50 Hz 2960 r/min, p.f. = 0.9, eff = 96%
- DE bearing: N316 C3; cylindrical roller bearing with steel cage
- NDE bearings: 6316 C3; deep groove ball bearing with steel cage.

1.5.2 Overall Vibration Measurements and Vibration Spectrum Analysis (VSA) – Motor Run Uncoupled and Coupled

Figure 1.60 shows the positions of the accelerometers and Table 1.12 presents the overall velocities during coupled and uncoupled runs.

The key features of the overall r.m.s. velocities presented in Table 1.12 with the motor running uncoupled compared to the motor driving the pump are as follows:

(i) DEV bearing housing: velocity dropped by a factor of 2
(ii) DEH bearing housing: velocity dropped by a factor of 1.8

Table 1.12 Overall r.m.s. velocity levels on the SCIM in mm/s r.m.s., to within ±10%; frequency span: 10–1000 Hz

Coupled and uncoupled runs			
Position	DEV	DEH	DEA
DE bearing housing	Coupled 3.0 mm/s Uncoupled 1.5 mm/s	Coupled 6.3 mm/s Uncoupled 3.5 mm/s	Coupled 2.7 mm/s Uncoupled 1.3 mm/s

**NDE on the bolt securing the fan cowl to the end frame.
Coupled: 10 mm/s; uncoupled: 4.7 mm/s.**

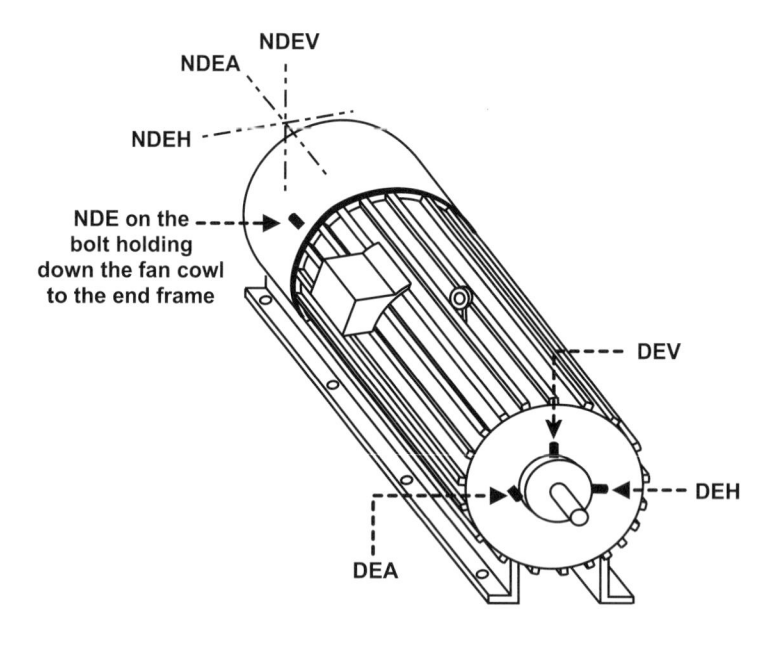

Figure 1.60 Positions of accelerometers.

(iii) DEA end frame: velocity dropped by a factor of 2
(iv) NDE end frame: velocity dropped by a factor of 2.

1.5.2.1 Vibration Spectra – Comparison between Coupled and Uncoupled Runs

Samples of vibration spectra for the DEH and NDE positions (see Figure 1.60) from the coupled and uncoupled runs are presented in Figures 1.61 to 1.64.

A comparison between Figures 1.61 and 1.62 shows that the velocities of the *1X* and *2X* components at the DEH position on the bearing housing have each dropped *by a factor of 2* during an uncoupled run compared to the coupled run.

1.5 Industrial Case History – Vibration Spectrum Analysis (VSA)

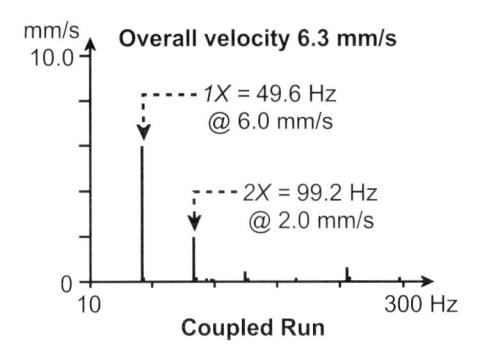

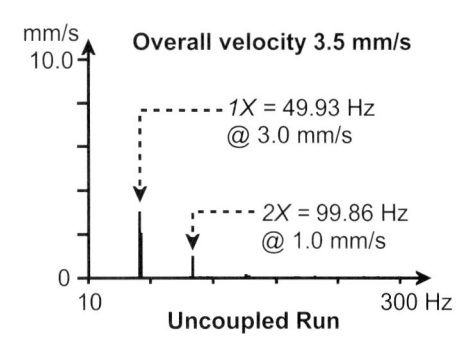

Figure 1.61 DEH bearing housing: velocity spectrum.
- Overall velocity 6.3 mm/s
- 1X = 49.6 Hz @ 6.0 mm/s
- 2X = 99.2 Hz @ 2.0 mm/s
- Coupled Run

Figure 1.62 DEH velocity (r.m.s.) spectrum.
- Overall velocity 3.5 mm/s
- 1X = 49.93 Hz @ 3.0 mm/s
- 2X = 99.86 Hz @ 1.0 mm/s
- Uncoupled Run

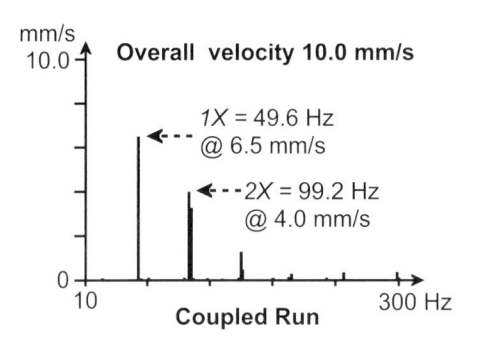

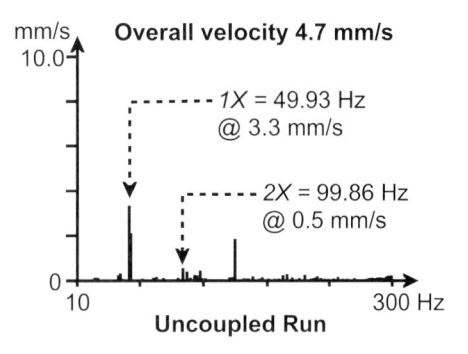

Figure 1.63 NDE velocity spectrum.
- Overall velocity 10.0 mm/s
- 1X = 49.6 Hz @ 6.5 mm/s
- 2X = 99.2 Hz @ 4.0 mm/s
- Coupled Run

Figure 1.64 NDE velocity spectrum.
- Overall velocity 4.7 mm/s
- 1X = 49.93 Hz @ 3.3 mm/s
- 2X = 99.86 Hz @ 0.5 mm/s
- Uncoupled Run

A comparison between Figures 1.63 and 1.64 shows that the velocity of the *2X* component at the NDE has *dropped by a factor of 8* during an uncoupled run compared to the coupled run.

The above comparisons indicate that there was an unacceptable level of shaft misalignment and possibly motor soft foot.

1.5.3 Overall R.M.S. Velocities on the Motor as a Function of Slackening and Tightening the Motor's Fixing Bolts During an Uncoupled Run – Soft Foot was Diagnosed

Figure 1.65 shows the positions and designations of the bolts holding down the motor. Table 1.13 presents the overall r.m.s. velocities during an uncoupled run for a range of bolt tightness conditions, starting with their tight positions and followed by combinations of tight and slack bolts.

The key observations from Table 1.13 are as follows:

(i) With B1 slackened and the other bolts tight the overall r.m.s. velocity at the DEH position dropped from 3.5 mm/s to 1.0 mm/s.

Table 1.13 Overall r.m.s. velocity levels on the SCIM in mm/s r.m.s., to within ±10%; frequency span: 10–1000 Hz

Slackened and tightened motor's fixing bolts – uncoupled run

Test condition uncoupled run	DEV	DEH	DEA	NDE on bolt as shown in Figure 1.38
Holding down bolts tight – in original positions	1.5 mm/s	3.5 mm/s	1.3 mm/s	4.7 mm/s
B1 slack, B2, B3, B4 tight	1.5 mm/s	1.0 mm/s	1.0 mm/s	3.4 mm/s
B2 slack, B1, B3, B4 tight	1.0 mm/s	3.0 mm/s	1.0 mm/s	4.8 mm/s
B3 slack, B1, B2, B4 tight	0.75 mm/s	1.3 mm/s	1.0 mm/s	2.8 mm/s
B4 slack, B1, B2, B3 tight	1.0 mm/s	3.0 mm/s	1.2 mm/s	3.38 mm/s

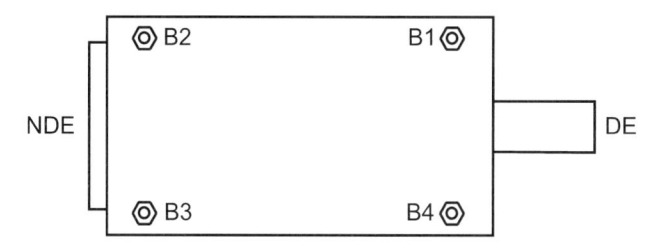

Figure 1.65 Positions of the motor's fixing bolts.

(ii) With B3 slackened and the other bolts tight the overall r.m.s. velocity at the DEH position dropped from 3.5 mm/s to 1.3 mm/s.
(iii) With B3 slackened and the other bolts tight the overall r.m.s. velocity at the DEV position dropped from 1.5 mm/s to 0.75 mm/s.
(iv) With B3 slackened and the other bolts tight the overall r.m.s. velocity at the NDE on the bolt holding down the fan cowl to the end frame dropped from 4.7 mm/s to 2.8 mm/s.
(v) The overall r.m.s. velocity at the DEA position on the end frame next to the DE bearing housing remained virtually constant at 1.3 mm/s ± 15% when the bolts were slackened and tightened in sequence.

These results provide the evidence that the motor mounting to the base had soft foot. Also, in the axial position (DEA) the velocity only varied by ±15% which further verified this prediction.

The bolts holding down the motor were sequentially slackened and tightened, and it was verified that the overall r.m.s. velocity at the drive end horizontal (DEH) position on the bearing housing dropped by a factor of 3.5 when one of the motor's fixing bolts was slackened. Similar trends occurred at the other positions and this was a clear indication that soft foot existed.

Table 1.14 Overall r.m.s. velocities on the SCIM in mm/s r.m.s., ±10%

Coupled run before and after shaft re-alignment and removal of soft foot

DEV bearing housing	DEH bearing housing	DEA DE end frame	NDE On bolt; see Figure 1.60
Before: 3.0 mm/s	**Before: 6.3 mm/s**	Before: 2.7 mm/s	**Before: 10.0 mm/s**
After: 1.3 mm/s	**After: 2.4 mm/s**	After: 1.2 mm/s	**After: 3.0 mm/s**

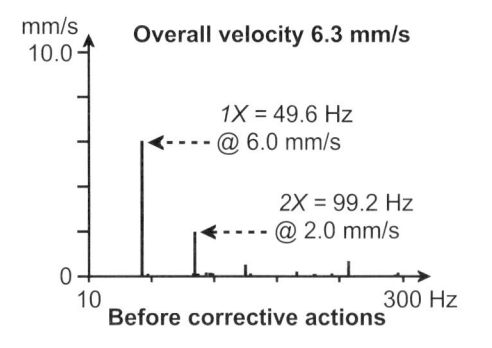

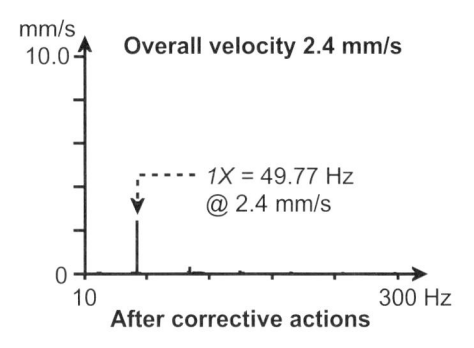

Figure 1.66 DEH bearing housing: velocity spectrum.

Overall velocity 6.3 mm/s
$1X = 49.6$ Hz @ 6.0 mm/s
$2X = 99.2$ Hz @ 2.0 mm/s
Before corrective actions

Figure 1.67 DEH velocity (r.m.s.) spectrum, on-site coupled run.

Overall velocity 2.4 mm/s
$1X = 49.77$ Hz @ 2.4 mm/s
After corrective actions

The rust was removed from the surfaces of both the base-plate and the motor's base and each of these surfaces were machined to ensure that they were clean and had flat surfaces.

New shims were installed under the motor's base and the motor and pump were coupled, then realigned using laser technology to achieve a maximum shaft misalignment in any of the three alignment planes of 0.05 mm (2 mils). No soft foot was indicated by the results presented in Table 1.14.

1.5.4 Overall Vibration Measurements and VSA – Comparison between before and after Removal of Misalignment and Soft Foot

Table 1.14 presents the overall r.m.s. velocities during a coupled run before and after shaft re-alignment and removal of motor soft foot.

The overall r.m.s. velocities have dropped by factors of between 2.3 and 3.3 at the DE and NDE positions after re-alignment and removal of motor soft foot, therefore the predictions from the measurements and VSA were proven to be valid.

Samples of velocity spectra **before and after corrective action** at the NDEH and NDE positions (see Figure 1.60) are presented in Figures 1.66 to 1.69.

The *1X* component from the original coupled run, which had a high shaft misalignment and motor soft foot has dropped due to corrective actions from

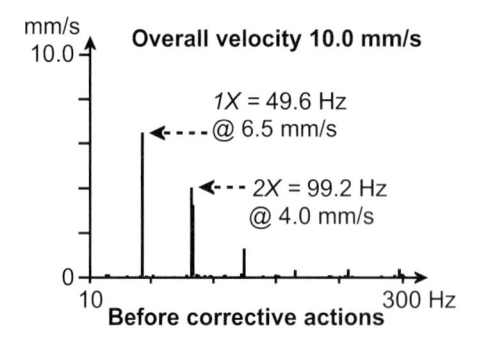

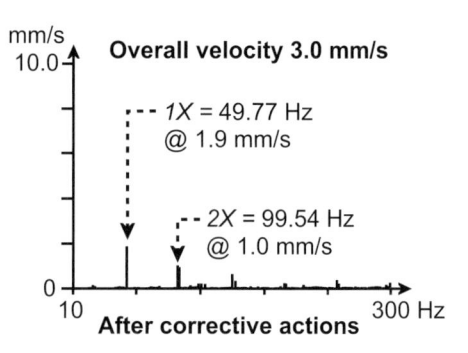

Figure 1.68 NDE velocity spectrum.

Figure 1.69 NDE velocity spectrum.

6 mm/s r.m.s. to 2.4 mm/s (a reduction factor of 2.5) and the 2X component dropped from 2 mm/s to only 0.12 mm/s r.m.s. (a massive reduction factor of 16.7).

1.5.5 Conclusions

Prior to carrying out the vibration measurements and VSA it was observed that the motor was rusty, as were the edges of the shims under the motor and the mounting base plate.

It might seem to the reader that using vibration monitoring to detect soft foot and consequential shaft misalignment was unnecessary. The motor could have simply been removed and corrective action taken to remove the rust, fit new shims, clean up the base-plate and then re-align the drive train. This is a rather naïve suggestion for the following reasons:

(i) The outer edges of the mounting shims for all the motors and base-plates were rusted, because this LNG site was next to the sea, and therefore in a salty and windy environment. The top and bottom faces of the shims are not visible.
(ii) The end user will certainly not shut down a strategic motor just because there is evidence of rust. The end user requires evidence via analysis of the vibration, while a motor is running, that soft foot and/or shaft misalignment is causing unacceptable levels of vibration.

The results presented in this case history verified that VSA was required to convince the end user to take the motor out of service for corrective action to remove the rust which was the root cause of the soft foot problem.

References and Further Reading

1.1 N. Tesla, A New System of Alternate Current Motors and Transformers, *Transactions of American Institute of Electrical Engineers*, V (10), 1888, pp. 308–27.
1.2 P. Dunsheath, *A History of Electrical Engineering*, Faber and Faber, first published in 1962.

References and Further Reading

1.3 Source: en.wikipedia.org/wiki/electricity consumption.
1.4 R. B. Randell, *Vibration-Based Condition Monitoring*, John Wiley, ISBN 978-0-470-74788-8, 2012.
1.5 A. Brant, *Noise and Vibration Analysis*, John Wiley, ISBN 978-0-470-74644-8, 2011.
1.6 R. Elsheman, *Basic Machinery Vibrations: An Introduction to Machine Testing, Analysis and Monitoring*, ASIN: B011YTFB9A, 1999.
1.7 J. I. Taylor, *The Vibration Analysis Handbook*, second edition, VCI, ISBN: 0-9640517-2-9, 2003.
1.8 J. S. Mitchell, *An Introduction to Machinery Analysis and Monitoring*, Penwell Publishing, Tulsa, Okla., 1981.
1.9 S. Timoshenko, *Vibration Problems in Engineering*, Wolfenden Press, ASIN: B0012J1C80, 4 Nov. 2008.
1.10 S. Goldman, *Vibration Spectrum Analysis – A Practical Approach*, Industrial Press Inc., ISBN-13: 978-0831130886, 1999.
1.11 A. W. Lees, *Vibration Problems in Machines – Diagnosis and Resolution*, CRC Press, ISBN-13: 978-1-138-89383-2, 2017.
1.12 J. P. Den Hartog, *Mechanical Vibrations*, fourth edition, McGraw Hill, New York, 1985.
1.13 R. A. Collacott, *Mechanical Fault Diagnosis and Condition Monitoring*, first edition, Chapman and Hall, ISBN-13: 978-94-009-5725-1, 1977.
1.14 D. J. Inman, *Engineering Vibration*, third edition, Prentice-Hall, Upper Saddle River, NJ, 2007.
1.15 R. E. D. Bishop, *The Mechanics of Vibration*, Cambridge University Press, Reissue Edition, ISBN-10: 110740245X, July 2011.
1.16 C. Scheffer and P. Girdhar, *Machinery Vibration Analysis & Predictive Maintenance*, Newnes, ISBN 0 7506 6275 1, 2004.
1.17 V. Kwok, *Machinery Vibration – Measurement and Analysis*, McGraw Hill Inc., ISBN: 0-07-071936-5, 1991.
1.18 J. Vance, F. Zeidan and B. Murphy, *Machinery Vibration and Rotor Dynamics*, John Wiley & Sons Inc., ISBN: 978-0-471-46213-2, 2010.
1.19 J. W. Cooley and J. W. Tukey, An Algorithm for Machine Calculation of Complex Fourier Series, *Mathematics of Computation*, 19 (90), 1965, pp. 297–301.
1.20 A. V. Oppenheim, *Papers on Digital Signal Processing*, MIT Press, Cambridge, Mass.
1.21 *Spectrum Analysis*: SKF.com; http://www.skf.com/binary/tcm:12-113997/CM5118%20EN%20Spectrum%20Analysis.pdf.
1.22 *Vibration Analysis Dictionary – Mobius Institute*: www.mobiusinstitute.com/site2/item.asp?LinkID=2002.
1.23 M. E. El-Hawary, *Principles of Electric Machines with Power Electronic Application*, Wiley–IEEE Press, July 2002.
1.24 M. Liwschitz-Garik and C. C. Whipple, *Electric Machinery Vol. II, A-C Machines*, Van Nostrand Company, first published Sept. 1946.
1.25 M. G. Say, *Alternating Current Machines*, fourth edition, ELBS and Pitman Publishing, 1976.
1.26 P. C. Sen, *Principle of Electrical Machines and Power Electronics*, second edition, John Wiley and Sons, 1997.
1.27 A. Chapman, *Electric Machinery Fundamentals*, McGraw Hill, 1985.
1.28 A. Hughes, *Electric Motor and Drives – Fundamentals, Types and Applications*, Butterworth-Heinemann, 1990.

1.29 G. Slemon, *Electric Machines and Drives*, Addison-Wesley Publishing Company Inc., 1992.
1.30 P. L. Alger, *Induction Machines – Their Behaviour and Uses*, Gordon and Breach Science Publications Inc., second edition, published by OPA Amsterdam, third printing with additions, 1995.
1.31 H. Vickers, *The Induction Motor*, Sir Isaac Pitman and Sons Ltd, London, first edition 1924, second edition 1953.
1.32 W. T. Thomson and I. Culbert, *Current Signature Analysis for Condition Monitoring of Cage Induction Motors*, IEEE-Press Wiley, ISBN: 978-1-119-02959-5, 2017.
1.33 P. L. Alger, Magnetic Noise in Poly-phase Induction Motors, *Transactions of the AIEE*, 73 (Part IIA), 1954, pp. 118–25.
1.34 M. J. Costello, Understanding the Vibration Forces in Induction Motors, *Proc., 19th Turbomachinery Symp.*, College Station, TX, Oct., 1989, pp. 179–83.
1.35 W. R. Findlay and R. R. Burke, Troubleshooting Motor Problems, *IEEE Transactions on Industry Applications*, 27, Nov./Dec. 1991, pp. 1204–13.
1.36 R. O. Eis, Electric Motor Vibration – Cause, Prevention and Cure, *IEEE Transactions on Industry Applications*, 1A11 (3), May/June 1975.
1.37 E. W. Sommers, Vibration in 2-pole Induction Motors Related to Slip Frequency, *IEEE Transactions on Power Applications*, 74 (3), Feb. 1955, pp. 69–72.
1.38 A. J. Ellison and C. J. Moore, Acoustic Noise and Vibration of Rotating Electric Machines, *Proceedings of the IEE*, 155 (11), Nov. 1968, pp. 1633–40.
1.39 F. T. Chapman, Production of Noise and Vibration by Certain Squirrel Cage Induction Motors, *IEE Journal*, 61, Dec. 1922, pp. 39–48.
1.40 W. R. Finley, M. M. Hodowanec and W. G. Holter, An Analytical Approach to Solving Motor Vibration Problems, *IEEE Transactions on Industry Applications*, 36 (5), Sept./Oct. 2000, pp. 1467–80.
1.41 J. Baumgardner, Vibration in Squirrel-Cage Induction Motors, *Proc. 18th Turbomachinery Symp.*, College Station, TX, Oct. 1989, pp. 179–83.
1.42 S. J. Yang, Low Noise Electric Motors, *Monographs in Electrical and Electronic Engineering*, IEE, Savoy Place, London, 1981.
1.43 W. J. Morrill, Harmonic Theory of Noise in Induction Motors, *Transactions of the AIEE*, 59 (8), Aug. 1940, pp. 474–80.
1.44 A. Arkkio, M. Antila, K. Pokki, A. Simon and E. Lantto, Electromagnetic Force on a Whirling Cage Rotor, *Proceedings of the IEE, Electrical Power Applications*, 147 (5), Sept. 2000, pp. 353–60.
1.45 A. Bonnett et al., Motor Vibes: Noise and Vibration Bibliographies and Abstracts, *IEEE Conference Record of 1993 Annual Pulp and Paper Industry Technical Conference*, 1993, pp. 184–205.
1.46 J. R. Cameron, W. T. Thomson and A. B. Dow, Vibration and Current Monitoring for Detecting Airgap Eccentricity in Large Induction Motors, *IEE Proceedings*, 133 (Part B, No. 3), May 1986.
1.47 D. G. Dorrell, W. T. Thomson and S. Roach, Analysis of Airgap Flux, Current and Vibration Signals as a Function of the Combination of Static and Dynamic Airgap Eccentricity in 3-Phase Induction Motors, *IEEE Transactions on Industry Applications*, 33 (1), Jan./Feb. 1997, pp 24–34.
1.48 W. T. Thomson, A Review of On-Line Condition Monitoring Techniques for Three-Phase Squirrel-Cage Induction Motors - Past, Present, and Future, *IEEE Symposium on*

Diagnostics for Electrical Machines, Power Electronics and Drives, Gijon, Spain, 1999, pp. 3–18 (opening keynote address).
1.49 W. T. Thomson and P. Orpin, Current and Vibration Monitoring for Fault Diagnosis and Root Cause Analysis of Induction Motors, *Proc. 31st Turbomachinery Symposium*, Texas, A&M University, USA, Sept. 2002.
1.50 W. T. Thomson, A. Barbour, C. Tassoni and F. Filippetti, An Appraisal of the MMF-Permeance Method and Finite Element Models to Study Static Airgap Eccentricity and its Diagnosis in Induction Machines, *Proc. ICEM'98*, Istanbul, 1998.
1.51 W. T. Thomson, J. R. Cameron and A. B. Dow, On-Line Diagnostics of Large Induction Motors, NATO ARW (by invitation only), Catholic University of Leuven, Belgium, August 1986, published in the NATO Api Series, Pub Martines Hijhoff, July 1988.
1.52 W. T. Thomson, N. D. Deans, R. A. Leonard and A. J. Milne, Monitoring Strategy for Discriminating Between Different Types of Rotor Cage Faults, *Proc. 18th Universities Power Engineering Conference Proceedings*, University of Surrey, April 1983.
1.53 W. T. Thomson, Vibration and Noise in Small-Power Electric Motors, MSc thesis by research, University of Strathclyde, Glasgow, Scotland, 1977.
1.54 R. A. Leonard and W. T. Thomson, Vibration and Stray Flux Monitoring for Unbalanced Supply and Inter-turn Winding Faults in Induction Motors, *Proc, Condition Monitoring Conference*, University College, Swansea, Wales, UK, 1983.
1.55 P. Tavner *et al.*, *Condition Monitoring of Rotating Electrical Machines*, The Institution of Engineering and Technology, ISBN: 978-0-86341-739-9, London, UK, 2008.
1.56 A. H. Bonnett, Root Cause Failure Analysis for AC Induction Motors in the Petroleum and Chemical Industry, *IEEE Petroleum and Chemical Industry Conference (PCIC), 2010 Record of Conference Papers Industry Applications Society 57th Annual Conference*, 2010, pp. 1–13.
1.57 A. H. Bonnett, Root Cause Methodology for Induction Motors; A Step-by-Step Guide to Examining Failure, *IEEE Industry Applications Magazine*, 18 (6), 2012, pp. 50–62.
1.58 http://new.abb.com/docs/librariesprovider53/about-downloads/motors_ebook.pdf?sfvrsn=4.
1.59 IEEE Committee Report, Report of Large Motor Reliability Survey of Industrial and Commercial Installations, *IEEE Transactions on Industry Applications*, 1A-21 (4), Parts I and II, July/Aug. 1985.
1.60 IEEE Committee Report, Report of Large Motor Reliability Survey of Industrial and Commercial Installations, *IEEE Transactions on Industry Applications*, 1A-23 (1), Part III, Jan./Feb. 1987.
1.61 O. V. Thorson and M. Dalva, A Survey of Faults on Induction Motors in Offshore Oil Industry, Petrochemical Industry, Gas Terminals and Oil Refineries, *IEEE Transactions on Industry Applications*, 31 (5), Sept./Oct. 1995.
1.62 https://www.efficientplantmag.com/2012/03/large-electric-motor-reliability-what-did-the-studies-really-say/ *Large Electric Motor Reliability: What Did The Studies Really Say?*, EP Editorial Staff 23 March 2012; *Efficient Plant Magazine* formerly *Maintenance Technology*: https://www.efficientplantmag.com/
1.63 J. I. Taylor and D. Wyndell Kirkland, *The Bearing Analysis Handbook*, ISBN: 0-9640517-3-7, Vibration Consultants, VCI USA, 2004.
1.64 J. Piotrowski, *Shaft Alignment Handbook*, third edition, CRC Press, ISBN: 13 978-1-57444-727-1, 2007.
1.65 J. S. Mitchell, *Introduction to Machinery Analysis and Monitoring*, second edition, Penwell Books, USA, ISBN 0-78714-401-3, 1993.

1.66 Pruftechnic, *A White Paper on a Practical Guide to Shaft Alignment*, distributed in the USA by Ludeca Inc. (www.ludeca.com), 2002, pp. 1–62.
1.67 ISO 1940-1:2003.
1.68 British Standard (BS) 60034-14: *Rotating Electrical Machines, Part 14 Mechanical Vibration of Certain Machines with Shaft Heights of 56 mm and Higher – Measurement, Evaluation and Limits of Vibration Severity*, Feb. 2004.
1.69 NEMA MG1: *Motors and Generators*, 2012.

Further Reading

1.70 A. H. Bonnett and G. C. Soukup, Rotor Failures in Squirrel Cage Induction Motors, *IEEE Transactions on Industry Applications*, IA-22 (6), 1986, pp. 1165–73.
1.71 A. H. Bonnett and G. C. Soukup, Analysis of Rotor Failures in Squirrel-Cage Induction Motors, *IEEE Transactions on Industry Applications*, 24 (6), 1988, pp. 1124–30.
1.72 A. H. Bonnett and G. C. Soukup, Cause and Analysis of Stator and Rotor Failures in Three-Phase Squirrel-Cage Induction Motors, *IEEE Transactions on Industry Applications*, 28 (4), 1992, pp. 921–37.
1.73 H. Barr, A. H. Bonnett and C. Yung, Understanding Vertical Motor Bearing Systems and Minimizing Their Failures, *IEEE Petroleum and Chemical Industry Conference*, 2007, pp. 1–9.

2 Rolling Element Bearings for Induction Motors

2.1 Basic Construction and Types of Rolling Element Bearings for Induction Motors

Abstract – The objectives are to describe the main types of rolling element bearings that are used in induction motors and these are as follows:

(i) To provide relevant information for Chapter 3, which is on the types of faults that occur in rolling element bearings.
(ii) To provide knowledge relevant to the 15 industrial case histories in Chapter 4.
(iii) To provide engineers responsible for the operation and maintenance of induction motors with knowledge on rolling element bearings.

This is not a book on the design, manufacture or in-depth operational characteristics of rolling element bearings; for more information, see references [2.1] to [2.12].

2.1.1 Basic Construction of a Deep Groove Ball Bearing

Rolling element bearings support the rotating rotor and keep the rotational friction to a minimum to allow smooth operation of induction motors. Figures 2.1 and 2.2 show respectively open type and sealed deep groove ball bearings.

Illustrations of different types of cages are shown in Figure 2.3 and the functions of a cage are as follows:

(i) To separate the rolling elements.
(ii) To keep the rolling elements evenly spaced for uniform load distribution.
(iii) To guide the rolling elements in the unloaded zone of the bearing.

The deep groove ball bearing is the most prevalent rolling element bearing used in induction motors – see references [2.1] and [2.2]. The term deep groove reflects the fact that the ring dimensions are relatively close to the dimensions of the balls in comparison to a cylindrical roller element bearing, which is described in Section 2.2. Different types of bearings have unique identification numbers which are universally used by bearing manufacturers. For example, the 63** series, means a single-row, deep groove ball bearing. Part numbers in this series typically range from 6300 up to 6340, which have bore diameters of 10 mm/0.393 inches and 200 mm/7.874 inches

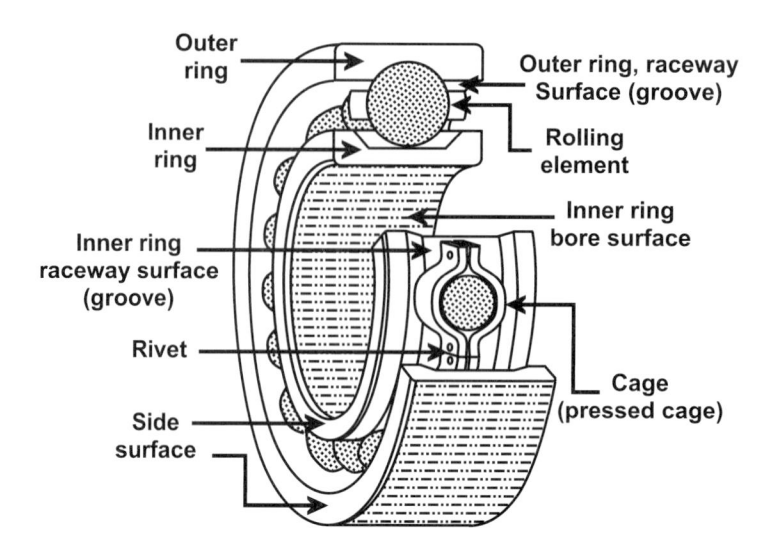

Figure 2.1 Illustration of a deep groove ball bearing.

Figure 2.2 Photograph of unshielded and shielded deep groove ball bearings, reproduced by kind permission of the NSK Corporation.

respectively. The first two numbers give the basic series for the part number, but further numbers are required to identify the bore size and clearance, C.

A deep groove ball bearing such as a 6316 C3 is defined as follows:

(i) The first two numbers give the basic series for the part number.
(ii) Multiply the pair formed by the third and fourth numbers by 5 to give the bore = 5 × 16 = 80 mm/3.15 inches.
(iii) Except with: 00 = 10 mm, 01 = 12 mm, 02 = 15 mm, 03 = 17 mm and 04 = 20 mm.

2.1 Basic Construction and Types of Rolling Element Bearings for Induction Motors

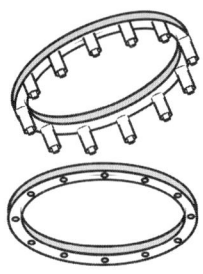

2 Piece Brass Cage

Pressed Steel Ribbon Cage

Figure 2.3 Commonly used types of cages in rolling element bearings.

(iv) C3 gives the internal clearance designation (see Section 2.1.1.2).
(v) If an M is included that means the bearing has a brass cage.

2.1.1.1 Bearing Load Zone in a Horizontal Motor

The rolling element bearings in a horizontal induction motor are normally subjected to a typical load zone as shown in Figure 2.4.

The rolling elements enter and depart the load zone and transmit the load on the rotor around the raceway as shown by the arrows. The load is normally a maximum at the 6 o'clock position and is proportional to the length of the arrows in Figure 2.4. Where there are no arrows in Figure 2.4 the rolling elements are considered to be in the unloaded zone of the bearing.

2.1.1.2 Bearing Clearances

Radial clearance is the internal radial movement of the outer ring with respect to the inner ring as shown in Figure 2.5.

Axial clearance is normally referred to as axial end-play and is the maximum relative movement of the inner ring with respect to the outer ring as shown in Figure 2.6. The inner ring is permanently fitted onto the rotor shaft of an induction motor and the outer ring is fitted into the bearing housing in the motor's end frame.

The internal radial clearances 'C' are specified from C1 (smallest) through to C5 (largest). The 'normal' clearance is CN, which is a range between C2 and C3. If the bearing clearance is not stated on the nameplate of a SCIM it is assumed to be a normal clearance. The majority of motor manufacturers normally fit a bearing with a C3 clearance in horizontal induction motors to allow for thermal expansion when the

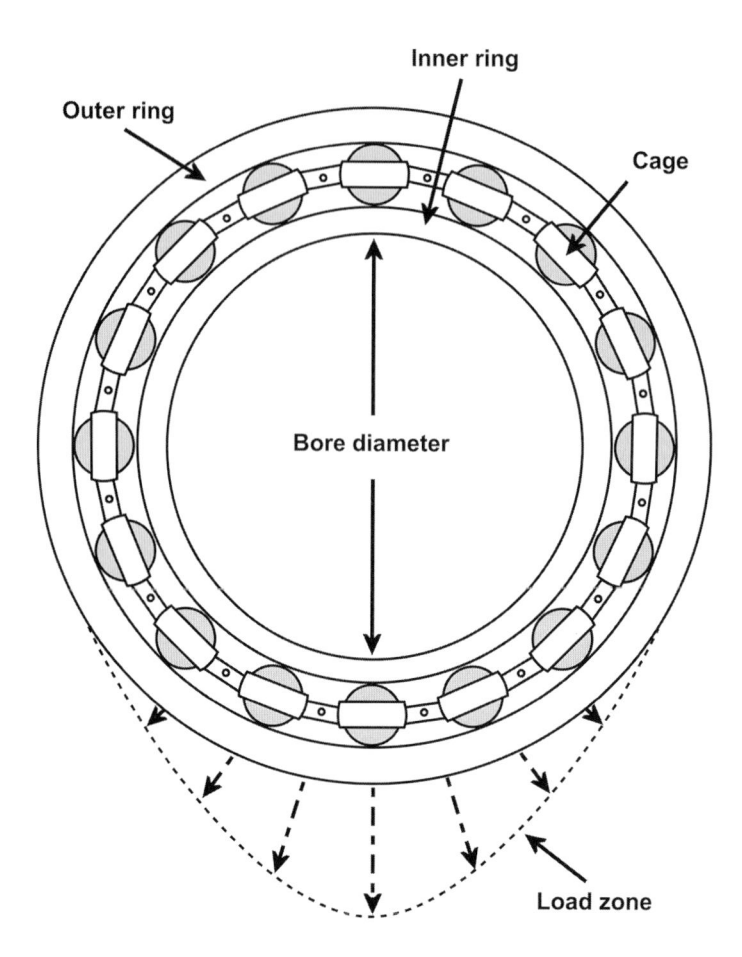

Figure 2.4. The arrows illustrate the load zone on a rolling element bearing in a horizontal machine; the lengths of the arrows are indicators of the load (stress) on those parts of the bearings.

motor reaches its operating temperature at full load. Table 2.1 presents typical radial clearances in a deep groove ball bearing.

2.1.1.3 Grease Lubrication of Rolling Element Bearings

Grease is normally used as the lubricant in a rolling element bearing and the reader is referred to a seminal textbook [2.13] titled *Grease Lubrication in Rolling Bearings* by Piet M. Lugt (John Wiley & Sons, 2003), which covers all there is to know about greasing of bearings and includes 640 references. The main function of grease is to prevent contact between the rotating parts within the bearing and this is achieved via the production of a thin oil film on the contact surfaces, and in rolling element bearings the grease has a number of additional advantages:

(i) reduces wear and friction
(ii) dissipates heat from friction

2.1 Basic Construction and Types of Rolling Element Bearings for Induction Motors 51

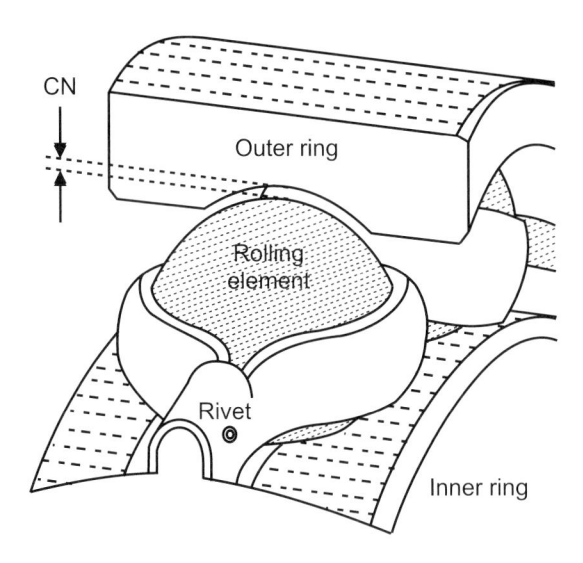

Figure 2.5 Illustration of radial bearing clearance.

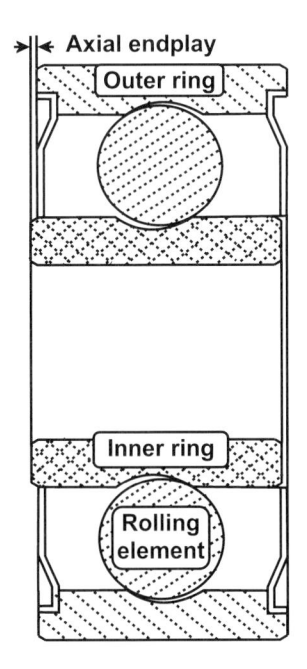

Figure 2.6 Illustration of axial/end play bearing clearance.

(iii) prevents rust
(iv) provides a form of seal between invasive particles and the bearing.

The National Lubricating Grease Institute (NLGI, founded in 1933 in the USA) [2.14], explains grease as the following:

Table 2.1 Internal radial clearances in deep groove ball bearings

CN is often referred to as the normal clearance for electric motors

Nominal bore diameter (mm)		\multicolumn{10}{c}{Clearance given is in μm (÷ 25.4 to obtain thou or mils)}
		C2		CN Normal		C3		C4		C5	
Over	Incl	min	max	min	max	min	max	min	max	min	max
10 only	0	0	7	4	11	2	13	8	23	14	29	20	37
10	18	0	9	4	11	3	18	11	25	18	33	25	45
18	24	0	10	5	12	5	20	13	28	20	36	28	48
24	30	1	11	5	12	5	20	13	28	23	41	30	53
30	40	1	11	9	17	6	20	15	33	28	46	40	64
40	50	1	11	9	17	6	23	18	36	30	51	45	73
50	65	1	15	12	22	8	28	23	43	38	61	55	90
65	80	1	15	12	22	10	30	25	51	46	71	65	105
80	100	1	18	18	30	12	36	30	58	53	84	75	120

Note: **Incl** means including.

Lubricating grease is a mixture of three main components: the lubricating fluid, performance enhancing additives and thickener.

The lubricating fluid can be petroleum-derived lubricating oil, any of various synthetic fluids or vegetable-based oil.

The additives are typically present in relatively low concentrations and are added to the grease to provide enhancement in one of multiple performance areas.

The thickener is what sets grease apart from liquid lubricants. This component gives the grease the property of consistency, making the product semi-solid rather than liquid.

The thickener is referred to as *soap* and is normally produced via a chemical reaction between a carboxylic acid and an alkaline earth metal hydroxide. Other thickeners such as polyurea and clays are also used. The purpose of a *soap* base is to retain the oil in suspension until it is removed by the rotating parts of the bearing and adheres to its surfaces. This action depletes the oil and effectiveness of the grease, which are functions of the operating hours temperature, load, speed and environment.

2.1.1.4 Main Operational Features of Deep Groove Ball Bearings

The theoretical and practical reasons for operational features can be highly complex and the reader is referred to the references. Only brief details of rolling element bearings, which are adequate for the application of vibration monitoring applied to induction motors, are presented in this book. The following characteristics of deep groove ball bearings are summarised as follows:

(i) Good radial load capability.
(ii) Axial load capability in both directions. Suitable for a locating (DE) and floating (NDE) bearing.

Table 2.2 Samples on the use of rolling element bearings in induction motors

3-Phase SCIM directly coupled to an incinerator fan – horizontal drive		
415 V, 225 kW, 369 A 1490 r/min, 50 Hz	**DE N324E M C3** **Cylindrical roller**	**NDE 6316 C3** **Deep groove ball**
3-Phase SCIM directly coupled to a reciprocating compressor in horizontal drive		
11,000 V, 270 kW, 20 A, 980 r/min, 50 Hz	**DE 6324M C3** **Deep groove ball**	**NDE 6319M C3** **Deep groove ball**
3-Phase SCIM directly coupled to a sulphate removal pump – horizontal drive		
6600 V, 685 kW, 69 A 3580 r/min, 60 Hz	**DE 6316 C3** **Deep groove ball**	**NDE 6316 C3** **Deep groove ball**
3-Phase SCIM directly coupled to a boiler feed pump – horizontal drive		
415 V, 110 kW, 176 A 2970 r/min, 50 Hz	**DE 6314 C3** **Deep groove ball**	**NDE 6314 C3** **Deep groove ball**
3-Phase SCIM directly coupled to a rich amine pump – horizontal drive		
415 V, 185 kW, 316 A 980 r/min, 50 Hz	**DE N324 C3** **Cylindrical roller**	**NDE 6316 C3** **Deep groove ball**
3-Phase SCIM directly coupled to a sea water lift pump – vertical drive		
11 kV 355 kW, 23 A 1792 r/min, 60 Hz	**DE 6324 C3** **Cylindrical roller**	**NDE 7322B** **Angular contact**
3-Phase SCIM directly coupled to a thruster propeller – vertical drive		
6.6 kV, 1139 kW, 118 A, 1186 r/min, 60 Hz	**DE 6324 C3** **Cylindrical roller**	**NDE 7322B** **Angular contact**

(iii) Limited suitability for compensation of misalignment.
(iv) Very suitable for high speeds.
(v) Dimensional stability up to typically 150 °C/302 °F (for O.D. >250 mm/9.84 inches) up to typically 200 °C/392 °F.

2.1.1.5 Typical Applications

Figure 2.7 illustrates a ball bearing arrangement in an induction motor and Table 2.2 gives examples of rolling element bearings fitted to 3-phase SCIMs.

2.2 Cylindrical Roller Element Bearings

A cylindrical roller element bearing such as an N324 C3 is defined as follows:

- The letter and first number give the basic series for the part number: in this case the N indicates the flange configuration and 3 the part number. There are different flange configurations such as NU and NJ, which means non-locating and semi-locating designs respectively.

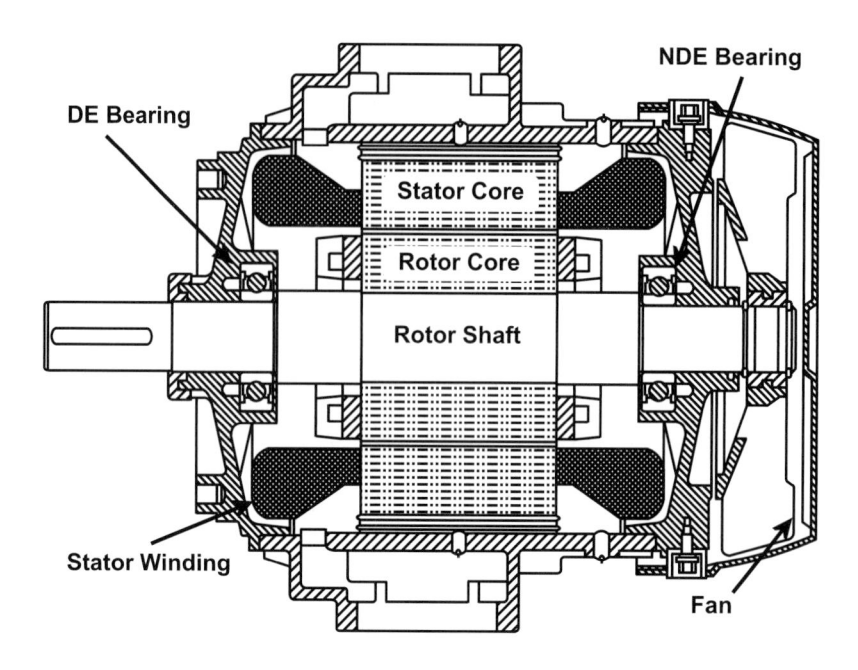

Figure 2.7 Illustration of a typical ball bearing arrangement in an induction motor.

- Multiply the pair formed by the third and fourth numbers by 5 to give the bore = 5 × 24 = 120 mm/4.724 inches.
- C3 gives the internal clearance designation.
- If an M is included this means a brass cage.
- If an E is included this means an extra capacity bearing.

Figure 2.8 illustrates the key constructional features of a cylindrical roller element bearing and a photograph is presented in Figure 2.9.

2.2.1 Main Operational Features of a Cylindrical Roller Element Bearing

The following characteristics of a single-row cylindrical roller element bearing are summarised as follows:

(i) high radial load capacity
(ii) very suitable for high speeds
(iii) high rigidity
(iv) can be selected to have axial load capability in both directions.

When an induction motor drives a mechanical load via a belt, the load on the DE bearing is greater than if it were directly coupled and a single-row cylindrical roller element bearing at the DE is used in preference to a single-row deep groove ball bearing.

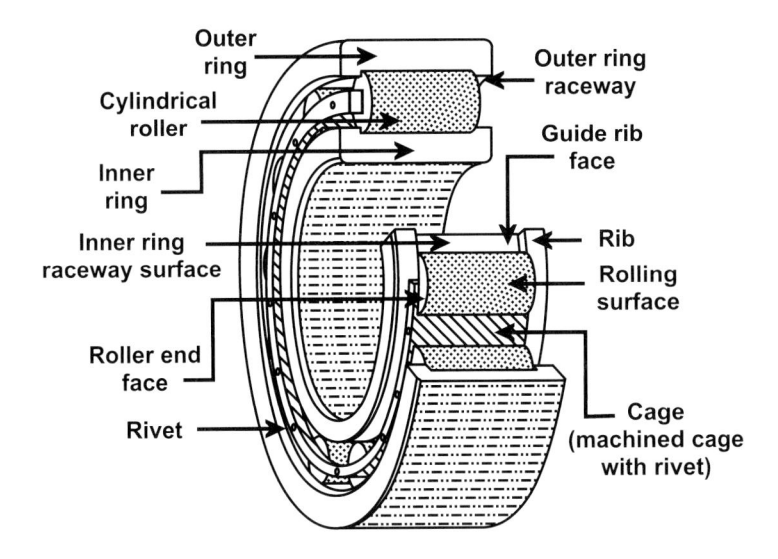

Figure 2.8 Illustration of a cylindrical roller element bearing.

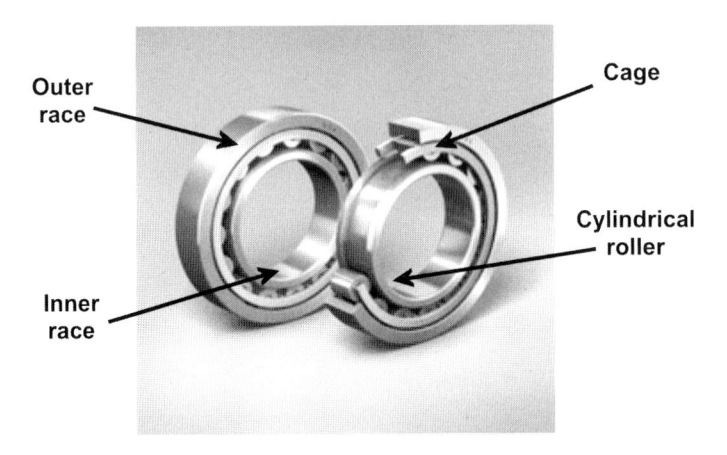

Figure 2.9 Photograph of a cylindrical roller element bearing, reproduced by kind permission of the NSK Corporation.

2.3 Angular Contact and 4-Point Contact Ball Bearings

2.3.1 Angular Contact Ball Bearings

The single-row angular contact bearing is often used at the NDE of vertically mounted induction motors to take the downward thrust and at the DE a cylindrical roller element bearing is used as a guide bearing. Figure 2.10 illustrates the construction of an angular contact ball bearing and a photograph is presented in Figure 2.11.

A bearing with a maximum contact angle up to 40° is available in contrast to a deep groove ball bearing where the contact angle (virtually a point) is nominally zero.

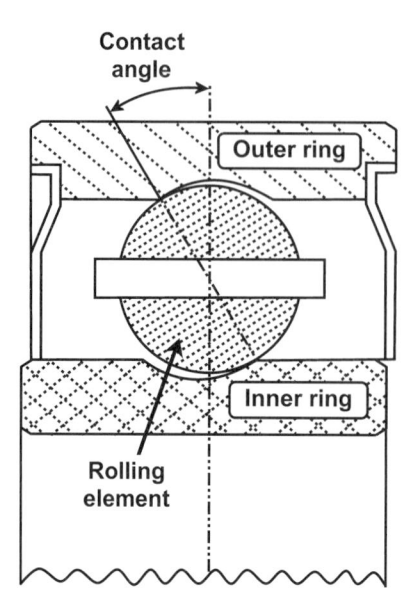

Figure 2.10 Illustration of an angular contact bearing.

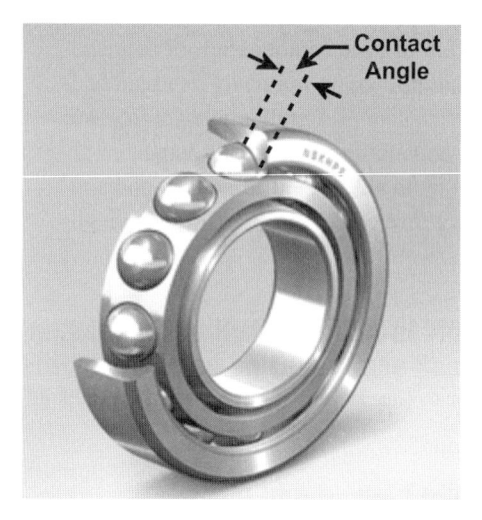

Figure 2.11 Photograph of an angular contact bearing, reproduced by kind permission of the NSK Corporation.

Clearly, for a given force, the stress (N/m^2) on the ball bearing in an angular contact bearing is much less compared to the same force applied to a deep groove ball bearing because the surface contact area in the latter is much smaller.

For example, a single-row angular contact bearing with the number 7322B is normally interpreted as follows:

2.3 Angular Contact and 4-Point Contact Ball Bearings

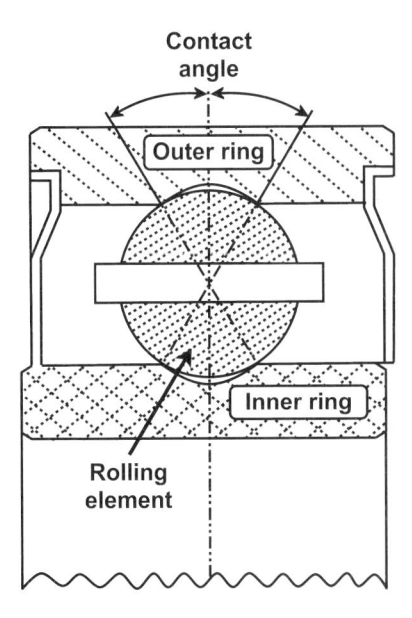

Figure 2.12 Illustration of a 4-point angular contact ball bearing.

(i) The first two numbers indicate that it is an angular contact bearing.
(ii) Multiply the pair formed by the third and fourth numbers by 5 to give the bore = $5 \times 22 = 110$ mm/4.33 inches.
(iii) An angular contact bearing often has a brass cage.

2.3.2 Main Operational Features of an Angular Contact Ball Bearing

The following characteristics of a single-row angular contact ball bearing are very briefly summarized as follows:

(i) Downward axial force (or thrust) can be supported in vertical SCIMs.
(ii) Can also support radial forces.

Two back-to-back angular contact bearings are often used in vertically mounted induction motors, which means downward and upward thrust in the drive train can be supported.

2.3.3 4-Point Angular Contact Ball Bearings – The QJ Series

A 4-point contact bearing is a novel design in that the ball path geometry results in double the number of contact points as shown in Figure 2.12 and a photograph is presented in Figure 2.12a. Note that it is a single-row ball bearing.

A typical 4-point angular contact ball bearing is a QJ316:

(i) The first two letters and the first number identify the bearing, and the pair (i.e. 16) of numbers is multiplied by 5 to give the bore diameter = 80 mm/3.15 inches.

Figure 2.12a Photograph of a 4-point angular contact bearing, reproduced by kind permission of the NSK Corporation.

The following characteristics of a 4-point angular contact ball bearing are very briefly summarized as follows:

(i) The bearing can cater for high axial loads in both directions and relatively small radial loads.
(ii) The contact angle is typically 30°.
(iii) This bearing requires considerably less space in the axial direction compared to double row designs or back-to-back angular contact 7*** bearings. The QJ bearing caters for downward and upward thrust in vertically mounted induction motor drives.

2.4   Miscellaneous Rolling Element Bearings

2.4.1   Tapered Roller Bearings

Figures 2.13a and b present schematic diagrams of tapered roller bearings with different contact angles [2.7].

This bearing has tapered rollers whose end surfaces make contact with the guide flange on the inner ring as shown in Figures 2.13a and b.

A tapered roller bearing has a high load-carrying capacity in the radial direction and in the axial direction. It is a highly durable bearing and is often used in back-to-back pairs so that axial forces can be supported in both directions. A photograph of a tapered roller element bearing is shown in Figure 2.13c.

The thrust rating capacity is dependent on the angularity of the rollers and as such the contact angle is given. For heavy thrust loads the contact angle is typically 28° to 30° [2.7].

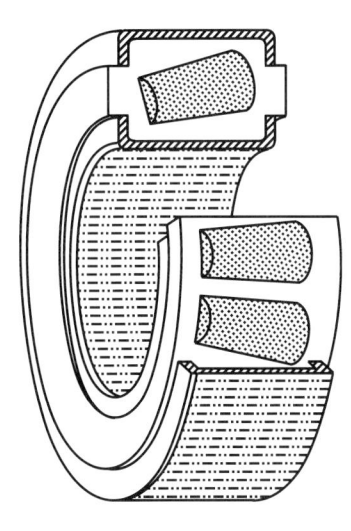

Figure 2.13a Contact angle, 10° to 16° [2.7].

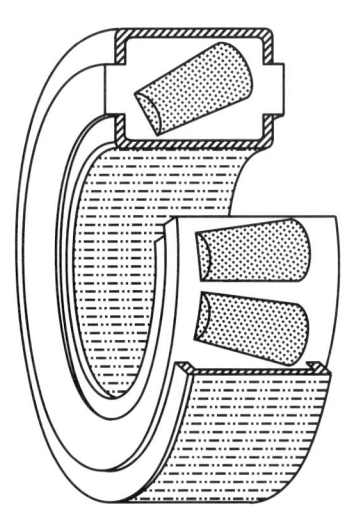

Figure 2.13b Contact angle, 28° to 30° [2.7].

2.4.2 Spherical Roller Bearings

These bearings are very suitable for induction motors which carry a heavy load and are often of the double row design as shown in Figures 2.13d and e. Both rows of rollers have a common spherical raceway on the outer ring and therefore the bearing is completely self-aligning [2.7]. The bearing has a very high load carrying capacity in both the radial and axial (thrust) directions. Because it has a unique self-alignment capability it can cope with misalignment between the shaft and bearing housing.

Figure 2.13c Photograph of a tapered roller element bearing, by kind permission of the NSK Corporation.

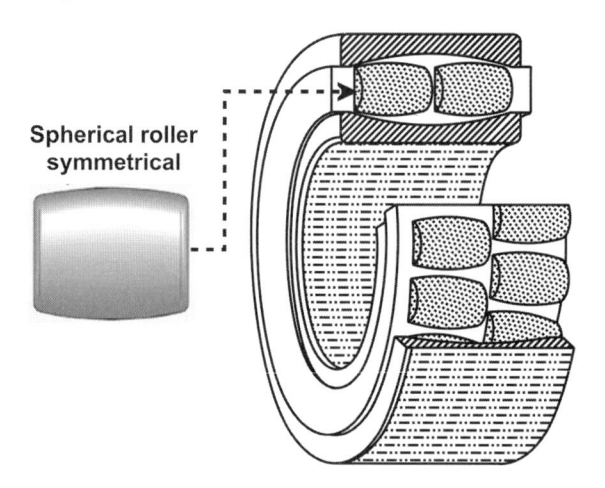

Figure 2.13d Illustration of spherical roller bearing.

Figure 2.13e Photograph of a spherical roller bearing, reproduced by kind permission of the NSK Corporation.

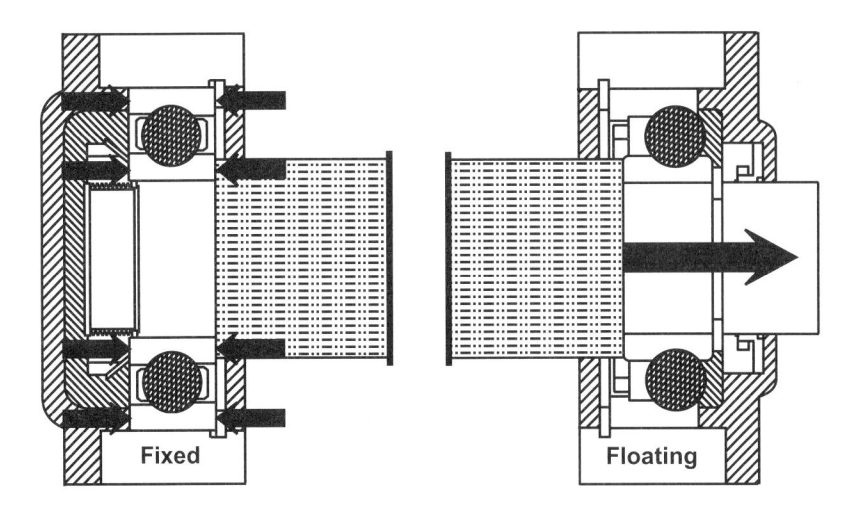

Figure 2.14 Fixed and floating bearing arrangements.

2.5 Bearing Arrangements

There are different types of bearing arrangement and only one is given as an illustration of their use as ball bearings in an induction motor. The reader should refer to the references for more examples. One of the bearings needs to be fixed and the other in this example is floating to allow for thermal expansion of the shaft–rotor assembly. The definitive functions of fixed and floating arrangements are as follows:

(i) A fixed bearing guides the shaft and rotor assembly and also supports axial forces.
(ii) A floating bearing compensates for changes in the bearing centre due to thermal growth.

Figure 2.14 illustrates fixed and floating bearing arrangements.

These are the main rolling element bearings which are used in the industrial case histories presented in Chapter 4. There is a vast selection of different types of rolling element bearings which is beyond the scope of this book, and, as already stated, this is not a book on bearings. The author considers that this chapter provides sufficient information to allow the reader to comprehend the subsequent chapters which deal with case histories on VSA to diagnose rolling element bearing faults before failures, which is the prime function of this book.

References

2.1 A. H. Bonnett and T. Albers, Motor Bearing Systems, *IEEE Industry Applications Magazine*, 8 (5), 2002, pp. 58–73.
2.2 A. H. Bonnett and T. Albers, Motor Bearing Systems for Forest Products Applications, *IEEE Conf., Record, Annual Pulp and Paper Industry Technical Conference*, 2001, pp. 96–110.

2.3 J. I. Taylor and D. Wyndell Kirkland, *The Bearing Analysis Handbook*, ISBN: 0-9640517-3-7, Vibration Consultants, VCI USA, 2004.

2.4 T. A. Harris and M. N. Kotzalas, *Essential Concepts of Bearing Technology*, fifth edition, CRC Press, ISBN: 13 978-0-8493-7183-7, 2007.

2.5 D. F. Wilcock and E. R. Booser, *Bearing Design and Application*, McGraw-Hill, New York, ASIN: B0011VRU2Q, 1957.

2.6 P. S. Haughton, *Ball and Roller Bearings*, Applied Science Publishers Ltd, London, 1985.

2.7 A. Palmgren, *Ball and Roller Bearing Engineering*, third edition, Burbank, Philadelphia, 1959.

2.8 R. K. Allen, *Roller Bearings*, Sir Isaac Pitman & Sons Ltd, London, 1945.

2.9 D. Dowson, *History of Tribology*, second edition, Longman, New York, 1999.

2.10 R. Juvinall and K. Marshek, *Fundamentals of Machine Component Design*, second edition, John Wiley, New York, 1991.

2.11 B. Hamrock, B. Jacobson and S. Schmid, *Fundamentals of Machine Elements*, McGraw-Hill, New York, 1999.

2.12 M. Spotts and T. Shoup, *Design of Machine Elements*, seventh edition, Prentice Hall, Englewood Cliffs, NJ, 1998.

2.13 P. M. Lugt, *Grease Lubrication in Roller Bearings*, John Wiley & Sons, Ltd, ISBN: 978-1-118-35391-2, 2013.

2.14 The National Lubricating Grease Institute, https://www.nlgi.org/glossary/l/

3 Types of Defects in Rolling Element Bearings

3.1 Bearing Life and Fatigue

Abstract – There are numerous publications with in-depth information on types and causes of defects in rolling element bearings, and the reader is referred to the sample in the list of references [3.1] to [3.26]. To support the industrial case histories in Chapters 4 to 8 this chapter presents a brief and general overview of the types and causes of defects which can occur in rolling element bearings. Figure 3.1 gives general guidelines [3.4] on the causes of rolling element bearing failures in induction motors.

An estimate of bearing life is highly complex and is in fact a statistical quantity when bearings are operating in industrial plant such as induction motors because degradation and failure is a function of many variables which in practice are difficult to quantify.

The International Standards Organization (ISO) publication ISO 281:2007 [3.1] covers the calculation of bearing ratings and life and the reader should refer to that standard for further information. In bearing technology, bearing life is referred to as a bearing's L_{10} *life* and in the ISO standard it is given as the life that 90% of a large group of nominally identical bearings can be expected to reach or exceed. In practice there are many variables which can affect the L_{10} *life* when nominally identical bearings are used in a wide variety of applications in industry.

3.1.1 Prediction of the L_{10} Life of a Rolling Element Bearing

A crucial factor is that the L_{10} *life* of a rolling element bearing is predicted on the assumption that normal material fatigue is the main cause of failure. Figure 3.2 is relative to the load relationship; for example, assume that L_{10} is 'x' years, then L_{50} is taken to be the average life because equal numbers of bearings have failed or survived, which is a factor of five or '5x' years [3.2]. Thus, for an industrial application with an L_{10} *life* of one year, 10% of the bearings could fail within the first year and 50% of the bearings could fail after five years.

The basic L_{10} *life* predictor is calculated using Equation 3.1 [3.2]:

64 Types of Defects in Rolling Element Bearings

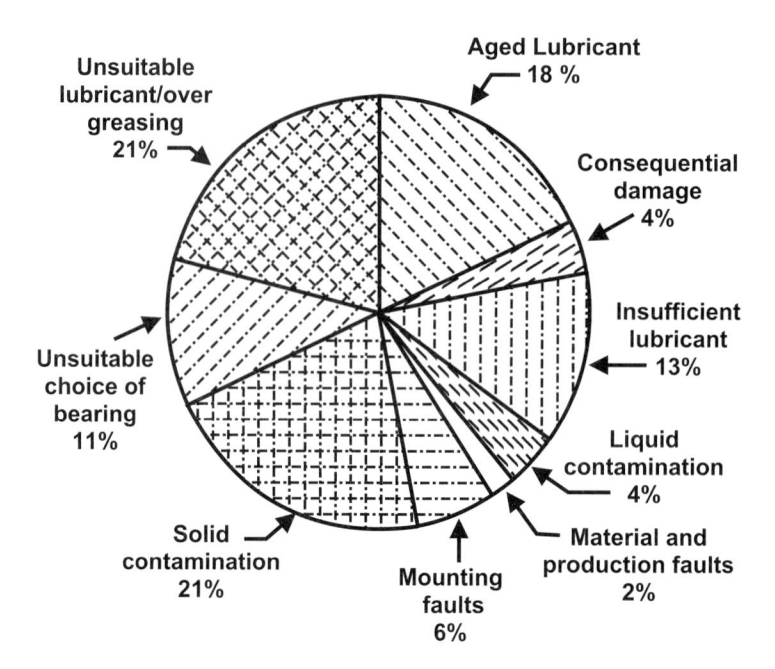

Figure 3.1 Typical causes of premature failures of roller element bearings in induction motors [3.4]. (A. H. Bonnett, Cause and Analysis of Bearing Failures in Electrical Motors, *IEEE Conf., Record of 39th PCIC Conference*, 1992, pp. 87–95.)

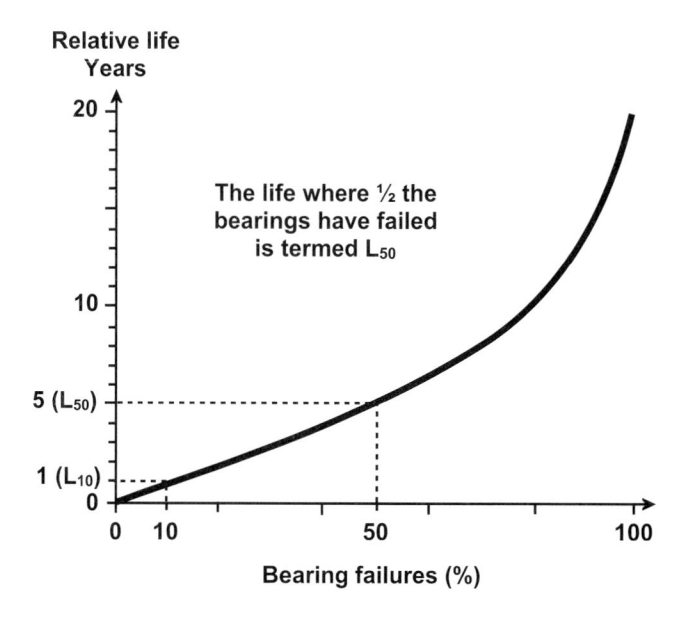

Figure 3.2 Relative bearing life versus bearing failures in % [3.2]. (A. H. Bonnett and T. Albers, Motor Bearing Systems, *IEEE Industry Applications Magazine*, 8 (5), 2002, pp. 58–73.)

$$L_{10} = \frac{10^6}{60n}\left(\frac{C}{P}\right)^a, \qquad (3.1)$$

where:

L_{10} = run life, in hours
C = bearing dynamic capacity, N or lbsf
P = equivalent bearing load, N or lbsf
n = speed, r/min
a = 3 for ball bearings
a = 10/3 for roller element bearings.

Equation 3.1 is only applicable within specific constraints such as design, size, lubrication, temperature, load and speed. Note that there is a limiting speed for each bearing, which is a function of size, type, lubrication and configuration. It is therefore obvious that the L_1 life given by Equation 3.1 is certainly not a *definitive* predictor of bearing life when they are installed in induction motors of different designs, which operate in a diverse range of industrial applications and environments with different planned maintenance strategies used by operators.

Bearing life is a function of numerous variables and an adjusted life rating can be expressed as follows:

$$L_{adj} = f(C, P, n, v_c, c_t) \qquad (3.2)$$

where:

L_{adj} = adjusted rating/run life, hours
v_c = lubrication viscosity, centiStokes or Pascal-second
c_t = level of contamination.

Note that lubrication viscosity v_c and amount of contamination c_t are interdependent and are normally expressed as a_{2-3} in Equation 3.2. It is this term which is highly variable and unpredictable to any degree of certainty.

The objective of the above information is to provide the reader with an appreciation of the factors that need to be considered with respect to bearing life in industrial plant and the requirement to select the correct grease for the application and prevent contamination.

3.1.2 Fatigue Failure

Bearings will eventually all fail by fatigue because this is the natural end to a bearing's life. Fatigue is normally defined as the fracture of the running surfaces and subsequent loss of small discrete particles of material from the rings or rolling elements and is often referred to as spalling or pitting. When fatigue has been initiated, it will spread with continued operation.

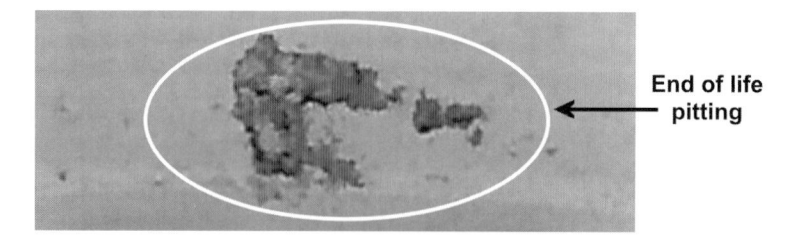

Figure 3.3 Illustration of bearing fatigue.

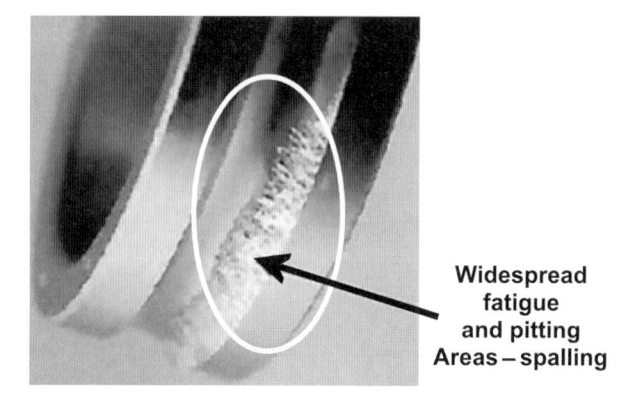

Figure 3.4 Illustration of effects of excessive loads on a bearing.

Figure 3.3 illustrates the classical case of fatigue and spalling. Fatigue usually starts in the most heavily stressed part of the bearing, for example, in the load zone, centred around 6 p.m. in a horizontal induction motor.

3.1.3 Excessive Loads

The symptoms of excessive loads on a bearing are characterized by heavy rolling element wear paths as shown in Figure 3.4 and evidence of overheating may exist.

3.1.4 False Brinelling

False brinelling in a bearing is the term that describes the result of damage to a stationary bearing caused by external vibration being transmitted to a *stationary* motor's bearings from adjacent running plant. Because the motor is stationary the external vibration leads to micro-motion between the rolling elements and raceways. There is no oil formation between the rolling elements (stationary) to prevent raceway wear when the bearing is not turning. The symptoms are often characterized by wear marks in an axial direction at each ball or roller's position in the load zone, around an arc centred at 6 p.m., as shown in Figures 3.5 and 3.6.

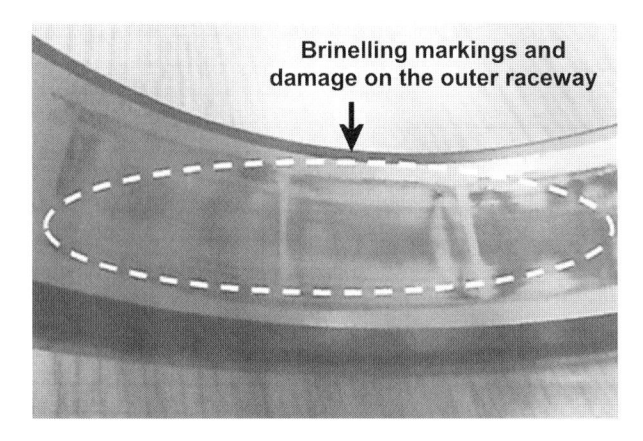

Figure 3.5 False brinelling on the outer raceway in the bearing's load zone.

Figure 3.6 Damaged rollers due to false brinelling.

3.1.5 True Brinelling

Permanent indents of a hard surface are normally referred to as true brinelling, a name derived from the Brinell scale of hardness. The symptoms of true brinelling are Brinell marks which appear as indentations in the raceways as shown in Figure 3.7, and severe Brinell marks can cause premature fatigue failure. Typical causes of true brinelling include:

(i) Static overload on the bearing.
(ii) Severe impact to the bearing, for example excessive impulse loads from rock crushers, etc.
(iii) Hammer usage to install the bearing.

3.1.6 Skidding or Slipping Tracks

'Skidding' is the sliding of the rolling elements on the inner ring caused by the bearing being lightly loaded in comparison to the bearing's dynamic loading capacity, causing

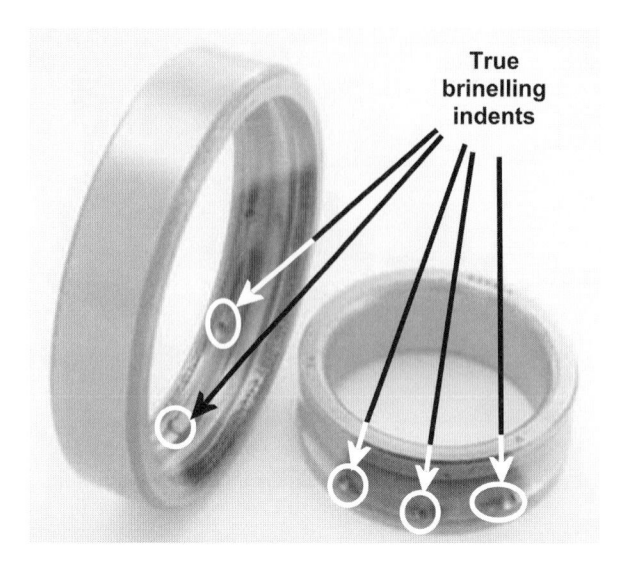

Figure 3.7 Indents caused by true brinelling.

ineffective lubrication. It is often observed (heard) when an induction motor is running uncoupled during a Factory Acceptance Test and normally disappears when the motor is installed on-site and is dynamically loaded while turning a mechanical load. Figure 3.8 illustrates the region of skidding

The symptoms of skidding are shown in Figures 3.9 and 3.10 and these include spotted smear marks or roughening of the rolling elements or raceways.

3.1.7 Contamination – Corrosion – Fretting

The symptoms are typically as follows:

(i) Red or brown staining and/or deposits on rolling elements and raceways as shown in Figure 3.11.
(ii) There can be an increase in the radial clearance.

A poor quality outer ring housing fit can result in fretting corrosion as shown in Figure 3.12. In this case the size of the load zone is revealed by the damage caused by fretting corrosion.

3.1.8 Shaft Currents

An electrical current flowing through bearings in a SCIM can cause bearings to fail – see references [3.27] to [3.31]. To prevent failures requires either insulation of the bearings to stop the flow of current or diversion of the current by grounding. Figure 3.13 is an example of fluting caused by bearing currents.

Causes of bearing currents include:

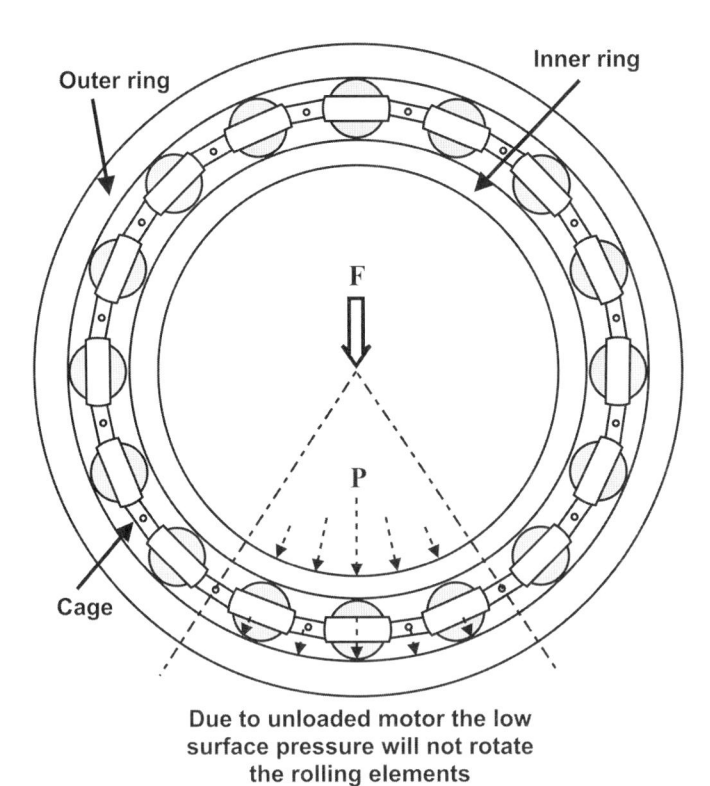

Figure 3.8 Region of skidding.

Labels in figure: Outer ring, Inner ring, Cage, F, P, "Due to unloaded motor the low surface pressure will not rotate the rolling elements"

Figure 3.9 Example of effects of skidding on rollers.

(i) If a time-varying axial flux exists along the shaft and through the bearings, a shaft voltage is generated via Faraday's law of electromagnetic induction. If the voltage is high enough to break down the oil film between the rollers and races a current can flow and, due to high temperatures, this creates small pits/fluting in the bearings. The axial flux is due to any magnetic asymmetry in the motor caused by,

Figure 3.10 Example of effects of skidding/slippage on the raceway of a cylindrical roller element bearing.

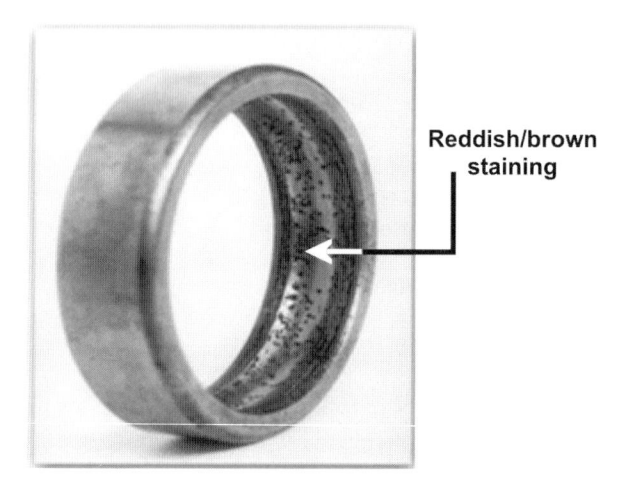

Figure 3.11 Illustration of contamination by, for example, water.

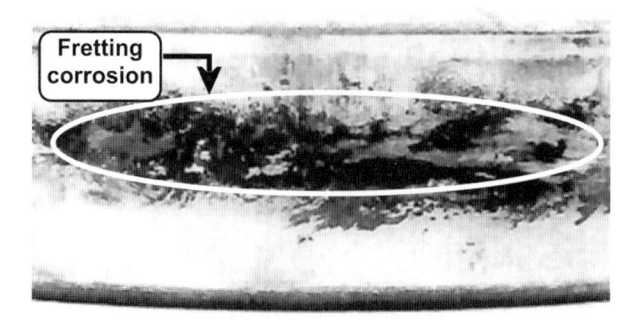

Figure 3.12 Illustration of fretting corrosion.

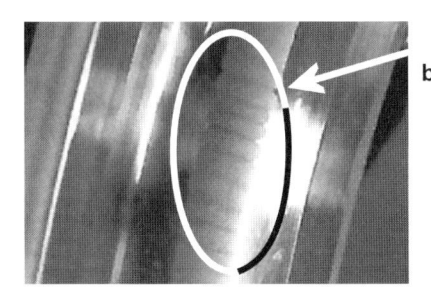

Figure 3.13 Illustration of fluting which can be caused by bearings currents.

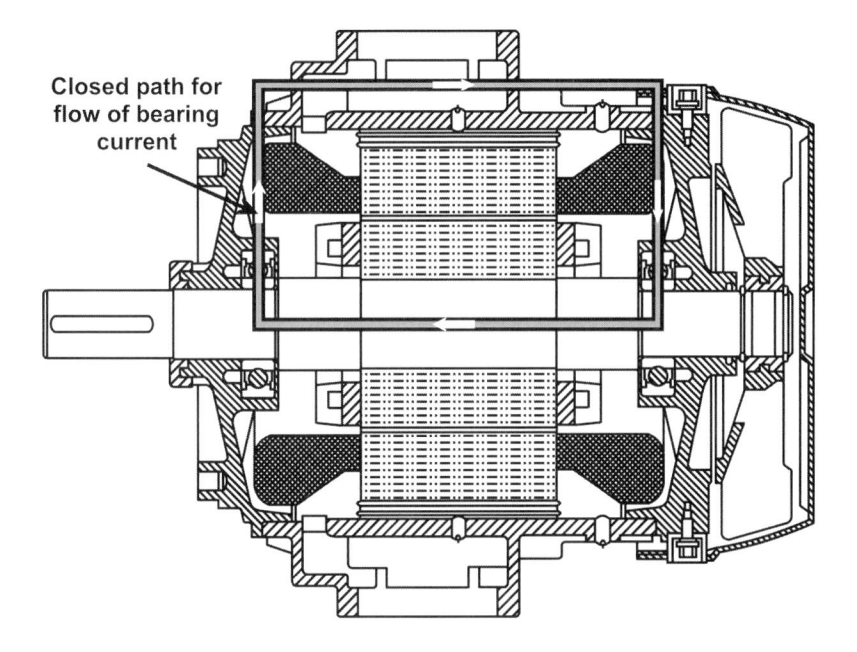

Figure 3.14 Illustration of the flow of bearing currents due to axial flux and induced shaft voltage.

for example, above normal (e.g. greater than 10%) airgap eccentricity or magnetic asymmetry in the magnetic steels used in the stator or rotor core. The asymmetry must be high enough to create an axial flux that generates a voltage which will break down the oil film. There is a closed path for the flow of current since the phenomenon is within the induction motor and not externally created.

Figure 3.14 illustrates the flow of current and insertion of insulation at the DE bearing to break that flow can solve the problem of bearing failures due to circulating bearing currents.

(ii) A different phenomenon is due to electrostatic discharges caused by friction, for example, charge is transferred from the driven mechanical load, which accumulates on the motor shaft through belts or via a shaft mounted fan operating

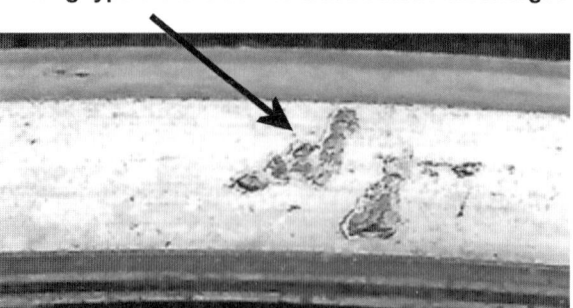

Pitting type craters due to electrostatic discharges

Figure 3.15 Electrostatic discharge currents which caused pitting as indicated by the arrow.

in very dry air. This can produce electrical discharges on the bearing surfaces and consequential pitting that can lead to failure. Electrostatic coupling can be produced from external power supplies, for example, from Pulse Width Modulated Inverters (PWMs), which have very high dv/dt.
(iii) This is inherent during switching of Gate Turn-Off (GTOs) and Insulated-Gate Bipolar Transistors (IGBTs), and is the major cause of electrical discharges. Figure 3.15 illustrates the effects of discharge current which caused pitting in a raceway.

Bearing manufacturers can now supply rolling element bearings with a ceramic layer coating around the surface of the outer ring [3.32], which can prevent bearing currents. The ceramic coating layer consists of an oxide-ceramic, which is plasma sprayed on to the outer ring and these bearings can be fitted if the end user is concerned about bearing currents, particularly in inverter-fed induction motors.

Also, shaft grounding via a carbon brush in contact with the shaft and earth can be used to divert bearing currents in large motors and generators, but in hostile environments the mechanism and contact pressure have to be regularly maintained to ensure a low resistance path to ground. Irrespective of the source of induced shaft voltage, when the value exceeds typically 500 mV r.m.s., it can produce large enough currents to permanently damage the bearings.

This chapter has presented a sample of the problems in roller element bearings, but it is emphasized that many other problems and failures can occur for example, due to:

(1) Excessive overheating caused by over greasing.
(2) Outer ring fracture.
(3) Misalignment during fitting of the bearings or excessive misalignment in the drive train, which causes severe reaction forces on the bearings.
(4) Tight bearing fits.
(5) Lubricant failure, for example due to under greasing, mixing of greases, etc.
(6) Effects of reverse loading/severe cyclic loads.

References

3.1 ISO 281:2007, *Rolling Bearings – Dynamic Load Ratings and Rating Life*, Geneva, ISO, 2007.
3.2 A. H. Bonnett and T. Albers, Motor Bearing Systems, *IEEE Industry Applications Magazine*, 8 (5), 2002, pp. 58–73.
3.3 A. H. Bonnett and T. Albers, Motor Bearing Systems for Forest Products Applications, *IEEE Conf., Record, Annual Pulp and Paper Industry Technical Conference*, 2001, pp. 96–110.
3.4 A. H. Bonnett, Cause and Analysis of Bearing Failures in Electrical Motors, *IEEE Conf., Record of 39th PCIC Conference*, 1992, pp. 87–95.
3.5 NTN Corporation, *Care and Maintenance of Bearings* http://www.ntnglobal.com/en/products/catalog/pdf/3017E.pdf.
3.6 NSK New Bearing Doctor www.mx.nsk.com/downloads/americas_bearing/BearingDoctor_for_web.pdf.
3.7 Bearing Defects and Damage http://www.zkl.cz/en/for-designers/11-bearing-defects-and-damage
3.8 Timkin Bearing Damage Analysis with Lubrication Reference Guide https://www.timken.com/pdf/5892_Bearing%20Damage%20Analysis%20Brochure.pdf.
3.9 F. R. Hutchings and P. M. Unterweiser, A Survey of the Causes of Failure of Rolling Bearings, from: *Failure Analysis: The British Engine Technical Reports*, Ed., American Society for Metals, 1981.
3.10 Z. Vintr and M. Vintr, An Assessment of Mean Time between Failures for a Group of Rolling Bearings, *IEEE Conference Record, 2011, International Conference on Quality, Risk Maintenance, and Safety Engineering*, 2011, pp. 177–81.
3.11 W. R. Finley and M. M. Howowanec, Optimal Induction Motor Bearing Selection, *Annual Record of the IEEE 2003 Annual Pulp and Paper Industry Conference*, 2003, pp. 29–41.
3.12 M. M. Howowanec, Evaluation of Antifriction Bearing Lubrication Methods on Motor Life Cycle Cost, *1996 IEEE Record of Annual Pulp and Paper Industry Conference*, 1996, pp. 170–7.
3.13 M. M. Howowanec, Satisfactory Motor Bearing Service Life: A Review of Often Overlooked Design Considerations, *IEEE Record of 1995 Annual Pulp and Paper Industry Conference*, 1995, pp. 177–184.
3.14 T. A. Harris and M. N. Kotzalas, *Essential Concepts of Bearing Technology*, fifth edition, CRC Press, ISBN: 13 978-0-8493-7183-7, 2007.
3.15 J. I. Taylor and D. Wyndell Kirkland, *The Bearing Analysis Handbook*, ISBN: 0-9640517-3-7, Vibration Consultants, VCI USA, 2004.
3.16 J. I. Taylor, *The Vibration Analysis Handbook*, second edition, VCI, ISBN: 0-9640517-2-9, 2003.
3.17 D. F. Wilcock and E. R. Booser, *Bearing Design and Application*, McGraw-Hill, New York, ASIN: B0011VRU2Q, 1957.
3.18 P. M. Lugt, *Grease Lubrication in Roller Bearings*, Wiley & Sons Ltd, ISBN:978-1-118-35391-2, 2013.
3.19 T. Tallian, Rolling Contact Failure Through Lubrication, *Journal of Institution of Mechanical Engineers, UK*, 182 (3A), 1987.
3.20 P. S. Haughton, *Ball and Roller Bearings*, Applied Science Publishers Ltd, London, 1985.

3.21 A. Palmgren, *Ball and Roller Bearing Engineering*, third edition, Burbank, Philadelphia, 1959.
3.22 R. K. Allen, *Roller Bearings*, Sir Isaac Pitman & Sons Ltd, London, 1945.
3.23 D. Dowson, *History of Tribology*, second edition, Longman, New York, 1999.
3.24 R. Juvinall and K. Marshek, *Fundamentals of Machine Component Design*, second edition, John Wiley, New York, 1991.
3.25 B. Hamrock, B. Jacobson and S. Schmid, *Fundamentals of Machine Elements*, McGraw-Hill, New York, 1999.
3.26 M. Spotts and T. Shoup, *Design of Machine Elements*, seventh edition, Prentice Hall, Englewood Cliffs, NJ, 1998.
3.27 Romero de Souza et al., Premature Wear and Recurring Bearing Failures in an Inverter-Driven Induction Motor – Part I: Investigation of the Problem, *IEEE Transactions on Industry Applications*, 51 (6), 2015, pp. 4861–7.
3.28 Romero de Souza et al., Premature Wear and Recurring Bearing Failures in an Inverter-Driven Induction Motor – Part II: The Proposed Solution, *IEEE Transactions on Industry Applications*, 51 (1), 2015, pp. 92–100.
3.29 S. Barker, Avoiding Premature Failure with Inverter Fed Induction Motors, *Power Engineering Journal*, 14 (4), 2000, pp. 182–9.
3.30 T. Hadden et al., A Review of Shaft Voltages and Bearing Currents in EV and HEV Motors, *42nd Annual Conference of the IEEE Industrial Electronics Society, IECON*, 2016, pp. 1578–83.
3.31 M. Asefi and J. Nazarzadeh, Survey on High Frequency Models of PWM Electric Dives for Shaft Voltage and Bearing Current Analysis, *IET Electrical Systems in Transportation*, 7 (3), 2017, pp. 179–89.
3.32 J. A. Oliver, G. Guerrero and J. Goldman, Ceramic Bearings for Electric Motors, *IEEE Industry Transactions Magazine*, 33 (6), 2017, pp. 14–20.

4 Introduction to Vibration Spectrum Analysis to Diagnose Faults in Rolling Element Bearings in Induction Motors

4.1 Summary

Abstract – This introductory chapter supports the application of VSA – see references [4.1] to [4.8] – which is used in the industrial case histories in Chapters 5 to 8, to diagnose the onset of faults in rolling element bearings before actual bearing failures occur. In the case histories, and whenever possible, the loop was closed between the prediction of bearing faults through to strip down of the motors and photographic evidence of the damaged bearings. Chapters 2 and 3 lead into this chapter because they cover the typical types of roller element bearings used in SCIMs and types of bearing faults.

Section 4.1.1 provides a typical example of the number of strategic LV SCIMs with rolling element bearings compared to HV, high-power, SCIMs with sleeve bearings in a large LNG processing plant. This section will also illustrate the vibration monitoring strategy applied to these motors by an end user. In Section 4.1.1.1 there is a discussion with photographic evidence of a catastrophic DE bearing failure in an LV, 185 kW/250 H.P., SCIM, culminating in the destruction of the DE end frame and a broken shaft.

The catastrophic destruction of a 185 kW/250 H.P. SCIM due to a bearing failure was a major health and safety violation and demonstrates the importance of reliable vibration monitoring to diagnose faults in rolling element bearings to prevent such a failure. Section 4.2.1 provides a brief overview of the different stages of bearing degradation and Section 4.2.2 presents the rationale for using VSA to diagnose faults in rolling element bearings. The case histories in this chapter have not been previously published.

4.1.1 Liquid Natural Gas (LNG) Processing Plant

This LNG processing plant had the following strategic SCIM drives:

- Ten 3-phase 11kV SCIMs, which have sleeve bearings with ratings from 930 kW/1247 H.P. to 2530 kW/3390 H.P., driving gas compressors and pumps.
- At the date of manufacture in 1992, these HV motors were fitted with permanent displacement probes to measure peak-to-peak shaft displacement. Alarm and trip settings set at 70 μm and 100 μm peak-to-peak respectively have prevented any

Figure 4.1 Photograph of catastrophic failure of an LV SCIM and positions of mounting studs on the DE end frame used by a condition monitoring vendor.

catastrophic bearing failures during the operational life (30 years) of these motors at the time of writing this book.
- Thirty-five strategic, LV, 415 V, SCIMs driving compressors, pumps and fans, with ratings from 75 kW/100 H.P. to 265 kW/355 H.P. These motors are fitted with rolling element bearings of different types, such as deep groove ball, cylindrical roller and angular contact bearings. None of these SCIMs has permanent accelerometers fitted for vibration monitoring.

The end user's strategy for vibration monitoring of strategic LV motors with rolling element bearings was, and still is, as follows:

(i) The end user sub-contracts a vibration monitoring vendor to carry out monthly vibration surveys of these 35 LV SCIMs using temporary accelerometers. As shown in Figure 4.1, these are mounted on permanent fixing studs and a hand-held vibration analyser collects the data.

(ii) The sub-contractor uses vibration envelope analysis to detect bearing faults, but the accelerometers used by the vendor are too large to mount on the bearing housings of any of the 35 LV SCIMs. They are therefore mounted on the outer periphery of the DE end frame as shown in Figure 4.1.

(iii) It is the author's opinion that, to reliably apply envelope analysis, the accelerometers must be mounted directly on the bearing housings, and this is also supported in reference [4.9]. A case history is presented in Section 8.2 which reports on the false diagnosis of a cage fault in a N324 C3 bearing in the DE of a SCIM by a condition monitoring (CM) sub-contractor. Envelope analysis was used, and the accelerometers were mounted on the outer periphery of the motor's flange and not on the bearing housing. The author was appointed as an

Figure 4.2 Catastrophic failure of the DE bearing.

Figure 4.3 Broken shaft, coupling spacer.

independent consultant by the oil company to carry out VSA before the motor was stripped down and after new bearings were fitted. There was no cage fault and VSA predicted the motor could have continued to run because there were no faults in the DE bearing.

4.1.1.1 Catastrophic Bearing Failure and Destruction of a SCIM

In 2006, there was a catastrophic failure of the DE bearing in one of the strategic LV SCIMs and photographs of the failed parts in the motor are shown in Figures 4.1, 4.2 and 4.3. The motor's nameplate data are as follows:

- Year of manufacture: 1992, serial number: T353Z
- Type of motor: SCIM (Squirrel Cage Induction Motor)
- 3-phase, 415 V, 185 kW, 305 A, 50 Hz, 2960 r/min, p.f. = 0.88, eff = 96%
- Connection: Δ stator winding, insulation class: F (temp. rise 80 °C)

- Rating: MC; Exn, temperature: T3; enclosure: IP55 TEFC
- DE: N316 C3 (a cylindrical roller element bearing with 13 rollers)
- NDE: 6316 C3 (a deep groove ball bearing with eight balls)
- Grease: Shell RA; greasing interval: 1800 run hours from motor data sheets.

4.2 Vibration Analysis to Diagnose Bearing Faults

4.2.1 Idealized Stages of Bearing Degradation and Vibration Analysis Techniques

Figure 4.4 illustrates the typical stages of bearing defects – see references [4.2] to [4.11] – from a very early stage (at the microscopic level), which commences with Zone A followed by B, C and D, which indicates a bearing failure is normally imminent.

4.2.1.1 Comments on the Application of Spike Energy and Envelope Analysis Techniques

The application of shock pulse measurements (SPM) or envelope analysis to diagnose the early inception of defects in rolling element bearings requires the accelerometer to be mounted directly on the bearing housing. However, the industrial reality is that in

Zone A	Zone B	Zone C	Zone D
Very early stages of defects – very high frequency content in the ultrasonic range	Early stages of defects can produce excitation pulses which can excite natural frequencies of bearing parts	Bearing defect frequencies are present, trend their magnitudes	When the spectrum is dominated by the 1X component and its multiple harmonics and the distinct bearing defect frequencies which were present in Zone C have disappeared, a serious bearing fault normally exists, and failure is imminent
Apply Spike Energy or SPM Analysis	Apply Envelope Analysis	Apply Vibration Spectrum Analysis (VSA)	
See references [4.2–4.8]	See references [4.2–4.8]	See references [4.2–4.8]	For example, see case history, Section 6.2

Decreasing Frequency, Hz ⟶

Figure 4.4 Various stages of degradation of a roller element bearing.

induction motors which have rolling element bearings the DE bearing housings may not be accessible due to practical restrictions such as the coupling guard. Furthermore at the NDE of induction motors which have external cooling fans and cowls the NDE bearing is certainly not accessible.

The author therefore does not use spike energy, SPM or envelope analysis techniques to assess the condition of rolling element bearings in induction motors.

End users are unimpressed with a false diagnosis that the bearing in a SCIM is seriously degraded when the motor repair shop subsequently reports that considerable operational life was still left in the bearings and there was negligible visual evidence of any flaws. End users do not want to have a microscopic inspection of the bearing just to prove that there are minor flaws. The end users require visual evidence that the bearing has a reduced operational life and action has been taken to prevent a catastrophic failure.

4.2.2 Vibration Spectrum Analysis (VSA) as Applied in the Industrial Case Histories

Vibration spectrum analysis (VSA) has been applied for many years and compared to the signal processing required for envelope analysis the application of VSA is easier to understand as will be verified in the case histories that follow in this chapter. This is borne out by the author's vast experience of meeting (during the past 40 years) numerous on-site maintenance personnel and mechanical and electrical engineers who are responsible for the operation, maintenance and condition monitoring of induction motors with rolling element bearings. When end users were asked if they understood the fundamental theory of envelope analysis and the associated signal processing, their answers have nearly always been:

No, with additional comments that they do not really understand the fundamental theory and signal processing used in envelope analysis.

In contrast, the end users are, in general, more comfortable with traditional VSA.

4.2.2.1 Rolling Element Bearing Defect Frequencies

The bearing defect frequencies due to bearing faults are a function of the bearing dimensions, as shown in Figure 4.5.

With radially loaded bearings the contact areas of the balls and raceways carry the largest loads hence fatigue failures normally involve these components. The ball spin frequency is produced by the rotation of each ball about its own centre. The frequency components due to an inner or outer race defect are generated when each ball passes over a defect. This occurs n_e times during a complete revolution of the raceway, where n_e is the number of rolling elements.

For completeness, the theoretical predictor equations for the bearing defect frequencies are presented in Equations 4.1 to 4.4. The reader is referred to an excellent book [4.3] that includes detailed theoretical derivations of the bearing defect frequencies. Due cognisance should be taken of the fact that these formulae are theoretical,

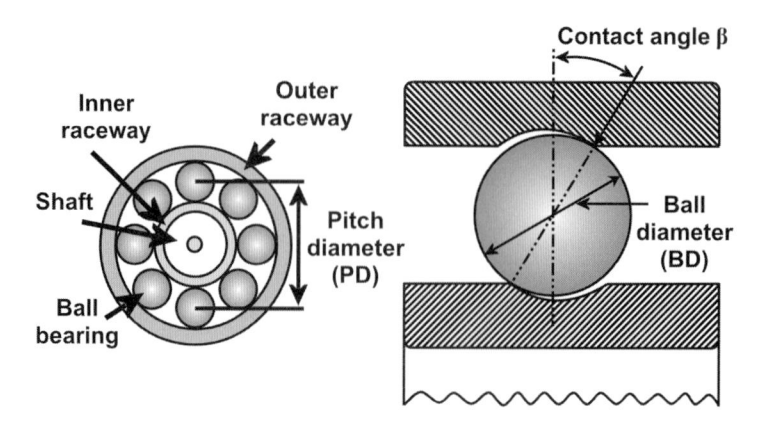

Figure 4.5 Illustrations of bearing nomenclature.

and predictions can differ from actual measurements due to for example, skidding (see Section 3.1.6) or the effects of high thrust loads.

Inner and Outer Race Defects

$$BPFI = \left(\frac{n_e}{2}\right)f_r\left[1 + \left(\frac{BD}{PD}\right)\cos(\beta)\right] \quad (4.1)$$

$$BPFO = \left(\frac{n_e}{2}\right)f_r\left[1 - \left(\frac{BD}{PD}\right)\cos(\beta)\right]; \quad (4.2)$$

$BPFI$ = ball pass frequency of the inner race and is the frequency at which the rolling elements contact a defect in the inner race,
$BPFO$ = ball pass frequency of the outer race and is the frequency at which the rolling elements make contact with a defect in the outer race.

Rolling Element Defect
A defective rolling element will be in contact with both the inner and outer raceways during each revolution hence the defect frequency will be double the spin frequency (BSF):

$$2BSF = \left(\frac{PD}{BD}\right)f_r\left[1 - \left\{\left(\frac{BD}{PD}\right)\cos(\beta)\right\}^2\right]; \quad (4.3)$$

$2BSF$ = rolling element defect frequency.

The defect on a rolling element will normally strike the outer and inner raceways during one revolution of the rolling element.

4.3 Flow Chart for Vibration Measurements

Abnormal Cage Operation

$$FTF = \left(\frac{f_r}{2}\right)\left[1 - \left(\frac{BD}{PD}\right)\cos(\beta)\right]; \tag{4.4}$$

FTF = the fundamental train frequency is the rotational speed frequency of the combined cage and rolling elements assembly. For example, FTF can be detected as a discrete frequency component due to internal looseness. It has also been reported [4.3] that FTF may modulate harmonics of $BPFI$, $BPFO$ and $2BSF$ which may show up as a difference frequency between these spectral components:
β = contact angle on raceways, degrees
PD = pitch diameter, mm
f_r = rotational speed frequency of the rotor, Hz
BD = ball diameter, mm
n_e = number of rolling elements, an integer.

4.3 Flow Chart for Vibration Measurements and VSA to Diagnose Bearing Defect Frequencies from Faulty Rolling Element Bearings in SCIMS

The reader should *refer to the preparatory guidelines* for vibration monitoring of induction motors presented in Section 1.1.3.1 and the *flow chart for vibration measurements* in Section 1.1.4.1 before proceeding to the following flow chart.

On-Site Information Gathering
Record the nameplate data directly from all motors to be tested. Determine if accelerometers can be mounted directly on the DE bearing housings (as shown in Figure 4.6) via a strong magnetic attachment. Or is only the DE end frame accessible due to the coupling guard covering the DE bearing housing?

Determine if the NDE bearing and NDE end frame are accessible. It is often the case that the fan cowl completely covers the NDE bearing and the end frame (as shown in Figure 4.6) and the cowl is bolted onto the outer periphery of the NDE end frame. The accelerometer can be mounted on this bolt, which has a direct transmission path through the end frame to the bearing housing. If permitted take photos of the motors.

⬇

Preparations for On-Site Vibration Measurements
Predict the bearing defect frequencies using reference [4.12]. Initially use the full-load rotational speed frequency from the full-load speed in r/min on the nameplate of each SCIM, which is given by: $1X = N_r/60$ Hz

(cont.)

Use the correct bearing numbers. If new bearings have been fitted in the past it may well be that extra capacity cylindrical rolling element bearings have been installed in place of standard ones.

For example, an induction motor nameplate specifies an N324 C3, but two major bearing manufacturers no longer provide an off the shelf, standard steel cage cylindrical roller bearing, which has been replaced by an extra capacity bearing N324E M C3. A steel cage version, of an N324 C3 has 12 rollers and the N324E C3 has 13 rollers. Therefore, the bearing defect frequencies will be different. If in doubt calculate the frequencies for both bearings.

⬇

On-Site Vibration Measurements

Ensure the correct accelerometer is selected so that it has a linear response in the range of the bearing defect frequencies and their harmonics and that it is securely mounted via permanent mounting studs or magnetic attachments.

The best mounting positions are directly on the bearing housings. If that is not possible then they should be mounted on the outer periphery of the DE end frame to measure the horizontal and vertical vibration.

At the NDE when a fan cowl is fitted (see Figure 4.6) the accelerometer should be mounted on a bolt securing the fan cowl. Measure, or record from an in-situ ammeter the operational current.

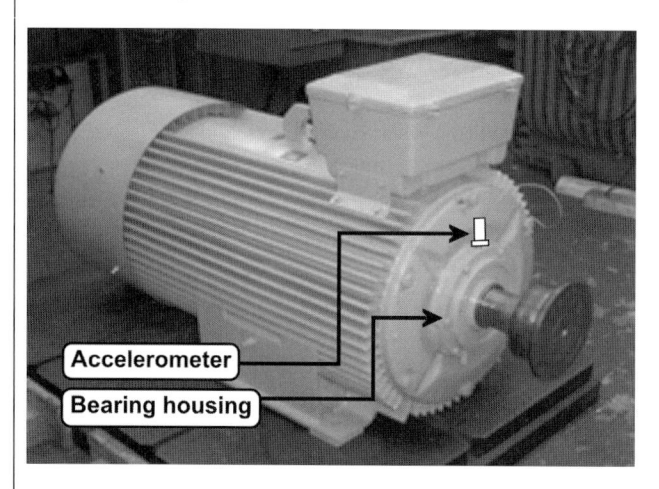

Figure 4.6 Photograph of a 415 V, 185 kW/250 H.P., SCIM.

Record the vibration velocity spectrum to accurately measure the r.m.s. velocity (or peak) and frequency of the *1X* component. If the measured *1X* component is

> (*cont.*)
>
> different from the full-load rotational speed frequency, then recalculate the bearing defect frequencies using the operational speed frequency.
>
> Measure the overall r.m.s. velocity, velocity spectra and r.m.s. velocity of each bearing defect frequency in the V, H and A directions and where possible, to be measured on the bearing housings.

4.4 Introductory Industrial Case History (1988) – Illustration of a VSA Procedure to Diagnose Bearing Defect Frequencies in SCIMs

Abstract – This is an introductory case history to illustrate a *VSA procedure*, used by the author to diagnose the bearing defect frequencies produced by faulty rolling element bearings operating in induction motors.

4.4.1 Stage One – Drive Train and Nameplate Data

Drive Train:
A SCIM driving a condensate extraction pump in an LNG processing plant has the following nameplate data:
- 3-phase, SCIM, 11 kV, 230 kW/300 H.P., 14 A
- 50 Hz, 2957 r/min, 0.9 p.f., eff = 95.8%
- DE and NDE bearings, 6317 M C3 deep groove ball bearings, M means brass cage.

4.4.2 Stage Two – Photographs of Motor and Positions of Accelerometers

The photographs presented in Figures 4.7 and 4.8 indicate where accelerometers could be mounted, and the next stage is to select a suitable accelerometer.

4.4.3 Stage Three – Select an Accelerometer

The accelerometer, which is shown in Figures 1.14, 1.15 and 1.16 in Chapter 1 with its magnetic attachment can be mounted at the positions shown in Figures 4.7 and 4.8. Its main features are:

- 100 mV/g ± 10%; linear frequency response: 2–15 kHz ± 5%
- Mounted base resonance nominally: 22 kHz ± 10%; weight: 25 grams/1.0 ounce.

4.4.4 Stage Four – Predict the Bearing Defect Frequencies at the Full-Load Rated Speed

The full-load speed from the nameplate = N_r = 2957 r/min.
Full-load slip $s_{f.l.}$ = (3000 – 2957)/3000 = 0.0143 or 1.43%.
The *1X* rotational speed frequency component = 2957/60 = 49.28 Hz.

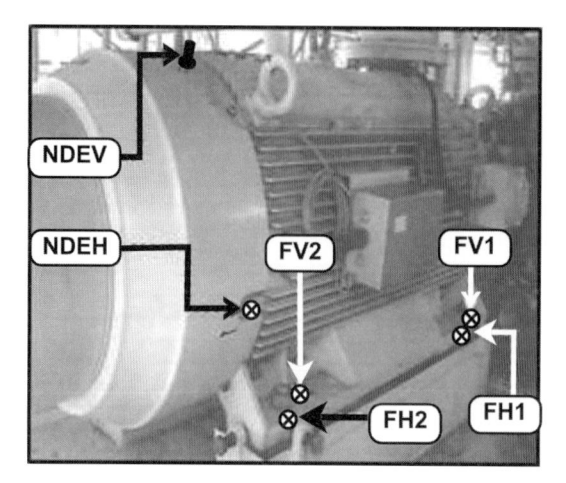

84 Introduction to Vibration Spectrum Analysis to Diagnose Faults

Figure 4.7 Positions of accelerometers on the NDE end frame and the motor's mounting base.

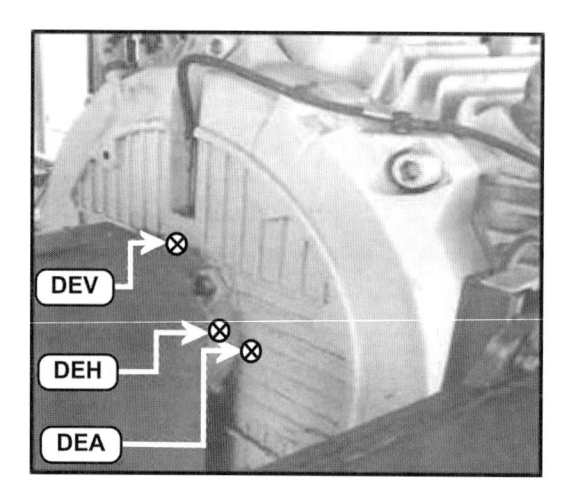

Figure 4.8 Positions of accelerometers at the DE on the DE bearing housing and end frame.

Table 4.1 presents the predicted bearing defect frequencies using the full-load nameplate speed. The predictions can be obtained via http://webtools.skf.com/BearingCalc/selectProduct.action;jsession (see reference [4.12]). Simply enter the bearing number (in this case 6317) and the shaft speed.

4.4.5 Overall R.M.S. Velocities and VSA to Detect the *1X* Frequency Component and the Bearing Frequencies using Velocity versus Frequency Spectra

First, if possible, record the motor's operating current.

Tag number D2501A = 10.8 A; full-load current = 14 amperes. The results are presented in Table 4.2 and the highest velocity was only 0.7 mm/s r.m.s. ±10% at the

4.4 Introductory Industrial Case History (1988)

Table 4.1 Predicted bearing defect frequencies for the DE and NDE bearings

Bearing type: 6317 M C3 deep groove ball bearing with a brass cage

Full-load speed: 2957 r/min (*1X* frequency component = 49.28 Hz)
BPFO: 152 Hz　　　　BPFI: 242 Hz　　　　FTF: 19 Hz　　　　2×BSF: 205 Hz

Table 4.2 Overall r.m.s. velocity levels in mm/s r.m.s., to within ±10%. Frequency span: 10–1000 Hz

Condensate export circulation pump		DM-2501A	8110004336.01/1
	Vertical mm/s r.m.s.	Horizontal mm/s r.m.s.	Axial mm/s r.m.s.
DE bearing housing	0.7	0.6	0.6
NDE end frame	0.6	0.66	Not accessible

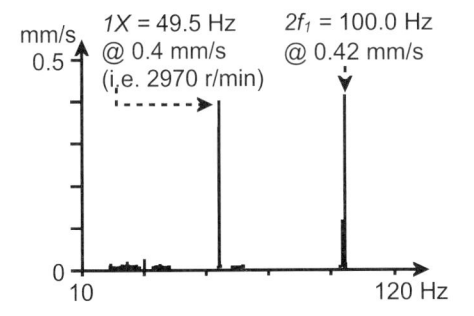

Figure 4.9 DEV velocity spectrum, 10–120 Hz, 0.0086 Hz/line.

Figure 4.10 DEV velocity spectrum, 10–500 Hz, 0.078 Hz/line.

DEV position with the motor running on load at a current of 10.8 amperes – the motor was running very smoothly indeed.

Figure 4.9 shows the velocity spectrum between 10 Hz and 120 Hz to exactly identify the *1X* frequency component which is 49.5 Hz (motor speed 2970 r/min).

The bearing defect frequencies in Table 4.1 were calculated using the full-load nameplate speed and can be corrected using the operational speed, for example *BPFO* = 152(2970/2957) = 153 Hz, this is a difference of only 0.65%. Therefore, the values in Table 4.1 can be used for the first pass VSA.

(i) Figures 4.10 to 4.14 show the r.m.s. velocity spectra between 10 to 500 Hz for the positions at the DE and NDE.
(ii) The objective was to present a VSA procedure to diagnose whether any of the bearing defect frequencies given in Table 4.1. were present.
(iii) Figure 4.11 and Figure 4.14 indicate that BPFO at 153.7 Hz (predicted at the operational speed = 153 Hz) at the DEH and NDEH positions exist but at very low velocities of 0.05 mm/s and 0.035 mm/s r.m.s. respectively.

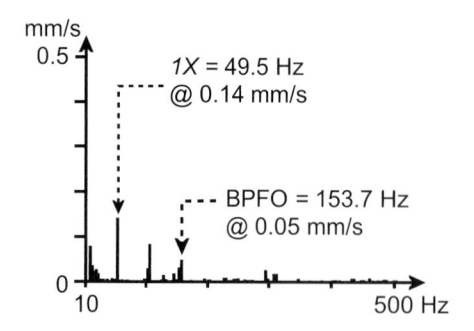

Figure 4.11 DEH velocity spectrum, 10–500 Hz, 0.078 Hz/line.

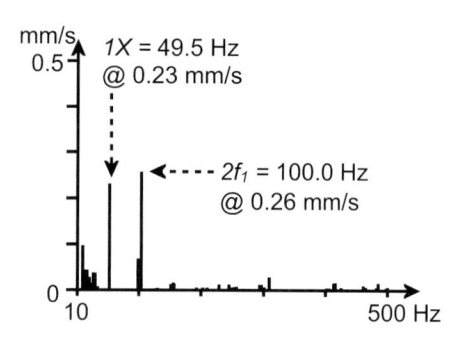

Figure 4.12 DEA velocity spectrum, 10–500 Hz, 0.078 Hz/line.

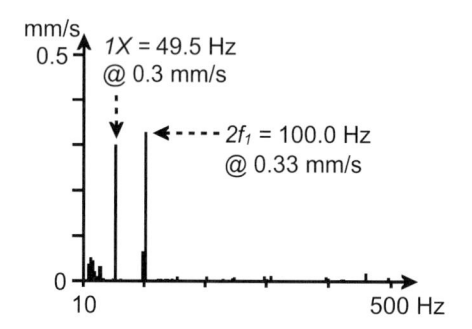

Figure 4.13 NDEV velocity spectrum, 10–500 Hz, 0.078 Hz/line.

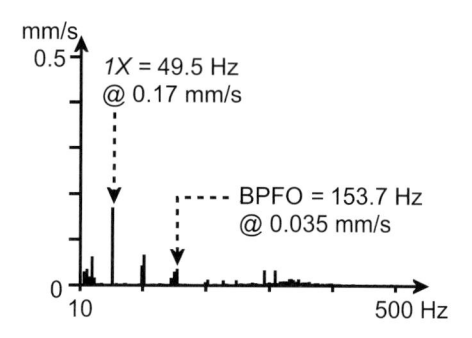

Figure 4.14 NDEH velocity spectrum, 10–500 Hz, 0.078 Hz/line.

(iv) These velocities can be trended because the results presented are the base-line measurements for this new motor.

(v) There was no evidence of any of the other bearing defect frequencies in the linear spectra presented.

4.4.5.1 VSA to Detect the Bearing Frequencies using dB versus Frequency Spectra

The dB scale is logarithmic and has a large dynamic range, for example, modern vibration analysers have the facility to display a dB spectrum, typically with a dynamic range of 80 dB, which is a factor of 10,000 in absolute velocity units. Table 4.3 gives the dB values and the factor it equates to in absolute units.

The decibel value can be calculated as

$$dB = 20 \log\left(\frac{v}{v_r}\right), \tag{4.5}$$

where:

v is the measured absolute velocity in mm/s.
v_r is the reference velocity in mm/s used to produce the dB spectrum.

Table 4.3 dB values and equivalent absolute factors

6 dB equates to a factor of 2	10 dB equates to a factor of 3.16	20 dB equates to a factor of 10
40 dB equates to a factor of 100	60 dB equates to a factor of 1000	80 dB equates to a factor of 10,000

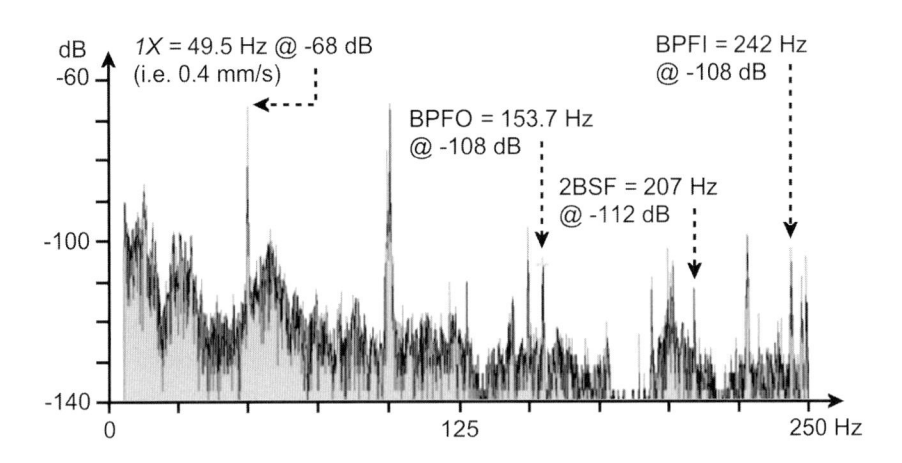

Figure 4.15 DEV vertical bearing housing dB velocity spectrum, 10–250 Hz, 0.078 Hz/line.

A dB versus frequency spectrum is now used to detect the bearing frequencies because much more information can be obtained as shown in Figure 4.15.

4.5 Conclusions

The following bearing frequencies are now very evident in the dB spectra of Figures 4.15 and 4.16, in comparison to the linear velocity spectra shown in Figures 4.10 and 4.11 which shows no bearing frequencies.

With respect to Figure 4.15:

BPFO = 153.7 Hz at −108 dB
BPFO is 40 dB smaller than the *1X* component, which means it is 100 times smaller; this is obtained using $\frac{1}{10^{-(40/20)}} = 100$
2BSF = 207 Hz at −112 dB
2BSF is 44 dB smaller than the *1X* component
BPFI = 242 Hz at 108 dB
BPFI is 40 dB smaller than the *1X* component.

The bearing frequencies are very low in magnitude, but they do exist and can be taken as base-line magnitudes at the DEV position on the bearing housing of this new motor.

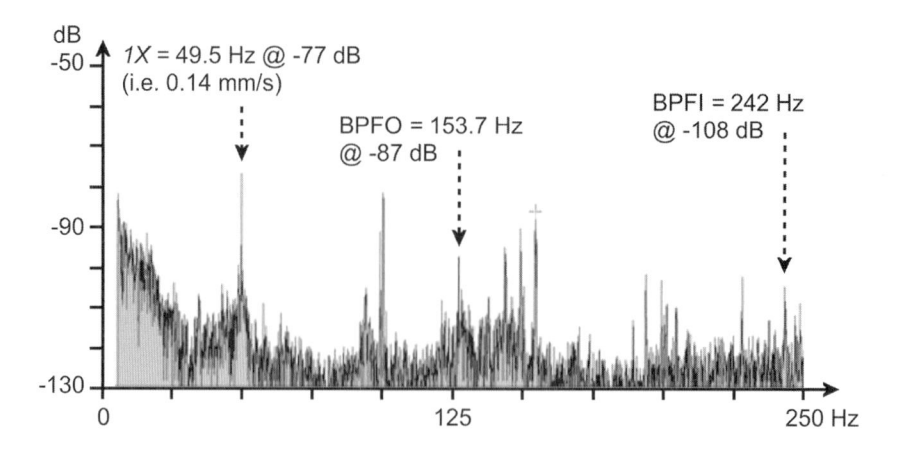

Figure 4.16 DEH horizontal bearing housing dB velocity spectrum, 10–250 Hz, 0.078 Hz/line.

(1) The dB scale is a much more powerful analysis tool compared to a linear velocity spectrum, as demonstrated in this case history, and vibration analysts are strongly encouraged to use the dB scale to obtain much more information.
(2) Unfortunately, there is a very strong reluctance by vibration analysts to use dB vibration spectra and, likewise, mechanical and electrical power engineers employed by end users find it alien.
(3) Overall vibration levels and linear spectra are normally presented in reports because vibration standards are usually given in linear absolute units of displacement, velocity or acceleration.
(4) Recall that acoustic engineers always use the dB scale and humans are perfectly happy to relate to that scale when assessing the level of acoustic noise perceived by the human ear in dBs.

References

4.1 S. A. McInerny and Y. Dai, Basic Vibration Signal Processing for Bearing Fault Detection, *IEEE Transactions on Education*, 46 (1), Feb. 2003, pp. 149–56.
4.2 *Spectrum Analysis*: SKF.com; http://www.skf.com/binary/tcm:12-113997/CM5118%20EN%20Spectrum%20Analysis.pdf
4.3 J. I. Taylor and D. Wyndell Kirkland, *The Bearing Analysis Handbook*, ISBN: 0-9640517-3-7, Vibration Consultants, VCI USA, 2004.
4.4 J. I. Taylor, *The Vibration Analysis Handbook*, second edition, Vibration Consultants, VCI, ISBN: 0-9640517-2-9, 2003.
4.5 R. B. Randell, *Vibration-Based Condition Monitoring*, John Wiley, ISBN 978-0-470-74788-8, 2012.
4.6 A. Brant, *Noise and Vibration Analysis*, John Wiley, ISBN 978-0-470-74644-8, 2011.
4.7 R. Elsheman, *Basic Machinery Vibrations: An Introduction to Machine Testing, Analysis and Monitoring*, ASIN: B011YTFB9A, 1999.

4.8 J. S. Mitchell, *An Introduction to Machinery Analysis and Monitoring*, Penwell Publishing, Tulsa, Okla., 1981.

4.9 Robert M. Jones, Enveloping for Bearing Analysis, *Sound and Vibration Journal*, 30 (2),1996, pp. 10–15.

4.10 Mig Xu, Spike Energy and its Application, *Shock and Vibration Digest*, 27 (3), May/June 1995, pp. 11–17.

4.11 T. Sundstrom, SPM White Paper: An Introduction to the SPM HD Method, pp. 1–36, SPM Instruments AB 1 Box 504 1 SE-645 25 Strangnas 1 Sweden.

4.12 NSK http://www.jp.nsk.com/app02/BearingGuide/m/html/en/BearingSearch.html.

5 Industrial Case Histories on VSA to Diagnose Cage Faults in Rolling Element Bearings of SCIMs

5.1 Introduction

Abstract – This case history reports on the application of VSA to diagnose a faulty cage and consequential damage to the balls and inner raceway in an angular contact bearing (7316 BE) at the NDE of a vertically mounted SCIM.

Photographs of the faulty bearing are presented. The fundamental cause of the defects was lack of grease caused by *not greasing* the NDE bearing at *intervals of run hours* recommended by the OEM of the motor. The end user did not have a proper quality control regime for the correct greasing of bearings in induction motors.

5.1.1 Motor Data and Positions for Vibration Measurements

The motor's nameplate provided the following data:

- 3–phase, SCIM, 415 V, 150 kW/200 H.P., 243 A
- 50 Hz, 2965 r/min, p.f. 0.89, eff. 96.5%
- Drive End (DE) bearing: NU316E M C3
- Non Drive End (NDE) bearing: 7316 BE. The NU316E M C3 is a cylindrical roller element bearing with an extra capacity rating E, a brass cage M and C3 clearance
- The 7316 BE is a single-row angular contact (of 40°) ball bearing which can sustain radial, axial or combined loads but the axial load can only be in one direction, which is downward in this case.

The SCIM drives a strategic LNG booster pump in a gas processing plant but permanent accelerometers were not installed on any of the strategic (75 kW and above), 415 V, SCIMs in the plant. Access to the NDE bearing or end frame, which contains the bearing was not possible due to the fan cowl.

It was not permitted to take photos in the LNG gas plant, but Figure 5.1 taken during a Factory Acceptance Test (FAT) illustrates where vibration measurements were taken at the NDE of the motor during on-site measurements.

The high mechanical stiffness of the end frame in the radial direction, which is outwards from the NDE bearing housing, can cause vibration from the bearing to

5.1 Introduction

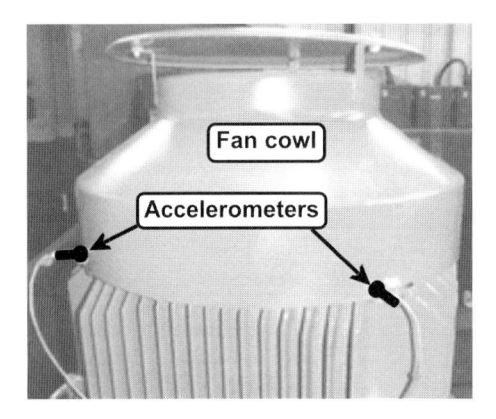

Figure 5.1 Positions of accelerometers at the NDE during a FAT, which is the same as the positions on-site.

Figure 5.2 Motor during a FAT with the fan cowl removed, which was not possible on-site.

be attenuated. Therefore, the diagnosis of the inception of bearing defects can be difficult.

Figure 5.2 shows the motor with the fan cowl removed during a FAT. The diagnosis had to be correct because the costs (approximately £10,000) to remove the motor have to cover the erection of scaffolding, provision of a crane to lift a vertical motor and the supply of mechanical and electrical personnel. This cost is normally greater than the cost to install a new set of bearings. An end user will quickly

Table 5.1 Overall r.m.s. velocities in mm/s (±10%); span 10–1000 Hz

DE$_{12}$	DE$_3$	DEA	NDE$_{12}$	NDE$_3$
Velocity r.m.s. 2.4 mm/s	Velocity r.m.s. 2.0 mm/s	Velocity r.m.s. 1.0 mm/s	Velocity r.m.s. 1.2 mm/s	Velocity r.m.s. 1.0 mm/s

Table 5.2 Predicted bearing frequencies at nominal full-load speed

NDE bearing type 7316BE
Nominal full-load speed 2965 r/min (49.42 Hz)
Contact angle: 40°
Number of balls: 16

Article I. BPFO	Article II. BPFI	Article III. FTF	2 × BSF
243 Hz	350 Hz	20.3 Hz	203.5 Hz

terminate the services of a condition monitoring company which is guilty of incorrect diagnosis.

5.1.2 Overall Velocity Levels and VSA

The overall r.m.s. levels are given in Table 5.1 and the predicted bearing defect frequencies for the NDE bearing at the full-load rated speed are presented in Table 5.2. The analysis focused on detecting the cage defect frequency (*FTF*) at the NDE$_{12}$ position and the subsequent trending of its magnitude during a period of 12 months (see Figure 5.3).

The reason was that in July 2010, the *FTF* component was evident, but no other bearing defect frequencies existed. The motor was not always running during the monthly vibration surveys.

The predicted *FTF* component was at the full-load speed of 2965 r/min because that is the nameplate speed and predictions were carried out before going on-site. The operational speed was not known until the measurements were analysed. The motor was not on full load and was operating on a reduced load, thus the speed was higher and from Figure 5.4 the *1X* rotational speed frequency component was 49.844 Hz, which is a rotor speed of 2990 r/min.

The predicted *FTF* was corrected for the higher speed, which is equal to 20.3(2990/2965) = 20.64 Hz. Interpretation of Figure 5.4 indicates there is a component at 20.47 Hz, which differs by only 0.8% from the predicted value of FTF at the operational speed.

It was concluded that this was the *FTF* at a magnitude of 0.58 mm/s, which is 84% of the *1X* component and 53% of the overall r.m.s. velocity. The severity of the cage

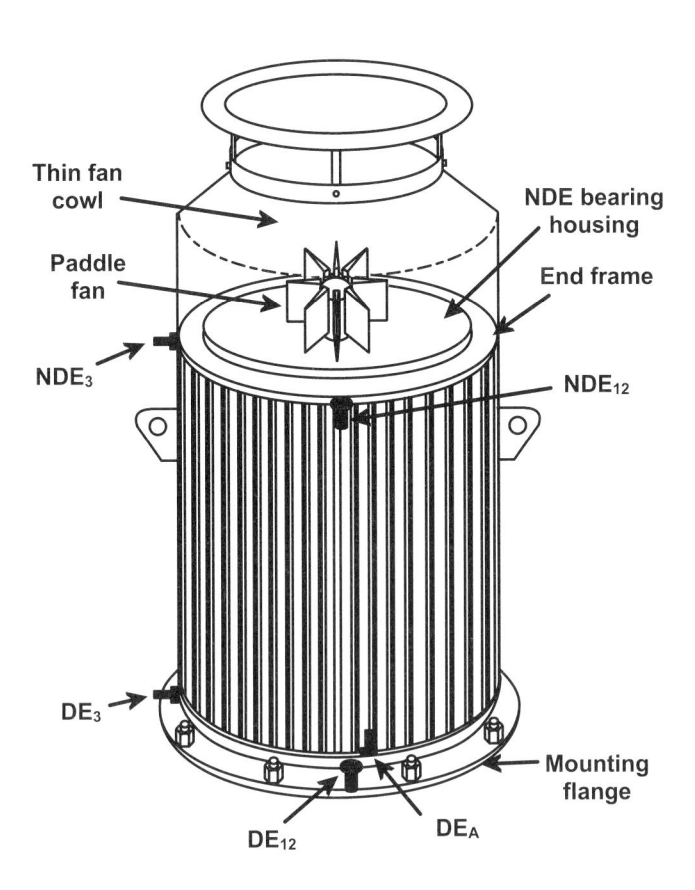

Figure 5.3 Schematic illustration of the motor to show the positions of accelerometers during on-site vibration measurements.

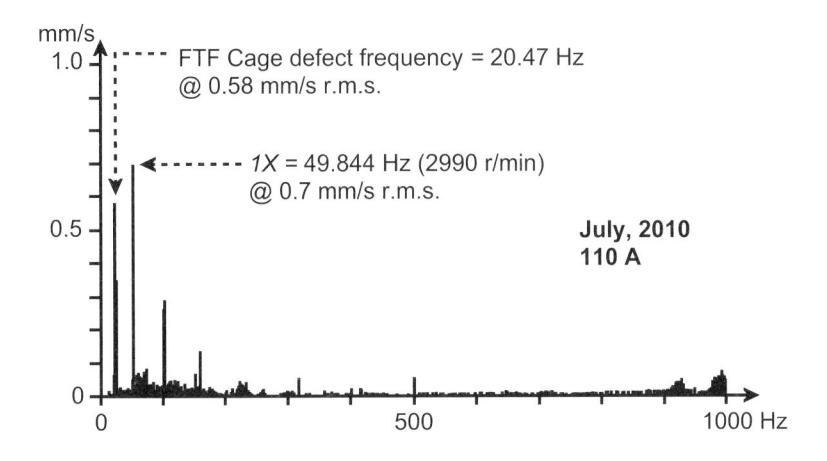

Figure 5.4 Velocity r.m.s. versus frequency for position NDE$_{12}$, 0.078 Hz/line.

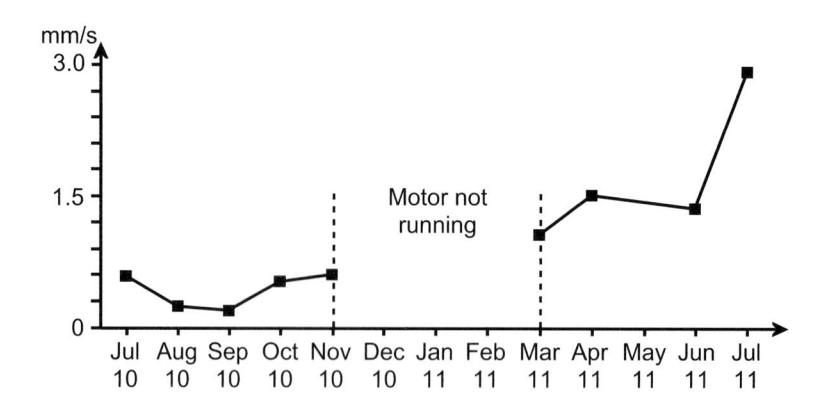

Figure 5.5 Trend graph of magnitude of cage defect frequency versus a 12 month period from July 2010 to July 2011.

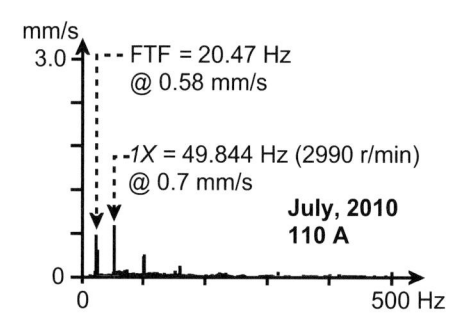

Figure 5.6a Velocity spectrum for position NDE$_{12}$, 0.078 Hz/line.

Figure 5.6 Velocity spectrum for position NDE$_{12}$, 0.078 Hz/line.

fault could not be estimated from a one-off measurement because the vibration was measured on the outer periphery of the NDE end frame.

Figure 5.5 presents the results of trending the *FTF* component during a year. Between March and July 2011 there was a doubling of the *FTF* component from 1.4 mm/s to 2.9 mm/s, and the rate of rise of *FTF* with respect to time is obvious in the plot.

The vibration spectra recorded in July 2010 and July 2011 are presented in Figures 5.6a and 5.6 to provide a direct comparison. The *FTF* component has increased by a factor of 5 in 12 months.

The author recommended that the motor should be shut down and removed from service as soon as possible. It was predicted that there was a cage defect and possibly secondary damage to the raceways.

5.1.3 Inspection of the NDE Bearing

The photographs shown in Figures 5.7 to 5.14 of Appendix 5A provide evidence to justify the following conclusions:

(i) The angular contact bearing was operating with no effective grease lubrication because the grease residue was in the form of consolidated soap.
(ii) The root cause of the faults in the bearing was lack of grease.
(iii) There was substantial wear in the cage pockets and inside diameter of the brass cage. Particles of brass debris from the damaged cage had accelerated the wear process and reduced the operating life.
(iv) There was also consequential damage to balls and the inner raceway.
(v) The remaining run life was dramatically reduced. If the motor had not been stopped, a failure was imminent and there was potential for a bearing collapse and a rotor to stator rub.

5.1.4 Conclusions

- Because the NDE bearing housing was inaccessible, the vibration in the V and H directions was measured on the outer periphery of the end frame. Therefore the vibration was attenuated and the cage frequency component was relatively small (0.58 mm/s) and thus the initial severity of the fault was masked.
- The magnitude of the *FTF* component was trended over a period of 12 months to determine this component's rate of rise, which was 2 mm/s r.m.s. during the last month. This was unacceptable and the motor was removed from service to prevent a catastrophic bearing failure.
- The end user was strongly advised to install permanent accelerometers directly on the DE and NDE bearing housing positions where access is impossible while the motor is running.
- The bearings should be re-greased at the run hours interval specified by the motor's OEM, and the correct volume and grade of grease must be inserted at re-greasing.
- The end user was advised to fit run hours counters on the induction motor drives to record the run hours for each motor so that a structured greasing regime could be implemented.
- Vibration monitoring is a valuable aid to prevent sudden and catastrophic failures but should not replace correct re-greasing of bearings.

Appendix 5A – Photos of the Faulty Bearing Parts

Figure 5.7 NDE bearing as removed from the SCIM.

Figure 5.8 Outer ring raceway.

Figure 5.9 Samples of the hardened soap thickener.

5.1 Introduction

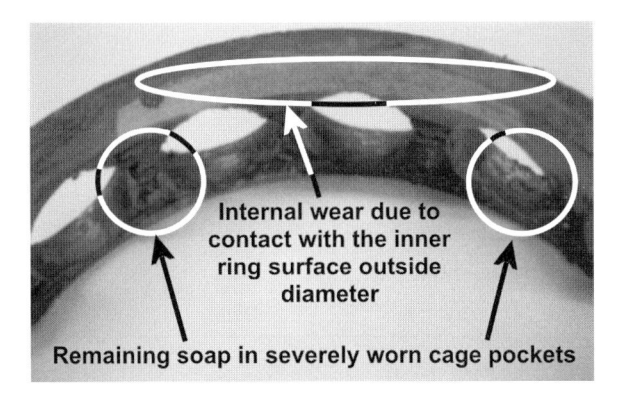

Figure 5.10 Internal wear in the brass cage.

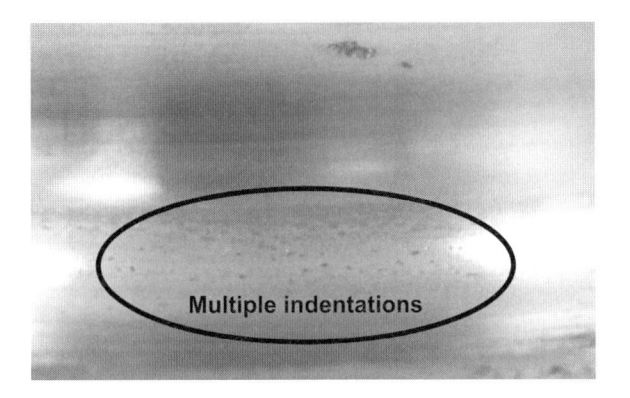

Figure 5.11 Inner ring raceway.

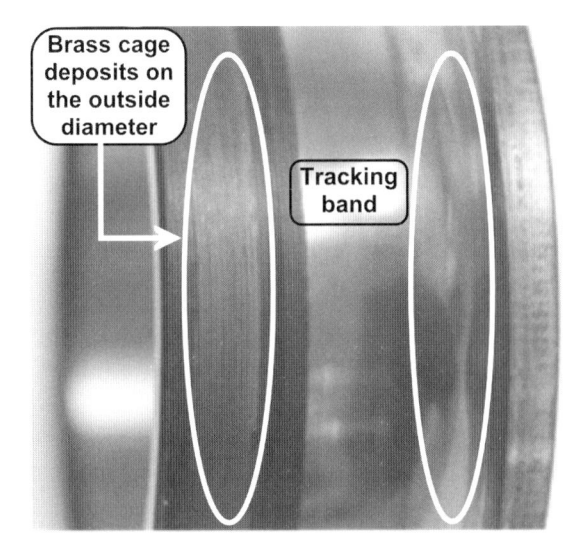

Figure 5.12 Inner ring raceway.

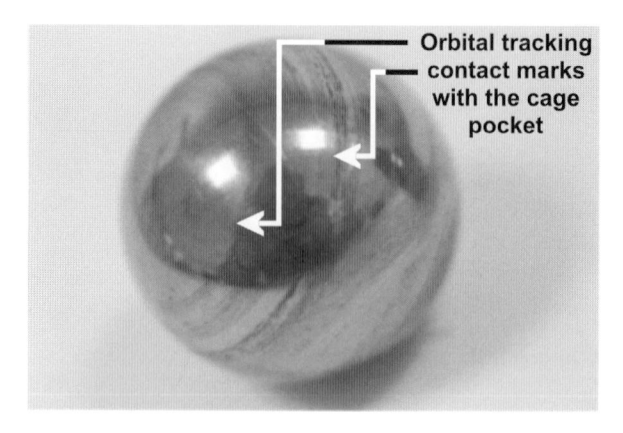

Figure 5.13 Sample of one of the balls.

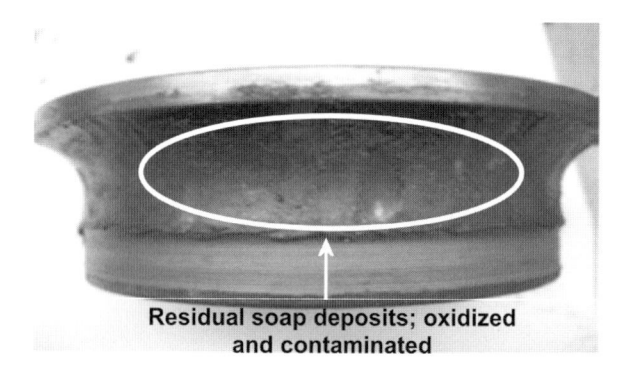

Figure 5.14 Inner ring raceway.

5.2 Industrial Case History – VSA Detected a Broken Cage in a Polyamide Cylindrical Roller Bearing in a 75 kW/100 H.P. SCIM

5.2.1 Introduction

Abstract – A SCIM that drives an Amine Transfer Pump at an LNG processing plant was removed from service in July 2007 for the following reasons:

(1) To repair the rusted frame via a *Thermally Sprayed Aluminium* (TSA) process.
(2) To refurbish the motor and fit new bearings.

The motor was installed in 1990 and was last overhauled in 2004. Prior to stripdown of the motor at the repair shop it was run uncoupled and vibration measurements were carried out by the author. The motor's nameplate details are as follows:

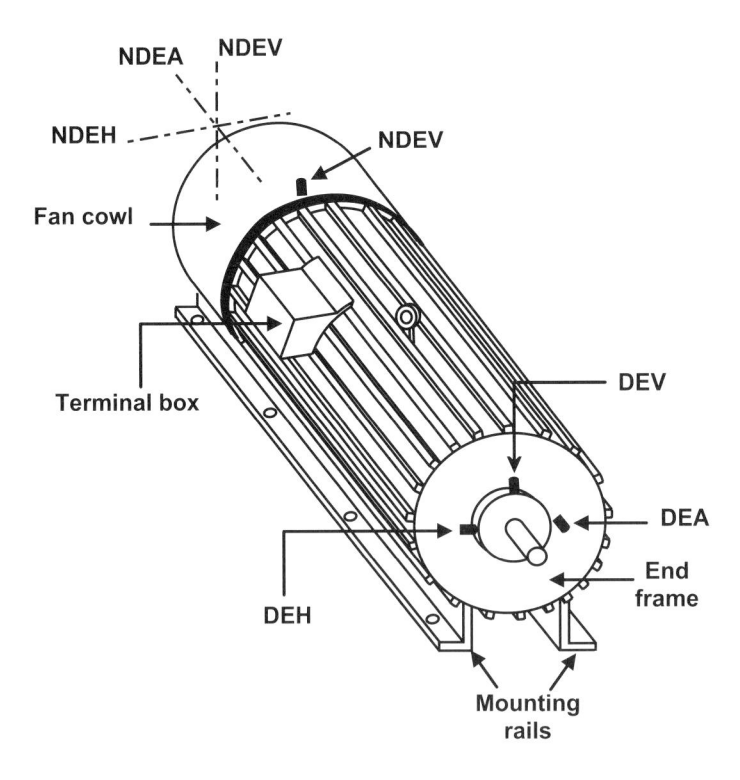

Figure 5.15a Positions of accelerometers.

- 3-phase SCIM, serial number V580-B2
- 75 kW, 415 V, 124 A, 2960 r.p.m., 50 Hz, p.f. = 0.88, eff = 95.6%
- Δ – connected
- Bearings – DE: N219-C3, NDE: 6219-C3.

5.2.2 Overall R.M.S. Velocities

The motor was mounted on a solid steel base-plate in the repair shop and supplied at rated voltage and frequency; the no-load current was 18 amperes (14% of $I_{f.l.}$). Vibration measurements were taken directly on the DE bearing housing and at the NDE, on a 12 o'clock positioned bolt, which secures the fan cowl to the NDE end frame. Figure 5.15a shows the positions where the vibration was measured on the motor.

The overall r.m.s. velocities are presented in Table 5.3 the levels were low with the highest being only 0.7 mm/s r.m.s. in the axial direction on the DE bearing housing. Note that when the motor was last overhauled the rotor was balanced to ISO G1.0 [5.1], hence the low levels of vibration.

Table 5.3 Overall r.m.s. velocities before motor strip down, span 10–1000 Hz

DEV bearing housing	DEH bearing housing	DEA bearing housing	NDEV on bolt securing fan cowl to NDE end frame
0.5 mm/s	0.65 mm/s	0.7 mm/s	0.7 mm/s

Table 5.4 Bearing defect frequencies

DE bearing type: N219 C3 cylindrical roller element bearing
No-load motor speed: 2998 r/min
Bearing frequencies

BPFO	BPFI	FTF	$2 \times BSF$
342 Hz	457 Hz	21 Hz	341 Hz

NDE bearing type: 6219 C3 deep groove ball bearing
No-load motor speed: 2998 r/min
Bearing frequencies

BPFO	BPFI	FTF	$2 \times BSF$
205 Hz	295 Hz	20.5 Hz	269 Hz

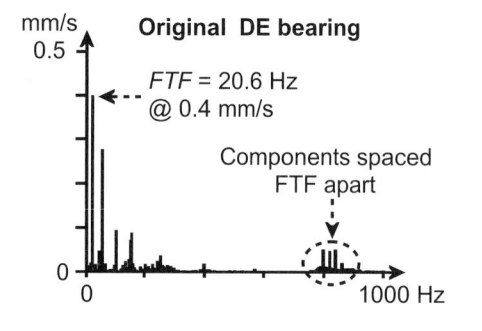

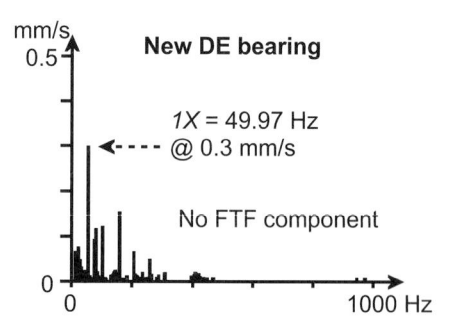

Figure 5.15 DEH velocity spectrum, 0.078 Hz/line.

Figure 5.16 DEH velocity spectrum, 0.078 Hz/line.

5.2.3 VSA Predicted a Broken Cage in the DE Bearing

The predicted bearing defect frequencies at the no-load speed are presented in Table 5.4.

5.2.3.1 Interpretation of Spectra before and after New Bearings were Fitted

(1) Figures 5.15 and 5.16 show that there was an *FTF* cage defect frequency component in the DEH and DEA velocity spectra from the DE bearing housing with magnitudes of 0.4 mm/s and 0.5 mm/s r.m.s. respectively. There were also high-frequency components, which were spaced *FTF* apart.

(2) The conclusion was that there was a cage defect and the motor was removed from service and the DE bearing inspected.

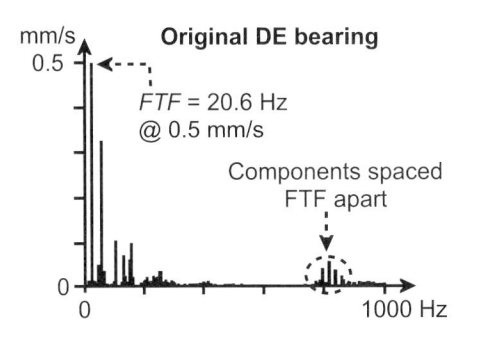

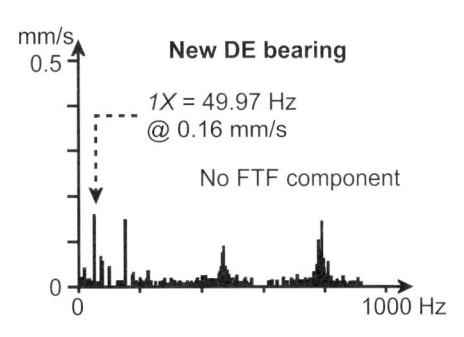

Figure 5.17 DEA velocity spectrum, 0.078 Hz/line.

Figure 5.18 DEA velocity spectrum, 0.078 Hz/line.

Figure 5.19 Broken polyamide cage in the N219 C3 DE bearing.

(3) A new set of steel cage bearings was fitted, the motor was run uncoupled, and the velocity spectra for the DEH and DEA positions are presented in Figures 5.17 and 5.18. There are no cage frequency components or harmonics.

5.2.4 Inspection of the Faulty Bearing

The motor was dismantled, and Figure 5.19 shows that the DE bearing had a broken polyamide cage, thus confirming the author's prediction that a serious cage fault existed.

At the previous refurbishment of the motor in 2004 the repair shop had fitted an N219 C3 with a polyamide cage, but it should have been a steel cage bearing as was originally fitted by the OEM of the SCIM. A new set of steel cage bearings was fitted.

5.2.5 Conclusions

(1) The overall r.m.s. velocities on the DE bearing housing were low (maximum of 0.7 mm/s r.m.s.), as shown in Table 5.3, and therefore this did not give any indication whatsoever that there was a problem with the bearing.

(2) However, VSA clearly detected that the *FTF* dominated the DEA velocity spectrum at a velocity of 0.5 mm/s r.m.s., which was 71% of the overall r.m.s. velocity.
(3) Overall r.m.s. velocities should not be used to assess the operational condition of rolling element bearings.

Reference

5.1 ISO 1940-1: 2003 Rotor Balancing.

6 Industrial Case Histories – VSA Detected Inner and Outer Race Faults in Rolling Element Bearings in SCIMS

6.1 Introduction

Abstract – Two vertically mounted SCIMs (A and B) were used to drive LNG pumps in an onshore LNG processing plant. The nameplates provided the following relevant data:

- 3-phase, SCIM, 3.3 kV, 385 kW/516 H.P.
- 83 A, 50 Hz, 2960 r/min., 0.88 p.f., eff 92.2%, star connected
- DE bearing: NU314 M C3
- NDE bearings: QJ312 and N312 M C3.

The motors were vertically mounted and therefore a QJ312 bearing was used at the NDE because it is a 4-point angular contact bearing which is designed to support axial loads in both directions. It has a relatively high mechanical stiffness with minimum deflection and has a split inner ring design to cater for the high speed.

The DE and NDE cylindrical roller element bearings are essentially guide bearings in this motor and drive train and the M means they have brass cages. There were no permanent accelerometers installed on any part of the drive train and the following points are strongly emphasised.

Owing to the construction of the motors it was impossible to measure vibration on the bearing housings and all measurements were taken on the outer frame of the motor as shown on the photograph in Figure 6.1 and on the schematic illustration in Figure 6.2.

To detect any bearing defect frequencies required the use of a logarithmic dB (amplitude) versus frequency spectrum because the mechanical stiffness characteristics between the bearings and access to measure vibration attenuated the actual vibration components due to bearing defects.

6.1.1 Overall Vibration Measurements

In January 2013, during a routine vibration survey of the two motors, the overall r.m.s. velocities were measured and are presented in Table 6.1. The highest was 6 mm/s r.m.s. at the DEH position on motor B. For a high-speed drive train, a value of 6 mm/s is acceptable because it was not an uncoupled run under FAT test conditions and at the DEH position the combination of motor and pump vibration is being measured.

Figure 6.1 Photograph of motor and motor–pump drive.

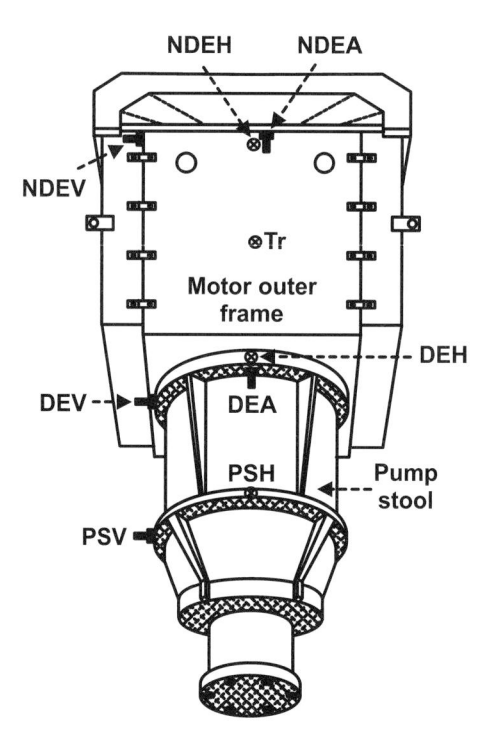

Figure 6.2 Schematic illustration of positions of vibration measurements.

Table 6.1 Overall velocities from motors A and B on 13th January 2013, coupled to pump

See **Figure 6.2** for positions of accelerometers

Motor A: NGL vertical pump motor: overall r.m.s. velocity ±10%
Motor operating current = 47 A; full-load current = 83 amperes

| DE motor flange | DEV 2.0 mm/s | DEH 1.0 mm/s | DEA 1.0 mm/s |
| NDE motor casing | NDEV 3.0 mm/s | NDEH 2.0 mm/s | NDEA 0.9 mm/s |

Motor B: NGL vertical pump motor: overall r.m.s. velocity ±10%
Motor operating current = 38 A; full-load current = 83 amperes

| DE motor flange | DEV 2.4 mm/s | DEH 6.0 mm/s | DEA 0.4 mm/s |
| NDE motor casing | NDEV 1.0 mm/s | NDEH 3.0 mm/s | NDEA 4.0 mm/s |

Table 6.2 Overall velocities from motor B on 7th May 2013

Motor C: Overall r.m.s. vibration in mm/s ±10%.
Frequency span: 10–1000 Hz
Coupled to pump motor operating current = 38 A

	Motor flange DE			Motor casing NDE				Pump stool	
	DEV	DEH	DEA	NDEV	NDEH	NDEA	Tr	PSV	PSH
16 Jan. 2013	2.4	6.0	0.4	1.0	3.0	4.0			
7 May 2013	5.4	13.0	1.5	1.4	10.7	6.0	8.5	5.0	1.6
Uncoupled no-load run motor operating current = 16 A									
7 May 2013	6.0	10.0	1.0	1.8	8.0	4.0	7.0		

Note that this was the first time the author had analysed vibration data from these motors and no historical records were available for trending vibration measurements. They were considered to be operational, but a re-survey was proposed for May 2013.

On 7th May 2013, the velocity at the NDEH position on motor B when driving the pump had increased from 3 mm/s to 10.7 mm/s and the results are presented in Table 6.2. The overall velocity on the 3.3 kV terminal box (Tr) was 8.5 mm/s. These levels were unacceptable.

The motor was uncoupled from the pump for a no-load run and the overall velocity at the NDEH position dropped from ***10.7 mm/s*** to ***8 mm/s***, verifying that the main source of the high vibration was from the motor and not the pump. A vibration spectrum analysis was applied to try to ascertain the source of the vibration problem.

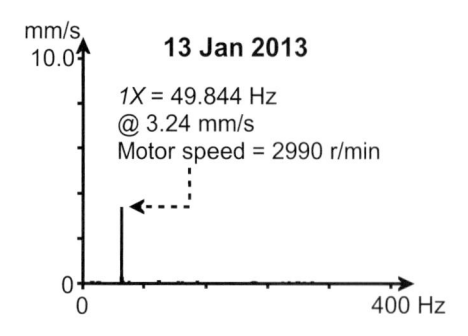

Figure 6.3 Motor B – vibration spectrum NDEH, coupled to the pump.

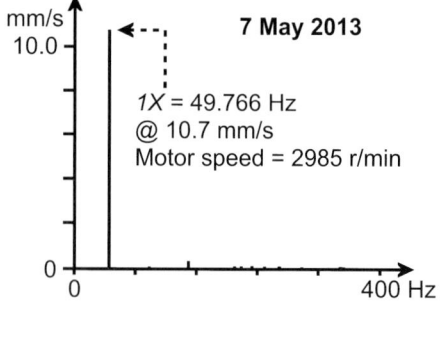

Figure 6.4 Motor B – vibration spectrum NDEH, coupled to the pump.

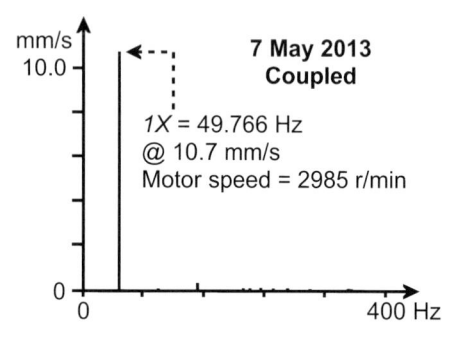

Figure 6.5 Motor B – vibration spectrum NDEH, coupled to the pump.

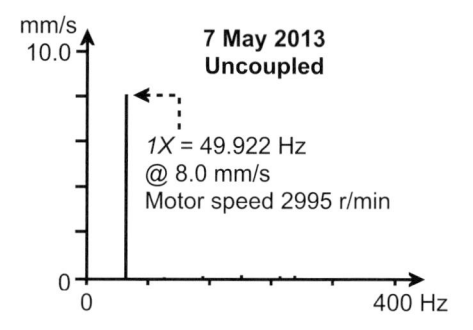

Figure 6.6 Motor B – vibration spectrum NDEH, uncoupled.

6.1.2 Vibration Spectrum Analysis

The routine survey of **13th January 2013** produced a vibration spectrum from the NDEH of motor B as shown in Figure 6.3 and this confirmed that the spectrum was dominated by the *1X* component at 3.24 mm/s, which was perfectly acceptable for a high-speed SCIM driving a centrifugal pump in an industrial site.

There were no components above 400 Hz and components on this linear spectrum between the *1X* component and 400 Hz were negligible. Figure 6.5 shows that four months later the *1X* component had increased from 3.24 mm/s to 10.7 mm/s ± 10% (as shown in Figure 6.4) which equates to a rate of rise of 1.9 mm/s per month. This was unacceptable. This increase may have been caused by an increase in mechanical imbalance or an increase in shaft misalignment in the drive train.

The pump was uncoupled, and a no-load run of the motor was carried out to ascertain the main source of the *1X* component. From Figure 6.6 the *1X* component at the NDEH position had dropped to 8 mm/s and it was clear that the high *1X* component was coming from the motor.

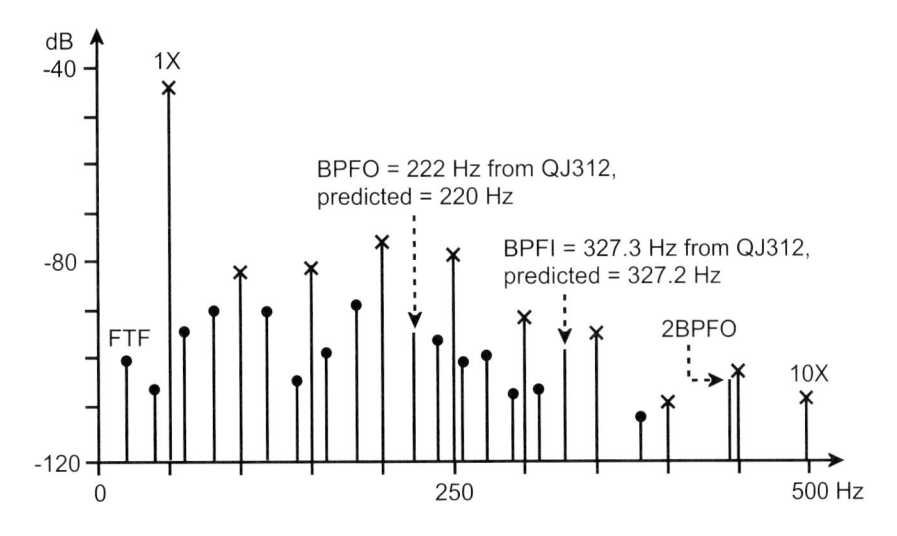

Figure 6.7 Motor B – uncoupled run, dB versus frequency spectrum at position NDEH on the motor.

Misalignment between the pump and the motor was therefore not the source of the problem but further analysis and interpretation were required.

6.1.3 Visual On-Site Inspection of the NDE of Motor B

A large volume of excess grease surrounded the motor's shaft and end frame assembly at the NDE at a height of approximately 100 mm (4 inches) next to the shaft and spreading out beyond the shaft in a radial direction onto the end frame for at least 150 mm (6 inches). This should not have been present, and by comparison, there was virtually no excess grease at the NDE of motor A.

The stainless-steel pipes between the motor's original grease pipe from the NDE down to nearly the bottom of the drive unit (for easy access for greasing) had five right-angle bends. The use of so many right-angle bends that had small radii of curvature was inadvisable. It was recommended that the number should be reduced and each given a much larger radius of curvature.

6.1.4 Interpretation of Logarithmic Spectrum and Predictions

The linear velocity spectra did not follow the classical vibration characteristics from faulty bearings. The linear velocity versus frequency spectra were dominated by the *1X* component and there appeared to be no multiple harmonics of *1X* from this motor and no evidence of bearing defect frequencies.

However, a logarithmic dB versus frequency spectrum of the vibration from the NDEH position on motor B when it was running uncoupled is shown in Figure 6.7.

Table 6.2a Predicted defect frequencies for a QJ312 bearing

NDE bearing information – bearing: QJ312			
Contact angle: 0	No. of balls: 13	Operational speed: 2995 r/min	
BPFO: 222.3 Hz	BPFI: 327.7 Hz	FTF: 20.2 Hz	$2 \times BSF$: 205.8 Hz

Figure 6.8 Damaged inner race of the QJ312 NDE bearing.

The predicted bearing defect frequencies for the QJ312 bearing at the NDE are presented in Table 6.2a.

The dB spectrum presented in Figure 6.7 shows the following:

(i) There are multiple harmonics of the *1X* component up to *10X*.
(ii) There are bearing defects at *BPFO*, *BPFI* and *FTF* (plus multiple harmonics of *FTF* were present).
(iii) It was therefore predicted that the QJ312 bearing was faulty.
(iv) Note that the spectrum was filtered to remove noise below 110 dB.

6.1.5 Conclusions

Based on the high (up to 13 mm/s r.m.s.) overall r.m.s. velocity measurements and the interpretation of the dB spectrum presented in Section 6.1.4, the motor was stopped and removed for an inspection of the bearings.

(i) A strip down and inspection verified the QJ312 bearing was faulty as is shown on the photograph in Figure 6.8.
(ii) The motor's nameplate stated an N314 M C3 rated bearing but an N312E M C3 bearing had been fitted at a previous overhaul. This means that the E rated bearing was overrated for the load on the bearing.
(iii) Figure 6.9 shows that the N312E M C3 bearing from the NDE had severely damaged rollers and this may have been initially caused by excessive skidding (see Section 3.1.6 for an explanation of skidding).

Figure 6.9 Damaged rollers in the N312E M C3 NDE bearing.

Figure 6.10 Large volume of grease inside the motor and on the end windings, etc.

(iv) Figure 6.10 is a photograph of the inside of the motor that shows excessive grease contamination of the end winding and blockage of ventilation, resulting in overheating which could escalate degradation of both bearings at the NDE.

6.2 Industrial Case History – VSA Diagnosed Outer Race and Ball Defects in a 7324 B Single-Row Angular Contact Ball Bearing in the NDE of a Vertical 1193 kW/1600 H.P. SCIM Driving a Thruster Propeller

6.2.1 Introduction

Abstract – Two vertically mounted SCIMs drive bow thrusters on a *Floating Production and Oil Off-Loading Ship* (FPSO). This is an interesting case history because the

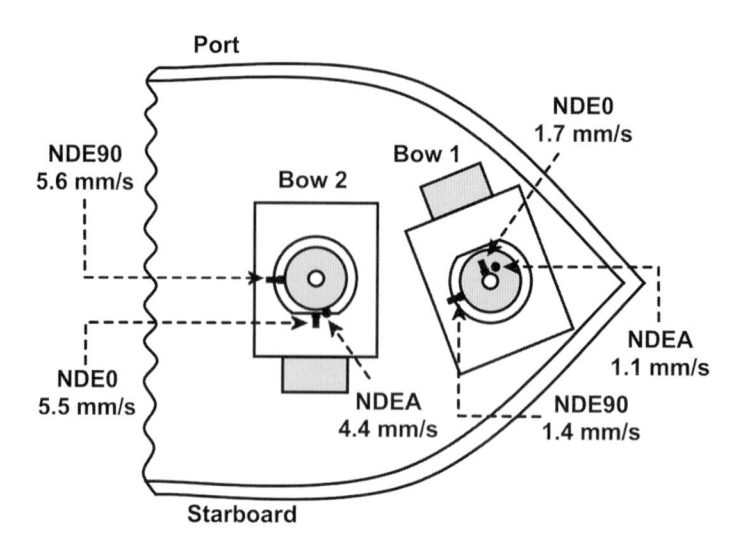

Figure 6.11 Positions of accelerometers and measured velocities (r.m.s.) at the NDE of Bow Thrusters 1 and 2.

bearing defect frequencies were not evident in the vibration spectrum although the faulty bearing was at an advanced stage of degradation.

However, a bearing with severe degradation is often characterised by a vibration velocity spectrum with pronounced multiple harmonics of the 1X fundamental rotational speed frequency ($N_r/60$ Hz) vibration component. This was indeed the case with the NDE bearing in one of the following motors. Overall velocities (r.m.s.) and vibration spectra are presented, and photos of the faulty bearing verified the diagnosis.

6.2.1.1 Motor Nameplate Data
- 3-phase, 6600 V, 1139 kW/1526 H.P., 118 A
- 60 Hz, 1186 r/min, 0.88 power factor, 96% efficient, star connected, SCIM
- Drive End (DE) bearing: 6324 C3. Non Drive End bearing: 7324 B
- The 6324 C3 is a deep groove ball bearing with a C3 clearance.

The 7324 B is a single-row angular contact ball bearing which can sustain radial, axial or combined loads with the axial load being in one direction. The B specifies a 40° contact angle to support the downward thrust.

6.2.2 Vibration Measurements and Overall Velocity Levels

Vibration measurements were taken at the DE and NDE of each motor and Figure 6.11 presents the overall r.m.s. velocities at each position. The photographs shown in Figures 6.12 and 6.13 give the positions of the accelerometers on the NDE of the motors driving Bow Thrusters 1 and 2.

6.2 Industrial Case History – VSA Diagnosed Outer Race and Ball Defects

Bow Thruster 1

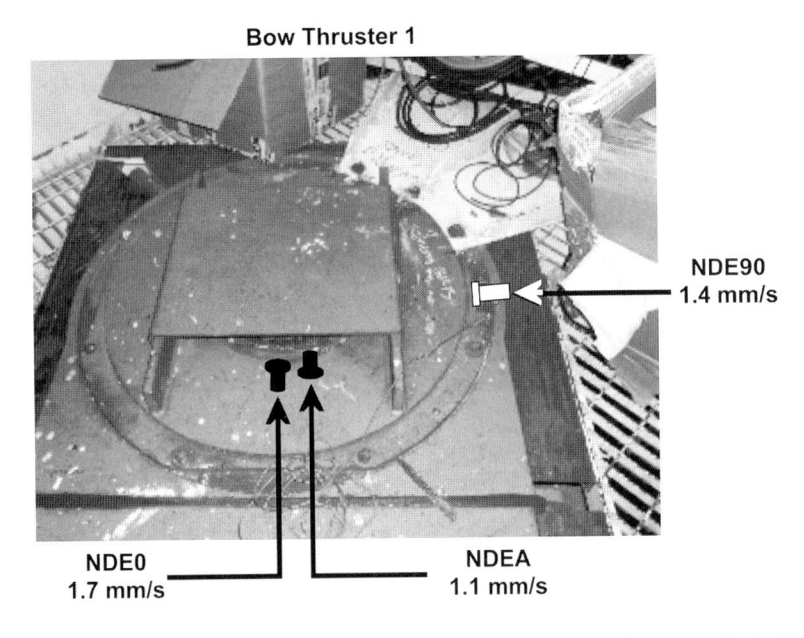

NDE90
1.4 mm/s

NDE0
1.7 mm/s

NDEA
1.1 mm/s

Figure 6.12 Photograph of actual positions of accelerometers and measured velocities (r.m.s.) at the NDE of Bow Thruster No. 1.

Bow Thruster 2

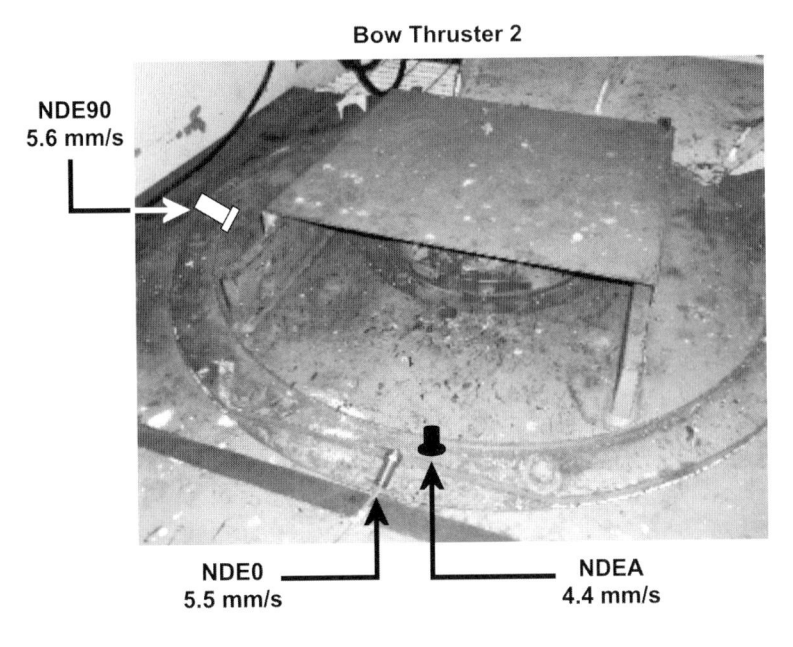

NDE90
5.6 mm/s

NDE0
5.5 mm/s

NDEA
4.4 mm/s

Figure 6.13 Photograph of actual positions of accelerometers and measured velocities (r.m.s.) at the NDE of Bow Thruster No. 2.

Industrial Case Histories – VSA Detected Inner and Outer Race Faults

Table 6.3 Overall r.m.s. velocity mm/s, span between 5.3 Hz and 1000 Hz

Accelerometer positions for Bow 1 (B1) and Bow 2 (B2) as shown in Figure 6.11

DE0 mm/s		DE90 mm/s		DEA mm/s		NDE0 mm/s		NDE90 mm/s		NDEA mm/s	
B1	B2	B1	B2	B1	B2	B1	B2	B1	B2	B1	B2
0.5	1.6	0.5	1.9	0.6	1.6	1.7	**5.5**	1.4	**5.6**	1.0	**4.4**

(i) The first step was to measure the overall vibration velocities r.m.s. (span of 5.3–1000 Hz) at the DE and NDE of each of the motors.

(ii) The lower frequency limit of 5.3 Hz was selected because the bearing cage frequency (FTF) is 7.7 Hz and 8.8 Hz for the DE and NDE bearings respectively (see Table 6.5 and Table 6.6 in Appendix 6A).

The overall velocities at the DE and NDE for both motors are presented in Table 6.3 and the key observations are as follows:

(a) Bow 1 DE motor: The overall velocities are low with the highest being only 0.6 mm/s r.m.s. in the axial direction (B1DEA).

(b) Bow 2 DE motor: The overall velocities are normal with the highest being 1.9 mm/s r.m.s. in the radial direction (B2DE90).

(c) Bow 1 NDE motor: The overall velocities are normal with the highest being 1.7 mm/s r.m.s. in the NDE0 direction (B1NDE0).

(d) Bow 2 NDE motor: The overall velocities r.m.s. were high at the NDE of Bow 2 in comparison to the velocities at the NDE of Bow 1. The velocity was 5.6 mm/s r.m.s. at B2NDE90 (recall not directly on the bearing housing) from Bow 2, which was four times greater than the velocity of 1.4 mm/s r.m.s. at B1NDE90 which was directly on the bearing housing of Bow 1.

In summary, the overall r.m.s. velocities at the positions on the NDE of Bow 2, namely, B2NDE0, B2NDE90 and B2NDEA, were 5.5 mm/s, 5.6 mm/s and 4.4 mm/s r.m.s. respectively, whereas at the NDE of Bow 1, namely, B1NDE0, B1NDE90 and B1NDEA, the velocities were only 1.7 mm/s, 1.4 mm/s and 1.0 mm/s r.m.s. respectively.

Note that from Figures 6.12 and 6.13 the positions of the accelerometers on the NDE of Bow 1 and Bow 2 were not the same, in fact the positions at Bow 2 were further away from the bearing housing than was the case with Bow 1. The reasons for this are as follows:

(i) During the vibration measurements there was a high acoustic noise level coming from the vicinity of the NDE bearing of Bow 2; this could best be described as a fast and repetitive chattering.

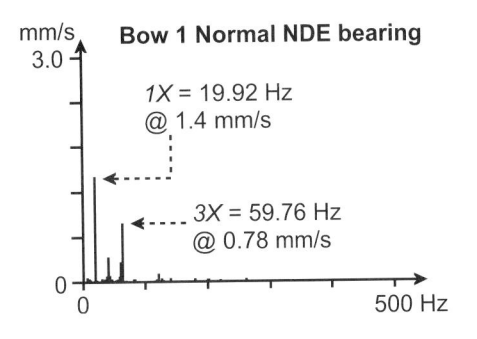

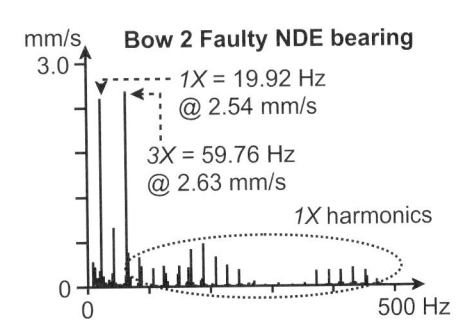

6.2 Industrial Case History – VSA Diagnosed Outer Race and Ball Defects

Figure 6.14 Velocity spectrum for position B1NDE0 on the bearing housing at the NDE.

Figure 6.15 Velocity spectrum for position B2NDE0 on the end frame at the NDE.

(ii) It was proposed that the noise was being produced from repetitive transient vibrations within the NDE bearing of Bow 2.

(iii) To obtain valid velocity (r.m.s.) vibration measurements the accelerometers were placed at the positions on the NDE assembly of Bow 2 as shown on Figure 6.13 because, when an attempt was made to measure the vibration directly on the actual NDE bearing housing, the same position as on Bow 1, the transient vibration pulses were very high.

(iv) This caused the vibration instrument to go into intermittent overload modes of very short duration so that r.m.s. velocity readings could not be obtained on the NDE bearing housing of Bow 2.

6.2.3 Vibration Spectrum Analysis

The high overall velocities (r.m.s.) measured on the main bearing end bracket and rotor support assembly (see Figure 6.13 for Bow 2 positions) at the NDE of Bow 2 compared to the levels directly measured on the bearing housing of Bow 1 indicated that there was a problem at the NDE of the motor driving thruster Bow 2. The cause and severity of the problem at the NDE of Bow 2 could not be established from overall velocity measurements (r.m.s.) and a vibration analysis was required.

The nominal full-load current of each motor was 118 amperes, and Bow 1 and 2 were operating at a reduced load of 60 and 63 amperes respectively. Therefore, the operational speed of each motor was higher than the nominal full-load speed of 1186 r/min. The vibration spectra at positions B1NDE0 from the NDE of the SCIMs driving Bow 1 and 2 are shown in Figures 6.14 and 6.15 respectively and a clear difference between them is evident.

6.2.3.1 Interpretation of Vibration Spectra

The fundamental rotational speed frequency of the motor is given by:

$$1X = N_r/60 \text{ Hz},$$

where N_r = rotor speed in r/min,

$1X = 19.77$ Hz at the full-load speed of 1186 r/min.
The measured $1X = 19.92$ Hz (1195 r/min).

The main features in the vibration spectra from the SCIMs driving Bow 1 and 2 are as follows:

(i) There are multiple harmonics of the $1X$ frequency component from Bow 2, up to the thirtieth harmonic, whereas from Bow 1 there are no harmonics of $1X$ above the third harmonic, $3X$.
(ii) The velocity (r.m.s.) of the $3X$ component at 2.63 mm/s from Bow 2 is 3.4 times greater than $3X$ at 0.78 mm/s from Bow 1.
(iii) No bearing defect frequencies were evident.

6.2.4 Inspection – Photos – Conclusions

(i) It was concluded that there was a serious fault in the NDE bearing of the SCIM driving Bow Thruster 2.
(ii) The motor was stopped, and the bearing was removed for inspection and the damage shown in Figures 6.16 and 6.17 verified the vibration analysis and prediction.
(iii) If the bearing had not been removed the next stage may well have been a bearing collapse, probably resulting in a rotor to stator rub and consequential stator core and winding damage, requiring a lengthy outage and an expensive motor repair.

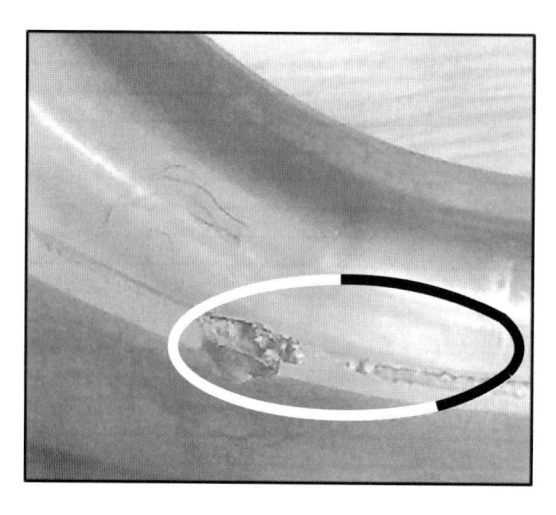

Figure 6.16 Damaged inner race way.

Figure 6.17 Sample of one of the damaged balls, all the balls were damaged.

Appendix 6A Prediction of Bearing Defect Frequencies

- *BPFO* = Ball pass frequency outer race
- *BPFI* = Ball pass frequency inner race
- *BSF* = Ball spin frequency
- *FTF* = Cage frequency

The predicted bearing defect frequencies were initially calculated using the full-load nameplate speed because that is the only information available prior to going on-site to take vibration measurements. They must be re-calculated with respect to the actual operational speed obtained from the measurement of the *1X* component. The measured speed using the *1X* component was 1195 r/min; therefore, all the frequency values in Tables 6.5 and 6.6 were multiplied by the ratio of 1195/1186.

6.3 Industrial Case History – VSA Diagnosed Outer Race Defects in a Rolling Element Bearing via Vibration Measurements on the Drive End Frame of a 160 kW/215 H.P. SCIM Driving a Boiler Forced Draft Fan

6.3.1 Summary

Abstract – This SCIM was driving a boiler forced draft fan in an LNG processing plant and only the end frames were available for assessing the operational condition of the bearings. The coupling guard and the fan cowl respectively prevented access to the DE and NDE bearing housings during the on-site vibration measurements. The overall r.m.s. velocities on the DE end frame were perfectly acceptable for a SCIM driving a

Table 6.5 Predicted bearing frequencies for the motor's DE bearing at the rated and nominal full-load speed on the nameplate

Bearing ID: 6324			Manufacturer: confidential			
No. of balls: 8			Motor speed at full load: 1186 r/min			

BPFO:		BPFI:		FTF:		$2 \times BSF$:
62 Hz		96 Hz		7.7 Hz		86.7 Hz

Harmonics	2	3	4	5	6	7
BPFO Hz	124	186	248	310	371	433
BPFI Hz	193	289	385	481	577	674
FTF Hz	15.5	23	31	38.6	46	54
$2 \times BSF$ Hz	173	260	347	434	520	607

Table 6.6 Predicted bearing frequencies for the motor's NDE bearing

Bearing ID: 7324B			Manufacturer: confidential			
No. of balls: 13			Motor speed at full load: 1186 r/min			

BPFO:		BPFI:		FTF:		$2 \times BSF$:
107 Hz		149 Hz		8 Hz		89 Hz

Harmonics	2	3	4	5	6	7
BPFO Hz	214	321	428	535	642	749
BPFI Hz	297	446	595	745	892	1041
FTF Hz	16	24	32	40	48	56
$2 \times BSF$ Hz	177	266	354	443	534	620

fan in an industrial plant, witnessed by the fact that the maximum velocity was 3.2 mm/s r.m.s. in the vertical direction at the DE.

However, the bearing defect frequency *BPFO* from the N319 C3 DE bearing, plus multiple harmonics, were evident in the vibration spectra on the DE end frame in the vertical, horizontal and axial positions. See Figure 6.18 for an illustration of the accelerometer positions, from which these spectra were obtained. Note that the taking of on-site photographs of rotating plant was not permitted in this LNG installation.

The r.m.s. velocities of the bearing defect frequency components, due to a bearing defect, when measured on a SCIM's end frame can be considerably less than the vibration measured directly on the bearing housing, because the bearing vibration can be attenuated by the mechanical stiffness of the end frame. That attenuation is also a function of frequency.

In this case there was considerable likelihood that the bearing vibration was greater than the measured value on the end frame. It was therefore predicted that there were defects in the outer race that merited the decision to stop the motor for an inspection of the DE roller element bearing. The removal and inspection of the motor had to be done

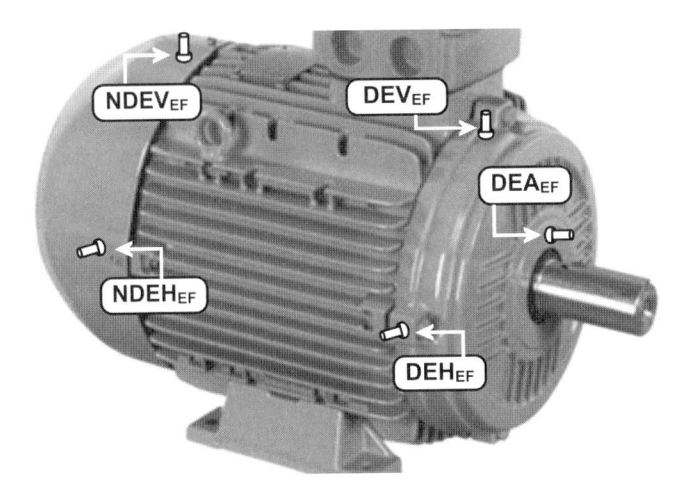

6.3 VSA Diagnosed Outer Race Defects in a Rolling Element Bearing 117

Figure 6.18 Illustration of positions of accelerometers on the DE end frame, and on the fan cowl at the NDE.

very quickly to minimize disruption to production. Fortunately, there was a planned outage of the LNG processing plant for five days, and the motor was removed from site and new bearings were fitted. The motor was refurbished with new bearings and re-commissioned within the time frame allocated, but there was no time for a vibration FAT in the repair workshop immediately after the repair.

Motor nameplate data are as follows:

- 3-phase, SCIM, 415 V 160 kW/210 H.P.
- 265 A, 50 Hz, 1480 r/min, 0.9 p.f.
- DE bearing: N219 C3
- NDE bearing: 6314 C3.

6.3.2 On-Site Vibration Measurements and Spectrum Analysis before New Bearings were Fitted

The overall r.m.s. velocities on the DE end frame during a routine vibration survey are presented in Table 6.7 and the maximum was 3.2 mm/s r.m.s. in the vertical direction which is certainly acceptable for a SCIM driving its mechanical load. Note that this was not a FAT vibration test during an uncoupled run.

It is important to point out that the levels on the actual DE bearing housing are very likely to be higher. The motor's operating current was 124 amperes (full-load current = 265 amperes).

The first step in the prediction of a defect in a roller bearing is to calculate the bearing defect frequencies for the synchronous speed of this SCIM, and these are presented in Table 6.8.

(1) The DEV_{EF} velocity spectrum in the vertical direction on the DE end frame between 10 Hz and 1000 Hz is presented in Figure 6.19 and this shows that

Table 6.7 DE end frame – overall r.m.s. velocities span 10–5000 Hz

The motor's operating current was 124 amperes

	On-site coupled run BEFORE THE MOTOR WAS REMOVED	On-site coupled run NEW BEARINGS
Accelerometer position	mm/s r.m.s.	mm/s r.m.s.
DE vertical DEV_{EF}	3.2	0.75
DE horizontal DEH_{EF}	2.0	0.9
DE axial DEA_{EF}	2.6	1.0

Table 6.8 Predicted bearing frequencies at the motor's synchronous speed

Bearing: N319 C3 **Contact angle: 0** **No. of balls: 8**		**Synchronous speed:** 1500 r/min (50 Hz)	
BPFO: 134 Hz	**BPFI:** 199 Hz	**FTF:** 10.3 Hz	**2 × BSF:** 137 Hz

BPFO and its harmonics are present. The fifth harmonic of *BPFO* was 1.7 mm/s r.m.s., which is 53% of the overall r.m.s. velocity of 3.2 mm/s.

(2) Note that the motor was on reduced load at an operating current of 124 amperes (full-load current is 256 amperes), therefore its speed is higher than the full-load speed.

(3) The ball pass outer race defect frequency (*BPFO*) was initially calculated at 1500 r/min and was 134 Hz, the *1X* component from the velocity spectrum in Figure 6.19 was 24.88 Hz, which is equivalent to a rotor speed of 1493 r/min. The predicted *BPFO* frequency based on the actual operating speed of the motor is 133.2 Hz and the measured *BPFO* was 133.4 Hz – a negligible difference.

(4) The motor was removed from service for an inspection of the bearings.

6.3.2.1 VSA after New Bearings were Fitted

New bearings were fitted and the rotor was mechanically re-balanced to ISO G1.0 in the refurbished motor (date of manufacture 1991) in comparison to G 2.5, which was the mechanical unbalance grade in the rotor of the motor removed for repair. Therefore, the *1X* component is reduced at each measurement position.

A comparison between Figures 6.19 and 6.20 clearly shows that with the new DE bearing there was not a *BPFO* component present and there were no harmonics of *BPFO*.

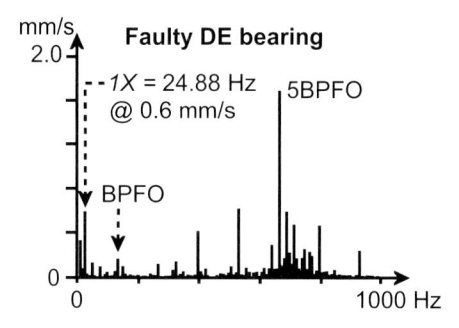

6.3 VSA Diagnosed Outer Race Defects in a Rolling Element Bearing

Figure 6.19 DEV$_{EF}$ vertical velocity spectrum, 10–1000 Hz, 0.078 Hz/line.

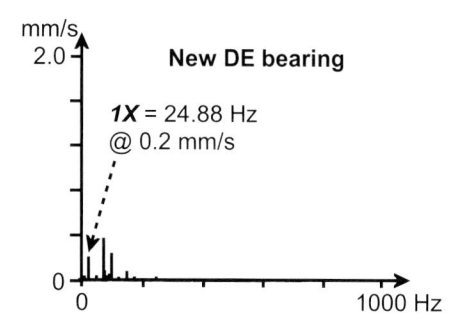

Figure 6.20 DEV$_{EF}$ vertical velocity spectrum, 10–1000 Hz, 0.078 Hz/line.

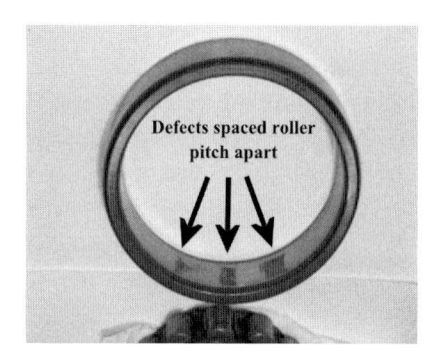

Figure 6.21 Inner race defects centred at 6.00 p.m. in the load zone and spaced apart at roller pitch.

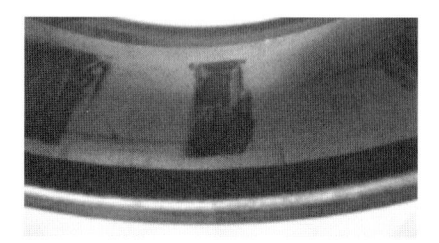

Figure 6.22 Zoom view of inner race defects centred at 6.00 p.m. and spaced apart at roller pitch.

6.3.3 Inspection of DE Bearing

The DE bearing was dismantled during the repairs and Figures 6.21 and 6.22 are photographs of the faults, which clearly show the defects in the outer race. The damage was in the load zone centred around 6.00 p.m. with defects spaced at roller pitch apart.

This suggests that the problem was caused by *false brinelling* and it was subsequently established that this was normally the standby SCIM, which had not operated for nine months prior to a routine vibration survey check. The motor had not been given any starts, for example once per month, or barred over to prevent false brinelling.

In industrial plants where production and income generation are the key drivers this non-turning of the rotor is very often the case because operators will not disrupt the operational process just to turn a motor to prevent false brinelling – that is the industrial reality and is based on the author's experience of 57 years.

6.3.4 Conclusions

(i) This case history proved that it is possible to diagnose a bearing defect via vibration analysis of vibration on the end frame of a SCIM even although the overall r.m.s. velocities on the end frame were perfectly acceptable.
(ii) For example, the highest overall velocity was 3.2 mm/s r.m.s. on the DE end frame; however, it was the presence of *BPFO* and its harmonics up to 1.7 mm/s that were the key factors for predicting a bearing fault.
(iii) The motor was removed from service and photographs of the faulty bearing confirmed the vibration analysis.
(iv) The bearing fault was caused by false brinelling because the motor had not operated for nine months prior to the routine vibration survey check. The motor had not been given any starts, for example once per month, or barred[1] over to prevent false brinelling.
(v) In industrial plants where production and income generation are the key drivers this non-turning of the rotor is often the case because operators will not disrupt the operational process just to turn a motor to prevent false brinelling.

6.4 Industrial Case History – VSA of Vibration Measured on the Outer Frame Predicted an Outer Race Bearing Defect in a Vertically Mounted 75 kW/100 H.P. SCIM

6.4.1 Summary

Abstract – Two identical, 75 kW, vertically mounted SCIMs were driving auxiliary sea water lift pumps (ASWLPs) on an offshore oil and gas production platform.

The DE and NDE bearing housings were inaccessible as shown in Figure 6.23. There were no permanent accelerometers on the drive trains and vibration measurements were taken at the positions shown in Figure 6.23 using small accelerometers mounted on strong magnets.

---

[1] Note: barred means manually turned via a bar connected to the motor shaft.

6.4 VSA of Vibration Measured on the Outer Frame Predicted an Outer Race Bearing Defect

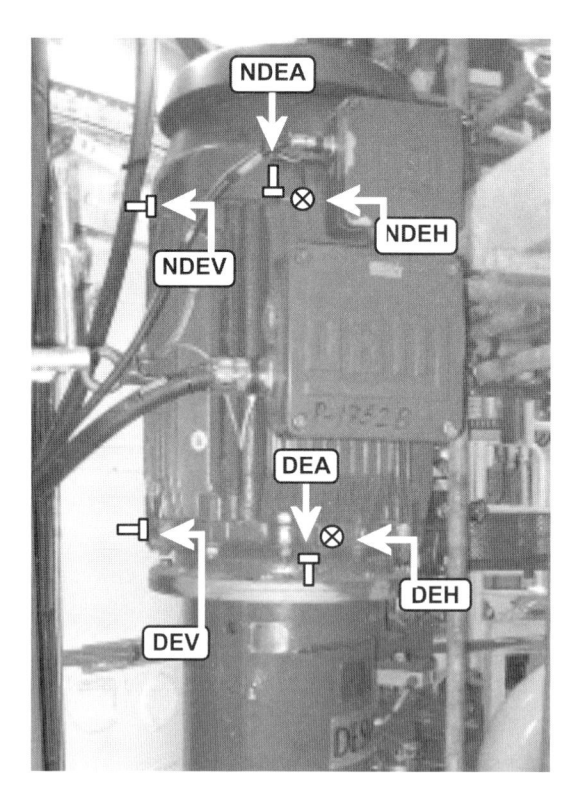

Figure 6.23 Positions of accelerometers on the vertical SCIM.

Motor nameplate data are as follows:

- 3-phase, SCIM, 440 V 75 kW/100 H.P.
- 117 A, 60 Hz, 3575 r/min, 0.9 p.f., eff = 93.5%
- DE bearing: 6314 C3
- NDE bearing: 6314 C3.

6.4.2 Overall Vibration Velocity Measurements

The overall r.m.s. velocities at all positions on both motors were acceptable for vertically mounted SCIMs driving pumps operating on an offshore oil and gas production platform because the highest was 4.4 mm/s r.m.s. at the NDEA in motor B as shown in Table 6.9.

These were the first on-site vibration measurements from these new motors, which had only been in operation for six months, and are therefore the base-line results for future comparisons. No FAT vibration results were available and, even during a FAT test, the bearing housings would not be accessible on these vertical SCIMs – this is obvious from the photograph of Figure 6.23. Hence comparisons between velocities on the outer frame and directly on the bearing housings cannot be achieved.

Industrial Case Histories – VSA Detected Inner and Outer Race Faults

Table 6.9 Overall r.m.s. velocities ±10% in mm/s and *BPFO* r.m.s. velocities; frequency span: 10–1000 Hz

Location	Position	Motor A mm/s	BPFO A mm/s	Motor A BPFO as a % of overall velocity	Motor B mm/s	BPFO B mm/s	Motor B BPFO as a % of overall velocity
NDE casing in line with bearing	NDEH	1.0	0.17	17 %	2.8	0.2	7 %
	NDEV	0.7	0.24	34 %	2.3	0.47	20 %
	NDEA	2.5	0.8	32 %	4.4	0.25	6 %
DE casing in line with bearing	DEH	1.6	0.33	21 %	2.3	0.16	7 %
	DEV	0.8	0.45	56 %	0.4	0.1	25 %
	DEA	2.7	0.85	32 %	4.3	0.16	3.8 %

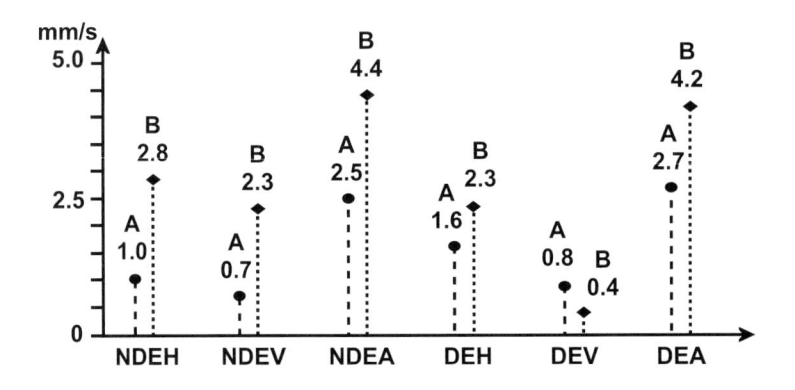

Figure 6.24 Overall r.m.s. velocities for motors A and B as a function of the positions shown in Figure 6.23.

Note that offshore oil and gas production operators in the North Sea (off the coast of Scotland) normally set the alarm and trip levels on SCIMs with rolling element bearings to 7 mm/s (0.28 inches/s) and 11 mm/s (0.43 inches/s) r.m.s. respectively. These levels are normally based on the measurements being taken directly on the bearing housings but the only option with this motor is to apply these alarm and trip levels to the measurement positions shown in Figure 6.23.

Graphical displays of the overall r.m.s. velocities for motors A and B as a function of the positions where the velocity was measured are presented in Figure 6.24 to provide a visual display compared to the tedious task of processing the numbers in Table 6.9.

6.4 VSA of Vibration Measured on the Outer Frame Predicted an Outer Race Bearing Defect

Table 6.10 Prediction of bearing defect frequencies DE and NDE bearing: 6314 C3 deep groove ball bearing

From the vibration spectra, the $1X$ component = 59.922 Hz, therefore the operating speed = 3595 r/min.

BPFO: 295 Hz	BPFI: 185Hz	FTF: 23 Hz	$2 \times BSF$: 245 Hz

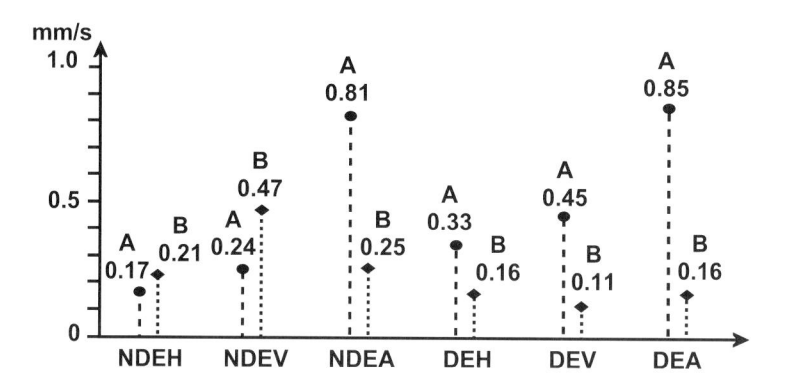

Figure 6.25 Velocities of *BPFO* for motors A and B.

The overall r.m.s. velocities on motors A and B are acceptable for vertical motors driving pumps on an offshore oil installation, with the highest being 4.4 mm/s r.m.s. on motor B.

6.4.3 Vibration Spectrum Analysis – Diagnosis of *BPFO*

The bearing defect frequencies are required for interpretation of the spectra and these are presented in Table 6.10. The predictions are based on the actual operating speed (3595 r/min) obtained from the measured $1X$ rotational speed frequency component at 59.922 Hz.

There were distinct differences between the velocity spectra from motors A and B; for example, the outer race defect frequency (*BPFO*) from motor A is 0.8 mm/s r.m.s. at the NDEA position whereas at the same position on motor B it is only 0.25 mm/s r.m.s.

The *BPFO* velocities for motors A and B as a function of the positions where the velocity was measured are presented in Figure 6.25.

In summary:

(i) At the NDEA position the velocity of *BPFO* from motor A was 3.24 times greater than that from motor B.
(ii) At the DEA position the velocity of *BPFO* from motor A was 5.3 times greater than that from motor B.

The *BPFO* velocities can often be less when measured on the outer frame of the motor than those measured directly on the bearing housings.

6.4.4 Conclusion and Recommendations

The spectrum analyses predicted that there was the initiation of outer race defects in the bearings in motor A and this is further substantiated by the following facts:

Motor A

(i) *BPFO*'s velocity was 56% of the overall r.m.s. velocity at the DEV position.
(ii) *BPFO*'s velocity was 33% of the overall r.m.s. velocity at the NDEA position.
(iii) *BPFO*'s velocity was 32% of the overall r.m.s. velocity at the DEA position.

For motor B the percentages are much less:

Motor B

(i) *BPFO*'s velocity was 25% of the overall r.m.s. velocity at the DEV position.
(ii) *BPFO*'s velocity was only 5.3% of the overall r.m.s. velocity at the NDEA axial position.
(iii) *BPFO*'s velocity was only 3.8% of the overall r.m.s. velocity at the DEA axial position.

The ratio (as a percentage) of *BPFO*'s velocity to the overall r.m.s. velocity is an important indicator of potential bearing problems, and the larger that percentage is, the more likely it is that a bearing fault exists.

The recommendation was to take vibration measurements every month and trend the overall r.m.s. velocities but with a focus on the velocities of *BPFO* from motor A. If the velocities of *BPFO* increase by a factor of 2 or greater in this particular case history,[2] from 0.85 mm/s r.m.s. and 0.8 mm/s r.m.s. up to 1.7 mm/s and 1.6 mm/s r.m.s. respectively at positions DEA and NDEA, then motor A should be stopped and the bearings inspected to prevent a sudden and catastrophic bearing failure and a consequential rotor to stator rub.

6.4.5 Vibration Analysis Predicted Misalignment in Drive Train B

Samples of the velocity spectra are presented in Figures 6.26 to 6.29, and these show that at the NDEH and DEH positions on motor B the velocities of the fundamental *1X* rotational speed frequency components are 3 and 1.8 times greater than those from motor A at the same positions.

A plot of the velocities of the *1X* component for motors A and B as a function of the positions where the velocity was measured is presented in Figure 6.30.

---

[2] Note: this **doubling factor** has proven to be an effective working guide used by the author for 50 years.

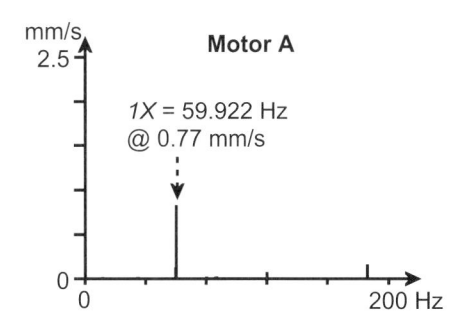

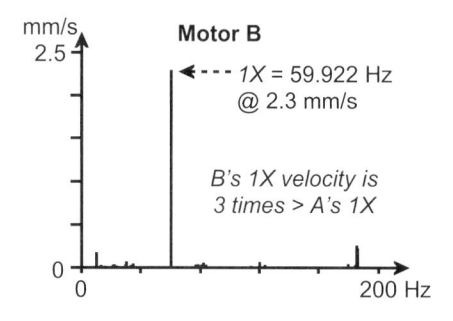

6.4 VSA of Vibration Measured on the Outer Frame Predicted an Outer Race Bearing Defect

Figure 6.26 Motor A NDEH spectrum of *1X*. (1X = 59.922 Hz @ 0.77 mm/s)

Figure 6.27 Motor B NDEH spectrum of *1X*. (1X = 59.922 Hz @ 2.3 mm/s; B's 1X velocity is 3 times > A's 1X)

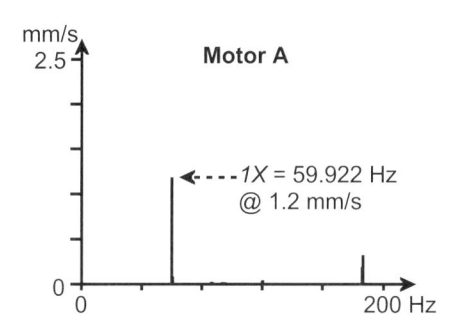

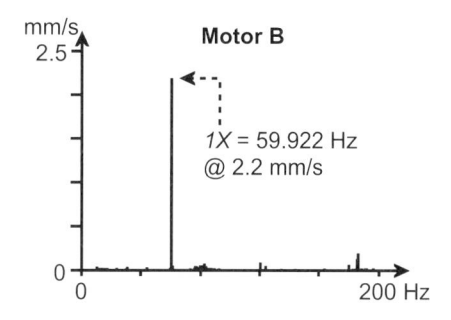

Figure 6.28 Motor A DEH spectrum of *1X*. (1X = 59.922 Hz @ 1.2 mm/s)

Figure 6.29 Motor B DEH spectrum of *1X*. (1X = 59.922 Hz @ 2.2 mm/s)

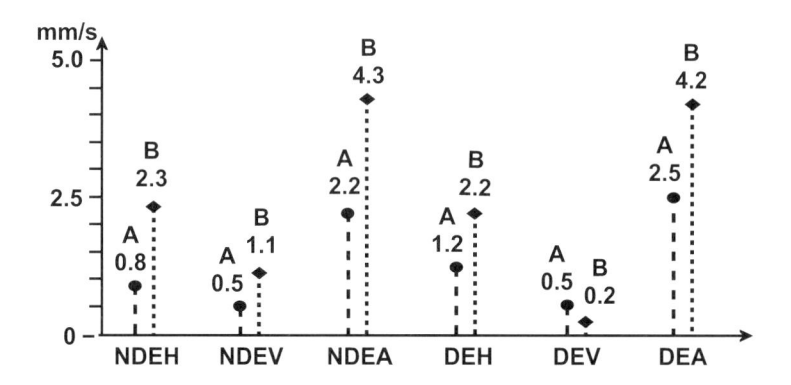

Position	A	B
NDEH	0.8	2.3
NDEV	0.5	1.1
NDEA	2.2	4.3
DEH	1.2	2.2
DEV	0.5	0.2
DEA	2.5	4.2

Figure 6.30 Vibration velocity levels of the *1X* component for motors A and B as a function of position.

6.4.5.1 Conclusions and Recommendations on Misalignment

Both rotors in each of these new motors (six months in operation) were balanced to ISO G1.0; consequently, it is very unlikely that the balance grade has increased in the rotor of motor B.

It was therefore concluded that the higher (up to a factor of 2.8) velocities of the $1X$ component in the NDEH, NDEA, DEH and DEA velocity spectra from motor B were due to a larger misalignment in the B drive train than that in the A drive train.

However, as already stated, the overall r.m.s. velocities from motor B are operationally acceptable but it was still recommended that the alignment in the B drive train should be checked at a planned maintenance outage.

7 Industrial Case Histories – VSA Diagnosed False Brinelling and Problems in Cylindrical Roller Bearings in SCIMs

7.1 Introduction

Abstract – This case history will show that *false brinelling*, which is caused by vibration being transmitted to a stationary motor's bearings from an adjacent running plant, was detected in a cylindrical roller element bearing even though the overall velocity levels were normal.

This was achieved by interpreting the velocity spectrum to detect the bearing defect frequencies. The SCIM was driving an Amine Transfer Pump in an LNG onshore plant and was an in-situ standby unit. There were lengthy periods of time during which the motor was neither started nor manually turned to prevent false brinelling.

Interpretation of the vibration spectra and photographs of the faulty bearing confirmed that the diagnosis was correct. The motor was removed and a refurbished spare motor was installed. The vibration analysis from the replacement motor established that no bearing defect frequencies were then present.

7.1.1 Motor Data and Overall Vibration Velocity Measurements

Relevant nameplate data:

- 3-phase, SCIM, 415 V, 75 kW/100 H.P.
- 126 A, 50 Hz, 2960 r/min, 0.88 p.f., eff 94%
- DE: NU219 C3; NDE: 6219 C3.

The positions of the accelerometers on the DE are illustrated in Figure 7.1 and Table 7.1 gives the overall velocities (r.m.s.) during a 12 month period.

The operational current was 75 amperes compared to the nominal full-load current of 126 amperes; therefore the motor was on reduced load and the operational speed was greater than the full-load speed of 2960 r/min.

Trends of the overall r.m.s. velocities on the DE bearing housing during a 12 month period are presented in Figure 7.2:

(i) The results in Table 7.1 confirm that all the overall r.m.s. velocities were normal. The highest velocity was 1.73 mm/s (r.m.s.) in the DEA direction and, based on overall r.m.s. velocity levels, this motor was fit for purpose.

Industrial Case Histories – VSA Diagnosed False Brinelling

Table 7.1 Overall velocity mm/s r.m.s. on the DE bearing housing

Time-line	DEV	DEH	DEA
December 2014	0.65	0.51	0.68
January 2015	0.46	0.53	0.57
February 2015	0.49	0.53	0.58
March 2015	0.65	0.66	0.63
April 2015	0.68	0.43	0.73
May 2015	0.72	0.47	0.77
June 2015	0.61	0.47	0.75
July 2015	0.52	0.64	0.59
August 2015	0.56	0.57	0.67
September 2015	0.49	0.73	0.58
October 2015	0.65	0.68	0.63
11 months	Average: 0.59	Average: 0.57	Average: 0.65
November 2015	0.98 63% higher than 11 month average	0.92 60% higher than 11 month average	**1.73** **166% higher than** **11 month average**

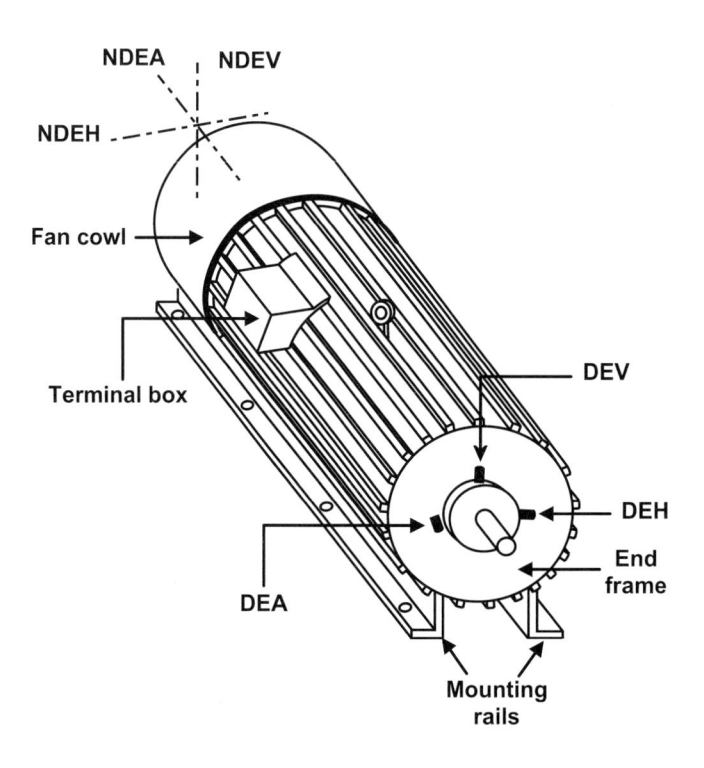

Figure 7.1 Schematic illustration of the positions of accelerometers at the DE.

7.1 Introduction

Table 7.2 Predicted bearing frequencies at nominal full-load speed

Bearing type NU219		Nominal full-load speed 2960 r/min (49.33 Hz)	
BPFO	BPFI	FTF	2 × BSF
338 Hz	451 Hz	21 Hz	337 Hz

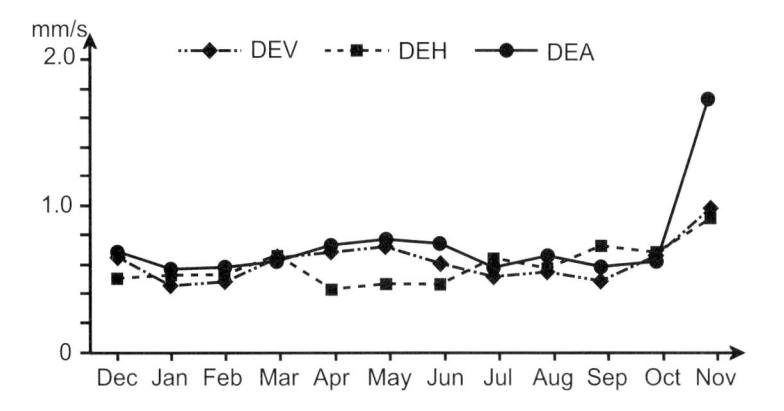

Figure 7.2 Trend graph for 12 months of DE overall velocities r.m.s. on the bearing housing.

(ii) After 12 months the velocity at the DEA position had increased by 166% compared to the average over the previous 11 months, and at the DEV and DEH the velocities had increased by 63% and 60% respectively.

(iii) The change in the DEA velocity between the eleventh and twelfth month is evident in the graph presented in Figure 7.2.

A vibration spectrum analysis was carried out to determine if bearing defect frequencies existed.

7.1.2 Diagnosis of Bearing Defect Frequencies

The predicted bearing defect frequencies at the full-load speed are given in Table 7.2, and Figure 7.3 shows the DEH velocity spectrum.

The *1X* component at 49.688 Hz corresponds to a rotor speed of $N_r = 2981$ r/min, therefore the bearing defect frequencies were re-calculated by multiplying the predicted frequencies at full load by (2981/2960), to give:

$BPFO = 340.5$ Hz; $BPFI = 454.4$ Hz; $FTF = 21.3$ Hz; $2 \times BSF = 339.6$ Hz.

A zoom analysis around *BPFO* produced the spectrum shown in Figure 7.4. Figure 7.5 presents a zoom spectrum to detect the *FTF* component.

7.1.2.1 Interpretation of the Velocity Spectra with the Faulty DE Bearing

(i) Figure 7.3 shows the *BPFO* component at 340.6 Hz (predicted 340.5 Hz) at a velocity of 0.23 mm/s r.m.s., which is due to an outer race defect. BPFO is 20% of the overall velocity at 1.0 mm/s r.m.s.

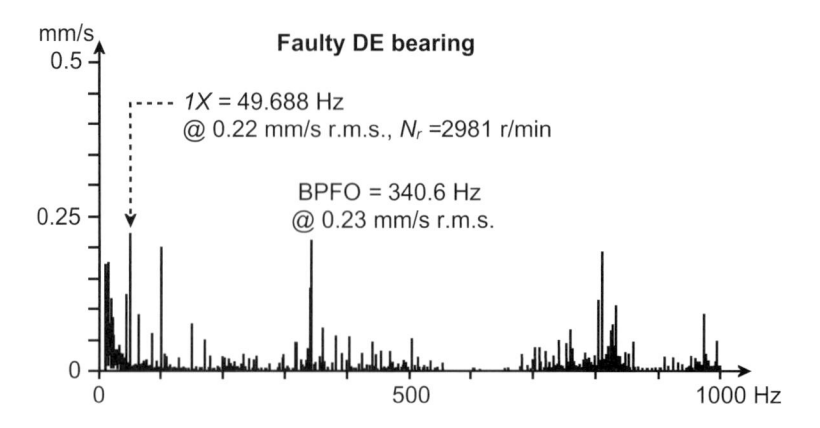

Figure 7.3 DEH velocity spectrum, 0.078 Hz/line.

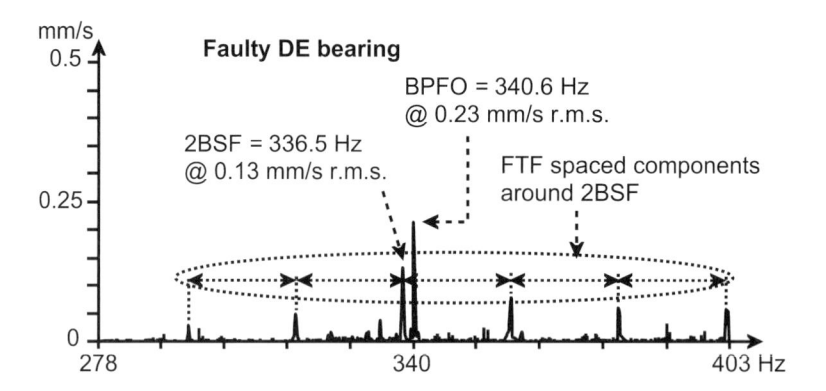

Figure 7.4 DEH velocity zoom spectrum, 0.078 Hz/line.

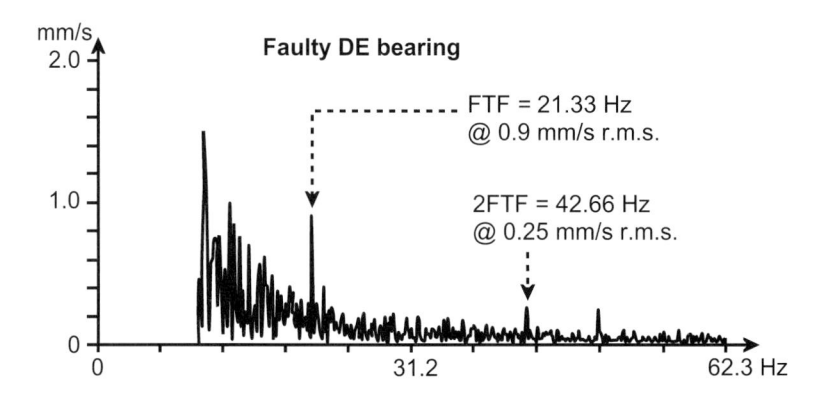

Figure 7.5 DEH velocity zoom spectrum, 0.078 Hz/line.

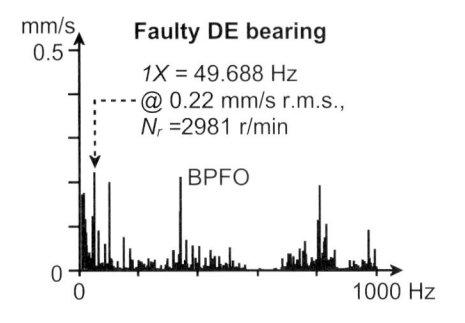

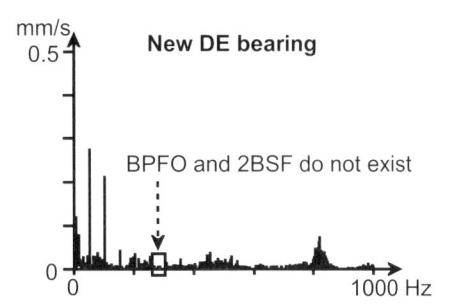

Figure 7.6 DEH velocity spectrum, 0.078 Hz/line.

Figure 7.7 DEH velocity spectrum, 0.078 Hz/line.

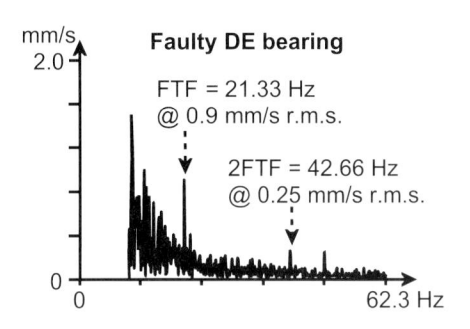

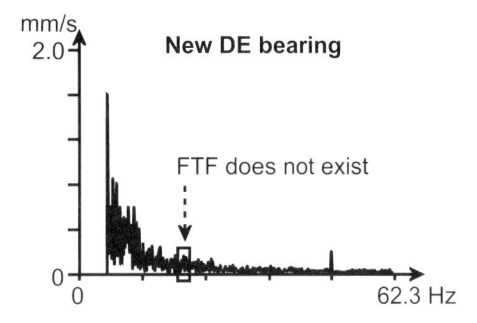

Figure 7.8 DEH velocity zoom spectrum, 0.078 Hz/line.

Figure 7.9 DEH velocity zoom spectrum, 0.078 Hz/line.

(ii) The zoom spectrum in Figure 7.4 shows $2 \times BSF$ at 336.5 Hz with a velocity of 0.13 mm/s r.m.s., which is due to defect/s on the rollers. It is amplitude modulated by the *FTF* component.

(iii) Figure 7.5 confirmed that there was a cage frequency component *FTF*, at 21.33 Hz (predicted 21.3 Hz) with a velocity of 0.9 mm/s r.m.s., which is 90% of the overall velocity.

7.1.2.2 Comparison between Vibration Spectra with the Faulty Bearing and a New DE Bearing

An identical spare motor was installed, which had new bearings previously fitted (see Figures 7.6 to 7.9). Figures 7.7 and 7.9 confirmed that there were no bearing defect frequencies present in the vibration spectra.

7.1.3 Conclusions

(i) Owing to the presence of three bearing defect frequencies, it was predicted that there would be a defect in the outer race and on several roller elements.

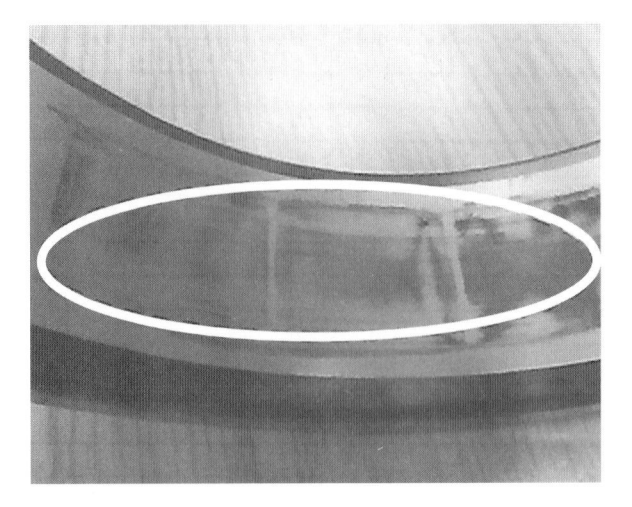

Figure 7.10 False brinelling on the outer raceway in the bearing's load zone.

Figure 7.11 Damaged rollers due to false brinelling.

(ii) This motor was an in-situ spare that typically had not run for six months and it was subjected to vibration from adjacent motors and pumps. It was therefore predicted that false brinelling would exist.
(iii) The DE bearing was removed in a motor repair shop and the photographs in Figures 7.10 and 7.11 confirmed that false brinelling was the problem, therefore the analysis and predictions were correct.

7.2 Industrial Case History – VSA Diagnosed Skidding from an Extra Capacity Cylindrical Roller Bearing (N234E M C3) in a 225 kW/300 H.P. SCIM

7.2.1 Historical Perspective and Summary

Abstract – The main goal of this case history is to verify, via vibration analysis, that *excessive skidding* of a brand new DE bearing during an uncoupled run was due to the repair shop fitting an over capacity cylindrical roller bearing. Skidding can be described as follows:

- The term *skidding* in a roller element bearing is best described as excessive and abnormal sliding of the rollers compared to normal operation where the rollers are rotating in the raceways.
- It is caused by insufficient traction forces between the rollers and raceways to overcome the inertial and drag forces.
- The typical causes of skidding in roller element bearings in SCIMs are during no-load (uncoupled) operation, when the dynamic load is insufficient to load the bearing, or when an overrated extra capacity bearing has been fitted.
- Skidding can result in shear stresses below the surfaces of the raceways, which can lead to premature failure, before the normal fatigue failure at the end of the bearing's operational life.

The industrial site was an LNG processing plant situated on the coastline of Scotland facing the North Sea.

The strategic LV SCIMs were 25 years old and their outer frames were severely corroded. A project to remove 35 motors for repairs and complete overhauls commenced in 2006, and the end user decided on the following procedures:

(i) The motors were to be run uncoupled at an electric motor repair shop and the vibration at the bearings was measured before strip-down.
(ii) The motors would be dismantled and all external parts, such as main frames and end frames, would be thoroughly cleaned and repaired, followed by a *Thermally Sprayed Aluminium* (TSA) process to provide future protection from rust.
(iii) The rotors would be balanced to ISO G1.0 [7.1], to reduce centrifugal forces and vibration on the bearings and transmitted vibration to other parts of the motor, because the OEM's original balance grade was G6.3 in 1981.
(iv) The repair shop was contracted to fit new bearings as per the nameplate designations.

The repaired motors were to be run uncoupled at rated voltage and frequency and the vibration recorded in accordance with the BS 60034-14 2004 vibration standard [7.2].

7.2.2 Vibration Measurements before and after New Bearings were Fitted

The motor's nameplate data were as follows, but note the p.f. and efficiency of this motor, which was manufactured in the UK in 1981, were not stated:

134 Industrial Case Histories – VSA Diagnosed False Brinelling

- 3-phase, SCIM, 415 V, 369 A, 225 kW, 1480 r/min, 50 Hz
- DE bearing: N324 C3, NDE bearing: 6316 C3.

The motor was mounted on a solid base-plate in an electric motor repair shop and supplied at rated voltage and frequency. The NDE fan cowl was removed and the positions of the accelerometers at the DE and NDE are shown in Figures 7.12 and 7.13 respectively. The no-load current was 60 amperes (i.e. 16% of $I_{f.l.}$).

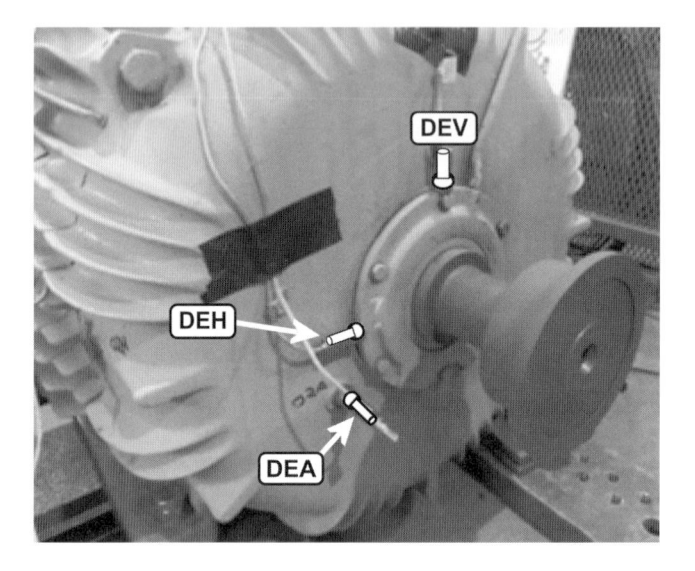

Figure 7.12 DE: positions of accelerometers.

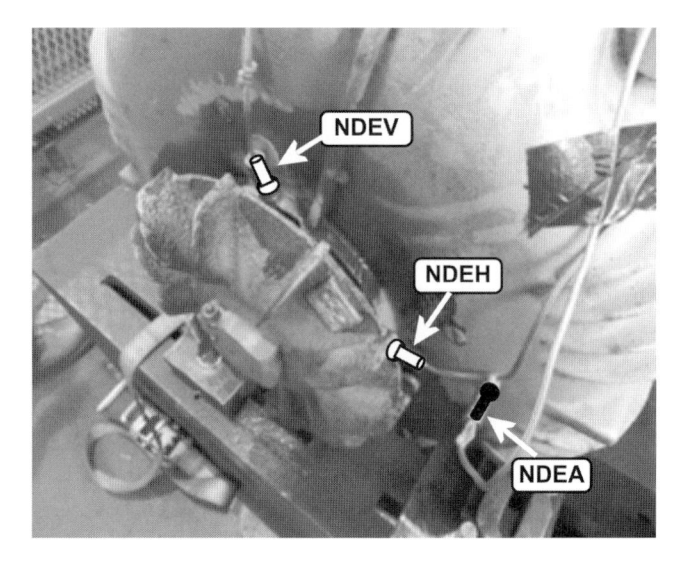

Figure 7.13 NDE: positions of accelerometers.

Table 7.3 Overall r.m.s. velocity levels in mm/s (±10%)

Prior to strip-down: rotor balance grade G6.3					
DEV velocity r.m.s.	DEH velocity r.m.s.	DEA velocity r.m.s.	NDEV velocity r.m.s.	NDEH velocity r.m.s.	NDEA velocity r.m.s.
2.7 mm/s	4.4 mm/s	3.2 mm/s	4.5 mm/s	1.7 mm/s	3.2 mm/s
New bearings: rotor balance grade G1.0					
DEV velocity r.m.s.	DEH velocity r.m.s.	DEA velocity r.m.s.	NDEV velocity r.m.s.	NDEH velocity r.m.s.	NDEA velocity r.m.s.
1.1 mm/s	0.7 mm/s	0.85 mm/s	0.55 mm/s	1.2 mm/s	0.5 mm/s

The overall r.m.s. velocity levels before and after the motor was overhauled and new bearings were fitted are presented in Table 7.3.

The main reason for the low levels (maximum 1.2 mm/s) after the motor was overhauled was that the rotor was re-balanced to G1.0 compared to G6.3 before the overhaul.

7.2.2.1 Comparison of Velocity Spectra before Strip-Down and after New Bearings were Fitted

Figures 7.14 and 7.15 show the DEV velocity spectra with the original DE bearing and after a new DE bearing was fitted.

Key observations:

(i) The *1X* component after balancing the rotor to G1.0 has dropped to 0.26 mm/s compared to 3.7 mm/s before overhaul due to the rotor balance being improved from G6.3 to G1.0.

(ii) However, the acoustic noise from skidding during the no-load run with the new DE bearing was very evident compared to the very low (virtually non-existent) acoustic noise from skidding prior to the overhaul.

(iii) There are no standards that cover unacceptable levels of skidding in roller element bearings. The author therefore could not state that the motor was unfit for purpose because the overall r.m.s. levels were low (maximum of 1.1 mm/s r.m.s. on the DE bearing housing) and the BS 60034-14 2004 vibration standard allows up to 2.3 mm/s r.m.s.

(iv) The electric motor repair shop put forward their view that skidding was normal during an uncoupled run and it should reduce when the SCIM is loaded and coupled to the mechanical load. The author agreed that this is normally the case but the noise level was excessive.

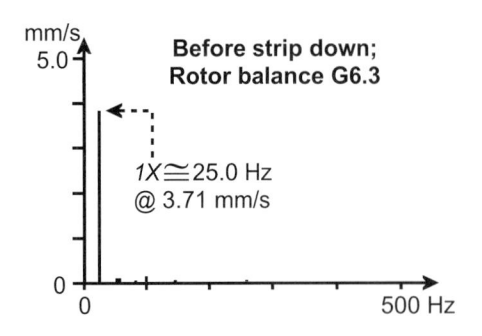

Figure 7.14 No-load uncoupled run prior to strip-down, vibration spectrum, DEV bearing housing.

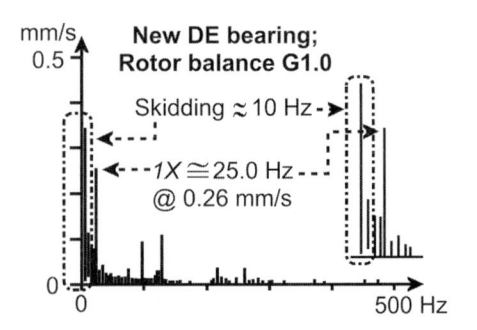

Figure 7.15 No-load uncoupled run with new DE bearing, vibration spectrum, DEV bearing housing.

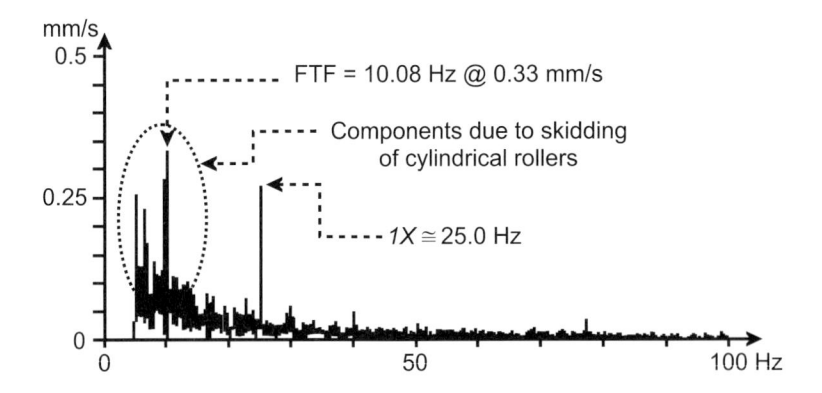

Figure 7.16 Vibration zoom spectrum, vertical DE bearing housing.

(v) The author probed the repair shop as to whether the new roller element bearing at the DE was the same number as shown on the nameplate because he considered the skidding noise to be too loud.

(vi) The repair shop then confirmed that they fitted an extra capacity bearing (N324E M C3) and that is the main reason why the skidding was excessive compared to that from the N324 C3 bearing that was in the motor before it was overhauled.

(vii) Also the centrifugal forces with the re-balanced rotor are less and therefore the dynamic loading on the bearing is reduced, which compounds the problem of skidding due to this overrated E (extra capacity) bearing.

7.2.2.2 Vibration Spectra as a Function of Skidding

A vibration zoom spectrum between 5 Hz and 100 Hz is shown in Figure 7.16. This indicates that the vibration components measured from the new DE bearing housing in the vertical position were due to skidding.

Table 7.4 Bearing defect frequencies

Bearing type: N324E M C3 cylindrical roller element bearing				
E: extra capacity		M: brass cage		C3: clearance
Contact angle 0°		Number of rollers: 13		Rotor speed: 1499 r/min

Bearing frequencies			
BPFO	BPFI	FTF	2 × BSF
130 Hz	194 Hz	10 Hz	121 Hz

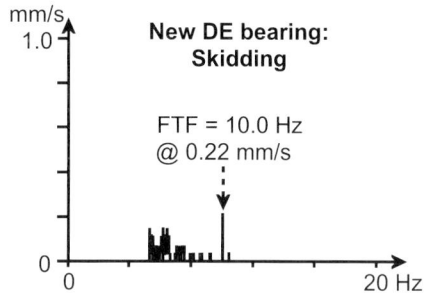

Figure 7.17 Vibration zoom spectrum, 5–20 Hz, vertical DE bearing housing.

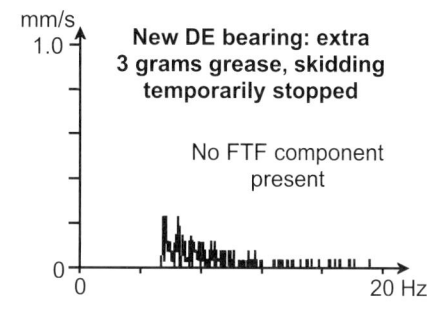

Figure 7.18 Vibration zoom spectrum, 5–20 Hz, vertical DE bearing housing.

There is also a component at 10.08 Hz which is very close to the predicted bearing cage frequency (*FTF*) of 10 Hz at a no-load speed of 1499 r/min as presented in Table 7.4.

A second zoom spectrum between 5 Hz and 20 Hz is shown in Figure 7.17, which was taken at a different time from the velocity spectrum presented between 5 Hz and 100 Hz in Figure 7.16; the component at 10 Hz (*FTF*) is still present in Figure 7.17 but at a lower velocity and the DE bearing was still emitting the classical skidding noise.

It is well known that when grease is inserted while the bearing is skidding the noise will temporarily disappear because a temporary oil film is created between the rollers and outer race. This film is quickly dissipated due to lack of loading on the motor during an uncoupled no-load run of a SCIM of this size and rating.

Figure 7.18 shows the velocity spectrum after the acoustic noise from skidding had disappeared immediately after 3 grams of grease were inserted and there were no components thus verifying that skidding was exciting the bearing cage frequency, *FTF*.

However, after the grease was thrown out due to minimum load on the bearing at no-load, the skidding returned, as expected.

7.3 Conclusions

Figures 7.19 and 7.20 show photographs of an N324E M C3 bearing; note how robust the brass cage is in this extra capacity bearing.

Figure 7.21 illustrates damage due to skidding.

The acoustic noise from skidding is a transient and not constant due to the skidding mechanism.

Inserting grease into the bearing will very temporarily reduce skidding noise.

The E rated extra capacity roller element bearing should not replace the non E rated bearing which is stated on the motor's nameplate. That was the choice selected by the OEM based on the design and operation of the motor.

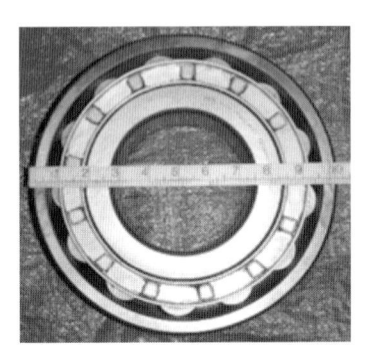

Figure 7.19 Photograph of an N324E M C3 cylindrical roller bearing – the rule indicates the overall diameter (254 mm/10 inches).

Figure 7.20 Photograph of an N324E M C3 to illustrate the rollers, cage and robustness of the bearing.

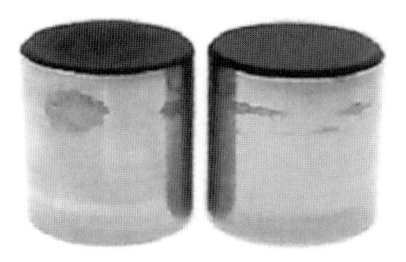

Figure 7.21 Illustration of damage to rollers due to skidding.

References

7.1 ISO 1940-1:Rotor Balancing, 2003.
7.2 BS 60034-14: *Rotating Electrical Machines, Part 14 Mechanical Vibration of Certain Machines with Shaft Heights of 56 mm and Higher – Measurement, Evaluation and Limits of Vibration Severity*, 2004.

8 Industrial Case Histories on VSA to Diagnose Miscellaneous Faults in Rolling Element Bearings in SCIMS

8.1 VSA Detected Corroded Deep Groove Ball Bearing in Vertically Mounted 1.5 kW/2 H.P. SCIMs

Abstract – At an LNG processing plant, four horizontally mounted axial fans are driven by 1.5 kW/2 H.P. SCIMs whose drive ends face upwards in a corrosive environment because the units are installed close to the sea. The DE end frame has channels that are designed so that the rotating shaft can throw off any accumulation of water to prevent it entering the bearing housing. This preventive design operates only when the motors are turning.

The motors only ever operate in pairs with a duty cycle of two weeks on and two weeks off, because only two are required to meet the cooling requirements. However, they still need to operate on a relatively regular basis to avoid the units being stationary for long periods of time with the consequent possibility of false brinelling. When the motors are stationary, the channel design described does not prevent the ingress of water.

New motors were lasting for only approximately 12 to 18 months from installation to failure, and removal of the faulty motors was based on their high acoustic noise, as perceived by the human ear, or when the DE bearing seized, but the faulty motors were never stripped down for inspection, the policy being to replace with new motors and discard the faulty ones. Maintenance personnel merely accepted that failures were an unavoidable consequence of the hostile operating environment.

The processing plant was thirty years old and the existing personnel did not know if the design of the replacement motors was identical to the original design supplied by the OEM of the fan drives. No vibration monitoring was applied to these small-power SCIMs. The costs to remove a faulty motor and replace it with a new one had, however, become unacceptable because it involved the following:

(i) cost of erecting and removing scaffolding
(ii) cost of time for maintenance personnel to remove a faulty motor and fit a new one
(iii) cost of a new replacement 1.5 kW/2 H.P. SCIM.

The maintenance manager at the plant decided to commission the author to carry out vibration measurements, identify the root cause of failures and propose a solution. The case history presents results from vibration measurements taken on the motors and

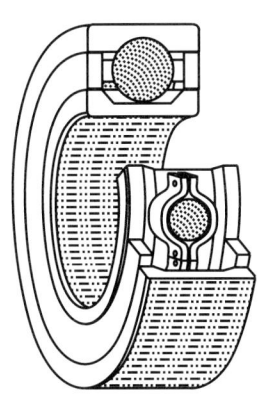

8.1 VSA Detected Corroded Deep Groove Ball Bearing in Vertically Mounted SCIMs

Figure 8.1 Schematic illustration of a deep a groove shielded ball bearing.

fabricated support structure together with spectrum analysis to determine when a motor should be removed for inspection followed by an RCFA (root cause failure analysis).

8.1.1 Motor Data and Description of Installation Layout

The motor's nameplate provided the following data:

- 3-phase, SCIM, 440 V, 1.5 kW/2 H.P., 3.9 A, 50 Hz, 930 r/min (i.e. 6-pole), delta connected, SCIM
- Drive End (DE) bearing: 6206 ZZ C3; (i.e. bore diameter 30 mm/1.18 inches)
- Non Drive End (NDE) bearing: 6205 ZZ C3; (i.e. bore diameter 25 mm/0.984 inches).

The 6205 ZZ C3 and 6206 ZZ C3 bearings are deep groove shielded ball bearings with a C3 clearance as shown in the photograph of Figure 2.2 and illustrated in Figure 8.1.

Figure 8.2 shows a photograph of one of the motors on its mounting bracket, which is welded to the 'H' beam support structure, and Figure 8.3 presents a photograph looking upwards towards the fan.

The main observations from the photographs in Figures 8.2 and 8.3 are as follows:

(i) There is extensive rust on the fan enclosure and 'H' support beams.
(ii) Each fan has four blades.
(iii) It is impossible to mount temporary accelerometers on either the DE, NDE bearing housings or on the end frames.
(iv) Owing to the external cooling channels there are a very limited number of positions on the motor's frame to mount accelerometers.

8.1.2 Vibration Measurements

The positions of temporary accelerometers, which were mounted via magnetic attachments, are shown in Figure 8.4. The diagnosis of bearing defect frequencies will be

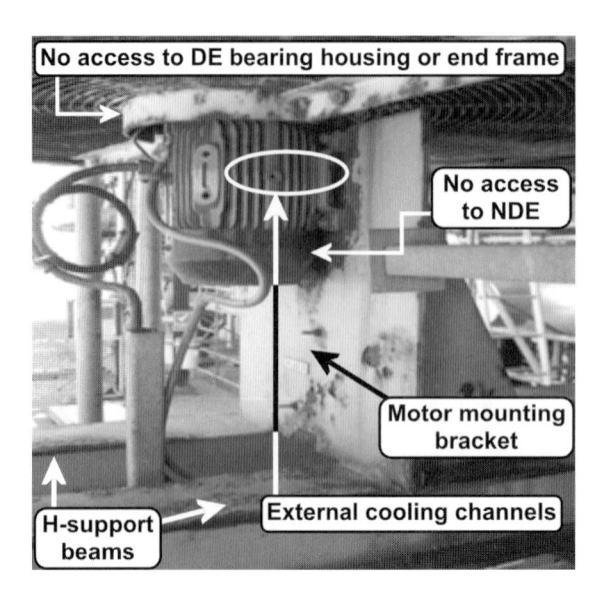

Figure 8.2 Photograph of motor, mounting bracket and 'H' beam support structure.

Figure 8.3 Photograph of motor, fan and its enclosure, which demonstrates the extensive corrosion.

difficult, because the magnitude of these components can be attenuated due to the mechanically stiff end frames in the radial direction, the stator core structure and the outer frame of the motor.

Figure 8.5 illustrates the 'H' beam support structure for mounting the motors to the vertical support brackets. It is obvious that there will be vibration transmitted through the 'H' beams between motors, and this further complicates the analysis of vibration.

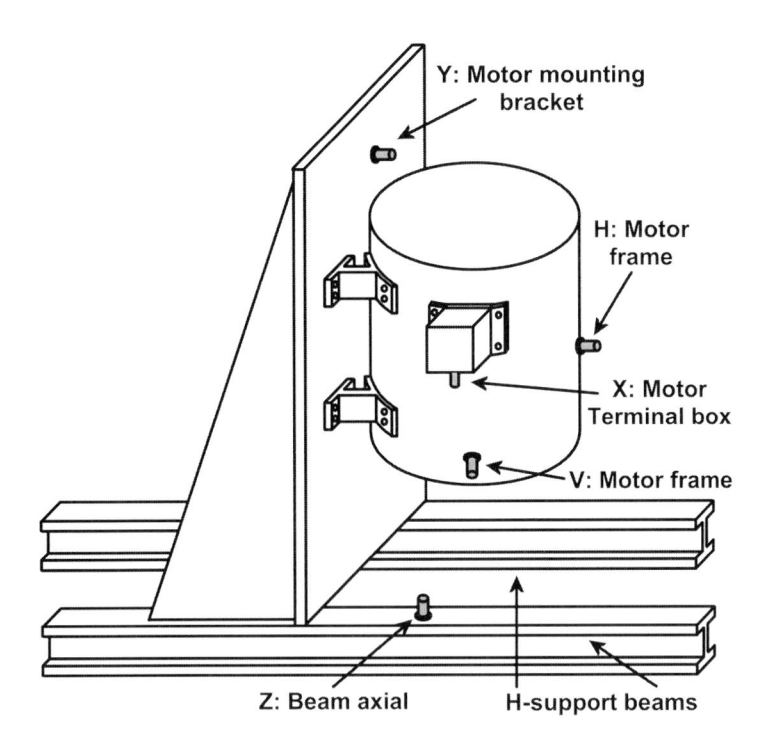

8.1 VSA Detected Corroded Deep Groove Ball Bearing in Vertically Mounted SCIMs 143

Figure 8.4 Schematic layout of accelerometer positions.

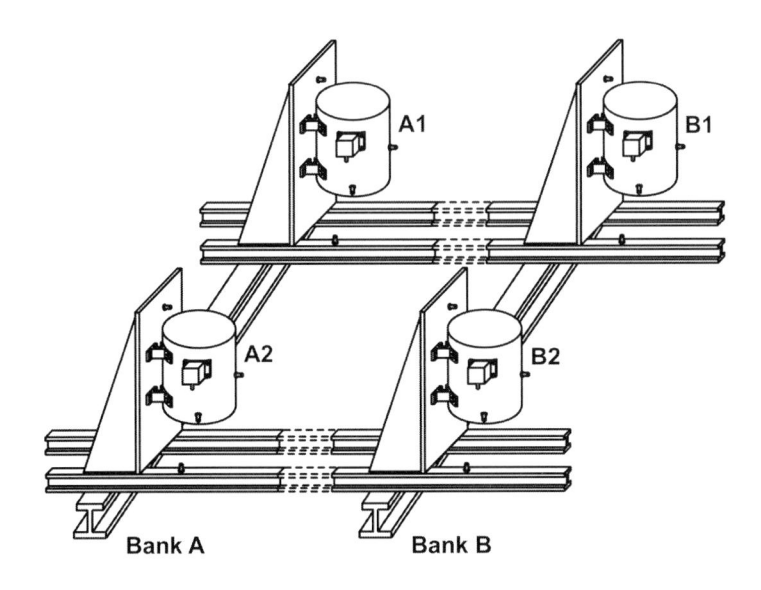

Figure 8.5 Schematic layout of 'H' beam support structure and positions of the four axial fan motors; Bank A: A1 and A2; Bank B: B1 and B2.

Because it was vital to obtain data on the vibration from each motor operating on its own, permission was given by the end user to run the units in various configurations, and vibration measurements were taken with the fan motors running in the following modes:

(i) Bank A: A1 and A2 running but with Bank B off – normal operation.
(ii) Bank B: B1 and B2 running but with Bank A off – normal operation.
(iii) A1, A2, B1, B2 – each motor running on its own – not normal operation.
(iv) Bank A and B on – all motors operating – not normal operation.

8.1.2.1 Overall Velocity Levels

Overall velocities (r.m.s.) were measured using a span of 10–1000 Hz and for the spectrum analysis the instrument was set to 12,800 lines, which gives a frequency resolution of 0.077 Hz/line.

Table 8.1 presents a full set of velocities (r.m.s.) immediately prior to one of the motors being removed. The actual acoustic noise perceived by on-site personnel and the author was used as a parameter for the selection of the quietest and noisiest motors when operating alone.

The key observations were as follows:

(i) Bank A: A1 running alone was perceived as the quietest motor.
(ii) Bank B: B2 running alone was perceived as the noisiest motor.

A comparison between the velocity levels from the quietest and noisiest motors is presented in Table 8.1.

(i) From Table 8.2 the velocity in the horizontal direction on the B2 motor's frame was 5.6 times larger than at the same position on motor A1.
(ii) There was a strong correlation between vibration levels and perceived acoustic noise from each of these motors operating alone.
(iii) Also, the vibration on the motor frames in the horizontal direction of A1 alone and with A1 and A2 operating together (normal) shows that the vibration on A1 has increased by a factor of 1.67, thus confirming vibration transmission from A2 to A1.

8.1.2.2 Vibration Spectrum Analysis and Interpretation

The vibration spectra for the problem motor (B2) and the healthy motor (A1) are presented in Figures 8.6 and 8.7. Note that the velocities of any components above 200 Hz were negligible.

An interpretation of the spectra indicates the following:

(i) There was a clear difference between the velocity spectra shown in Figures 8.6 and 8.7 for the noisy (B2) and quiet (A1) motors respectively.
(ii) The predicted bearing defect frequencies are given in Table 8.2 for the actual speeds of motors B2 and A1.

8.1 VSA Detected Corroded Deep Groove Ball Bearing in Vertically Mounted SCIMs

Table 8.1 Overall velocities levels in mm/s (within ±10%), 10–1000 Hz

Vibration measurements on motor A1				
Accelerometer position	A bank on B bank off	B bank on A bank off	All on	A1 alone
Motor frame V	1.4	0.9	1.6	0.6
Motor frame H	1.5	1.2	2.0	0.9
Motor terminal box A	2.0	0.8	1.6	1.4
Mounting bracket H	1.4	1.2	1.7	0.65
H beam A	1.2	0.8	1.3	0.4
Vibration measurements on motor A2				
Accelerometer position	A bank on B bank off	B bank on A bank off	All on	A2 alone
Motor frame V	2.4	1.2	2.5	2.7
Motor frame H	3.3	3.2	3.8	3.0
Motor terminal box A	6.0	1.8	5.2	6.3
Mounting bracket H	3.0	2.7	4.6	2.8
H beam A	2.0	1.5	2.3	2.0
Vibration measurements on motor B1				
Accelerometer position	A bank on A bank off	B bank on A bank off	All on	B1 alone
Motor frame V	0.8	2.3	3.0	0.5
Motor frame H	0.95	2.0	3.7	3.2
Motor terminal box A	1.0	2.8	4.3	4.0
Mounting bracket H	1.0	2.0	2.5	2.8
H beam A	0.6	1.5	1.8	0.55
Vibration measurements on motor B2				
Accelerometer position	A bank on B bank off	B bank on A bank off	All on	B2 alone
Motor frame V	1.0	3.8	4.4	3.3
Motor frame H	4.0	4.0	5.0	5.0
Motor terminal box A	1.2	5.5	8.6	6.6
Mounting bracket H	2.7	4.2	5.3	5.0
H beam A	1.2	2.4	2.9	2.7

Table 8.2 Bearing defect frequencies

DE bearing: 6206 ZZ C3		NDE bearing: 6205 ZZ C3	
Contact angle: 0°	No. of balls: 9	Motor speed: 956 r/min; 1X = 15.94 Hz	
BPFO: 56.9 Hz	BPFI: 86.6 Hz	FTF: 6.3 Hz	2 × BSF: 73.6 Hz

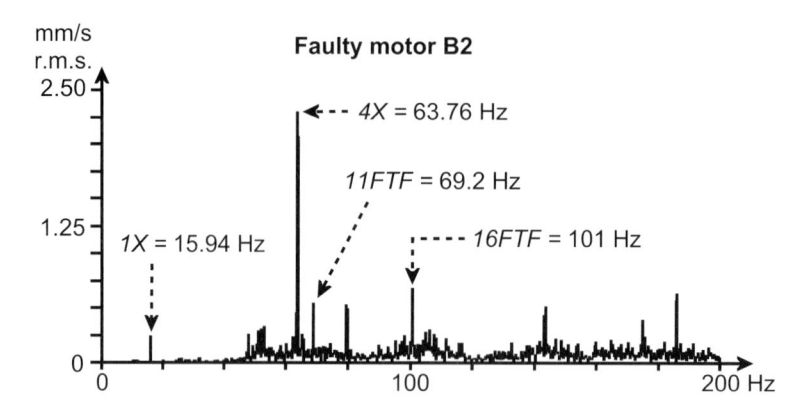

Industrial Case Histories on VSA to Diagnose Miscellaneous Faults

Figure 8.6 Faulty motor B2, DE horizontal velocity r.m.s. versus frequency spectrum.

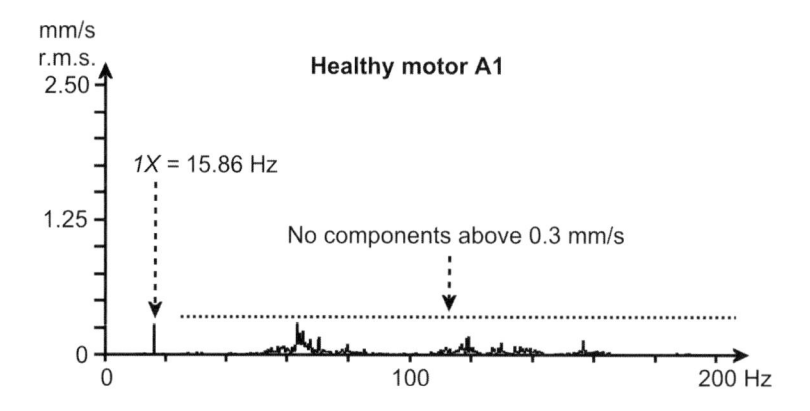

Figure 8.7 Healthy motor A1, DE horizontal velocity r.m.s. versus frequency spectrum.

In the healthy motor there are no components higher than 0.3 mm/s r.m.s., whereas in the spectrum from the problem motor there are in fact harmonics of the cage defect frequency at 11*FTF* and 16*FTF* at 0.6 mm/s and 0.7 mm/s r.m.s. respectively.

(iii) The fan blades passing frequency at *4X* was 2.3 mm/s r.m.s. in the problem motor (B2), which will cause higher centrifugal forces on the DE bearing, compared to *4X* at 0.3 mm/s in the healthy motor (A1).

(iv) This suggests that the mechanical unbalance in the fan of the faulty motor was much greater (value unknown) than in the healthy motor's fan.

(v) It was also noted that, when a strong wind was imposed on the fan in the opposite direction to its rotation, the *4X* component increased in velocity.

(vi) It was therefore recommended to remove motor B2 for a full strip-down and inspection of the DE bearing.

8.1 VSA Detected Corroded Deep Groove Ball Bearing in Vertically Mounted SCIMs 147

Figure 8.8 Motor before strip down – recall it is vertically mounted with the drive end upwards.

Figure 8.9 Close-up of the labyrinth seal set up.

8.1.3 Inspection of the Motor and DE Bearing

Figures 8.8 and 8.9 show photographs of end views of the motor's DE to illustrate the channels and labyrinth seal arrangement, which ensure that any accumulation of water is thrown off the bearing provided that the shaft is rotating.

The motors remained stationary for 26 weeks per annum and this was therefore the cause of the ingress of water because the seal system cannot function when the motor is stationary. Figures 8.10 and 8.11 verify the highly corroded and contaminated condition of the DE bearing.

Figure 8.10 Corroded and contaminated DE bearing.

Figure 8.11 Heavily corroded and contaminated DE bearing – the face shown is towards the DE end frame.

8.1.4 Proposed New Motor Design

The bearing seals must prevent the ingress of water during all operating and stationary conditions, and special features include:

(i) There must be radial and axial sealing at the DE.
(ii) The rotor shaft must be stainless steel.
(iii) Fan cowl and fixings to be stainless steel.

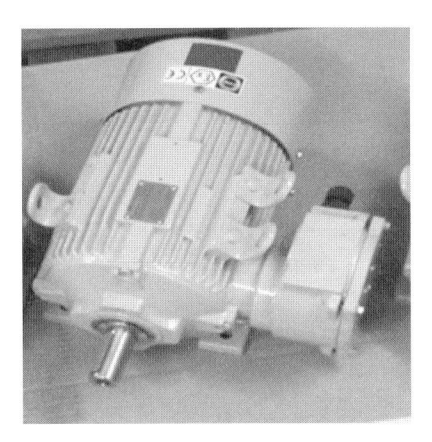

8.1 VSA Detected Corroded Deep Groove Ball Bearing in Vertically Mounted SCIMs

Figure 8.12 Photograph of new motor.

Figure 8.13 Close-up photograph of the DE of the new motor.

Motor rating:

- 3-phase SCIM, 1.5 kW/2 H.P., 440 V($\pm$10%), 3.9 A, 50 Hz, 945 r/min
- $\cos\Phi$ = 0.75, η = IE1–77. DIN 60034; PTB 09 ATEX 1011 X; IP66
- Type: DB100L, vertical foot mounted drive shaft upwards IMV6
- DE and NDE bearings 6206 2ZR C3.

Photographs of the new motor are shown in Figures 8.12 and 8.13, and Figure 8.14 illustrates the special seal system at the DE.

8.1.4.1 Overall Vibration Levels and VSA of New Motor Design

Table 8.3 presents a comparison between overall velocities r.m.s. from the old motor design with a faulty DE bearing and the new motor design in tag slot B2. Note that the fan was re-balanced.

Table 8.3 Overall velocities levels in mm/s (within ±10%); span: 10–1000 Hz

Original noisy motor and new motor design

Accelerometer position	Noisy motor: only B2 running mm/s r.m.s.	New motor design: only B2 running mm/s r.m.s.
Motor frame V	3.3	0.7
Motor frame H	5	2
Motor terminal box A1	6.6	2.2

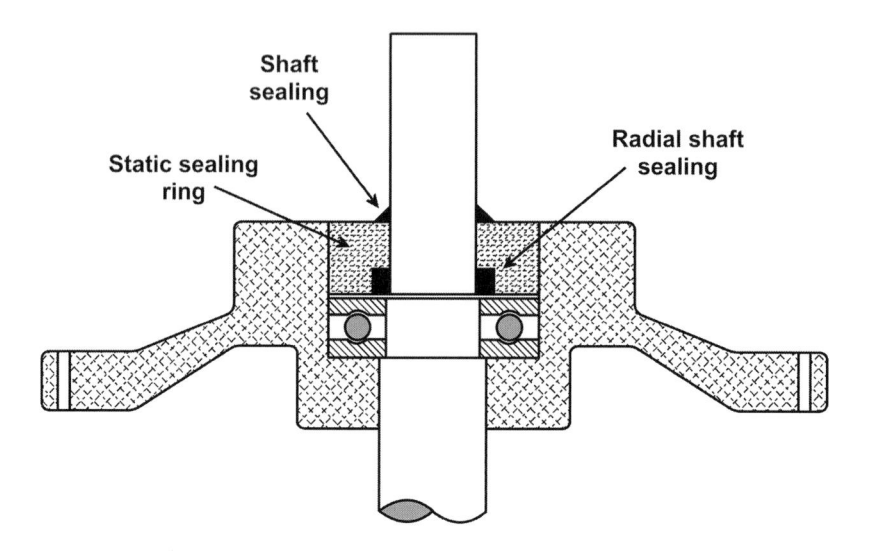

Figure 8.14 Illustration of double sealing system to prevent the ingress of water or any other contamination.

Table 8.3 shows that, by comparison to the overall velocities produced by the original faulty motor, there was a large reduction of those from the new design. Both motors were in tag slot B2 when the measurements were taken.

The quantitative reductions were as follows:

Motor frame V: Reduction in velocity by a factor of 4.6.
Motor frame H: Reduction in velocity by a factor of 2.5.
Motor terminal box A1: Reduction in velocity by a factor of 3.

The velocity spectra from the faulty motor and new motor design are presented in Figures 8.15 and 8.16.

8.1.5 Conclusions

(i) There was evidence of harmonics of *FTF* as shown in Figure 8.15 in the original motor with the faulty DE bearing. The spectrum in Figure 8.16 shows that there are no bearing defect frequencies from the new motor design.

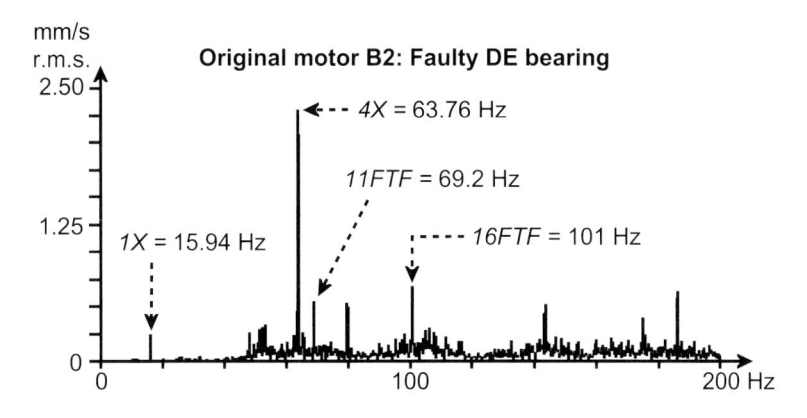

8.1 VSA Detected Corroded Deep Groove Ball Bearing in Vertically Mounted SCIMs

Figure 8.15 Faulty motor in slot B2, DE horizontal velocity r.m.s. versus frequency spectrum.

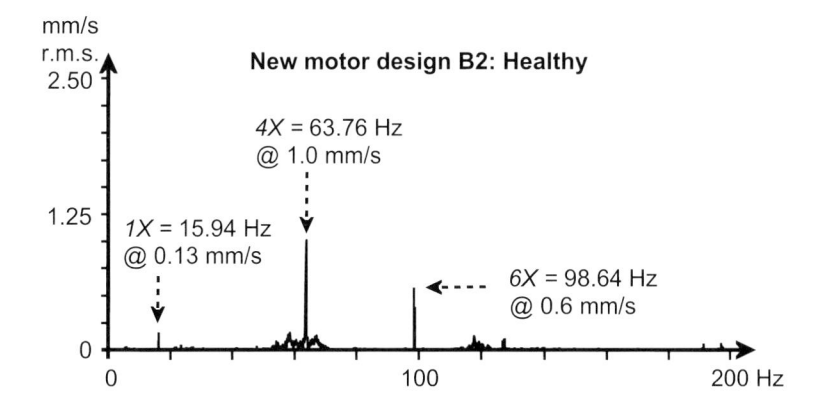

Figure 8.16 New motor in slot B2, DE horizontal velocity r.m.s. versus frequency spectrum.

(ii) The *4X* vibration component on the original motor was 2.3 mm/s which was due to mechanical unbalance of the external fan; whereas with the new motor and the fan re-balanced it dropped to 1.0 mm/s.

(iii) The original axial fan motors operated on a two week on/off regime; thus for 26 weeks they were stationary in a hostile environment because the LNG plant is on the coast directly next to the North Sea off Scotland.

(iv) Figures 8.8 and 8.9 show photographs of end views of the original motor's DE to illustrate the channels and labyrinth seal arrangement, which is supposed to prevent any accumulation of water entering the bearing. This design only functioned properly provided that the shaft was rotating, but for 26 weeks (annually) it was stationary.

(v) The fundamental root cause of bearing seizures at the DE of the motors was caused by the ingress of water with consequential corrosion and contamination of the DE bearings.

(vi) A new design of motor was proposed to prevent the ingress of water and is presented in Section 8.1.5. The design has a double bearing seal assembly and has proven to be successful because no DE bearing failures have been reported since 2015. This is in comparison to the original motor designs which had regular DE bearing failures of eight motors every 12 to 18 months.

8.2 Industrial Case History – Envelope Analysis used by a Vibration Sub-Contractor Produced a False Diagnosis of a Cage Fault in a Cylindrical Roller Element Bearing in a 225 kW/300 H.P. SCIM

8.2.1 Introduction

Abstract – An oil company that operated an offshore oil production platform in the North Sea off the coast of Scotland sub-contracted a vibration condition monitoring (CM) vendor to carry out monthly vibration measurements on rotating equipment.

The sub-contractor used envelope analysis to assess the operational condition of rolling element bearings in the LV (440 V, 60 Hz) SCIMs. On a previous occasion, the sub-contractor had predicted a bearing cage defect in one of the SCIMs, but this was proven to be false after strip down and examination of the DE bearing by an independent motor repair workshop.

The sub-contractor subsequently reported another cage fault in the DE bearing of a different SCIM, and this is the subject of this case history. The oil company decided to carry out a quality assurance (QA) test case on the sub-contractor's prediction of another cage fault and commissioned the author to carry out an investigation with the following objectives:

(i) With the motor running uncoupled in a motor repair workshop, carry out independent vibration measurements on the motor prior to strip-down for overhaul.
(ii) To analyse the vibration prior to strip-down of the motor and predict if either a cage fault or any other bearing fault existed in the DE bearing.
(iii) To obtain a preliminary report from the repair shop on the condition of the bearings.

The author predicted, by using VSA, that there was not a cage fault in the DE cylindrical roller element bearing, and subsequent examination of the bearing confirmed that there were no bearing defects whatsoever and only slight and normal wear as reported by the repair workshop.

Figure 8.17 Photograph of the motor on the repair shop's base-plate.

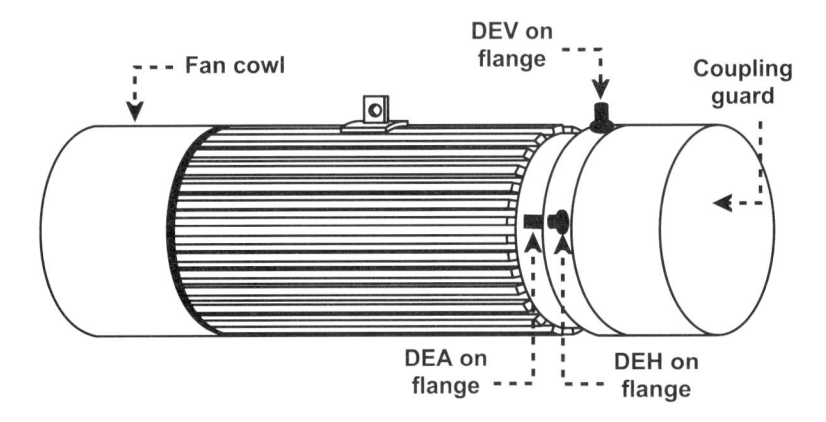

Figure 8.18 Illustration of motor to show coupling guard covering the DE bearing and the cowl over the NDE.

8.2.2 Nameplate Data and Construction of the Motor

- 3-phase SCIM, 440 V, 225 kW/300 H.P., 60 Hz, 352 A, 1775 r.p.m.
- Ins class: F, temp rise: 80 °C
- Eff = 93.2%, 0.9 p.f. Δ – connected
- Bearings: DE N324-C3, NDE 6316-C3.

A photograph of the motor mounted on the repair shop's base-plate is shown in Figure 8.17. It is a flange mounted motor, which drives a centrifugal pump. and when it is installed on the offshore oil production platform the coupling guard is bolted to the mounting flange at the DE, as illustrated in Figure 8.18, such that access to the DE bearing housing is impossible.

Table 8.4 Overall vibration measurements by a condition monitoring sub-contractor at the DE

Accelerometer	Measured quantity	Instrument reading
DEH on motor flange	Overall r.m.s. velocity	1.5 mm/s r.m.s.
	Enveloped acceleration	4.0 g Env
	Acceleration	1.0 g
DEV on motor flange	Overall r.m.s. velocity	1.4 mm/s r.m.s.
DEA on motor flange	Overall r.m.s. velocity	1.5 mm/s r.m.s.

A SCIM of this voltage and power rating will normally be cooled by an external fan, which is covered by a fan cowl, and therefore access to the NDE bearing housing was impossible, as shown in Figure 8.17.

The CM sub-contractor applied vibration envelope analysis to assess the condition of rolling element bearings in the motor shown in Figure 8.17 by measuring the vibration on the outer periphery of the DE flange and on the fan cowl in line with the end frame at the NDE. The author considers it was unwise to use envelope analysis, because the vibration on the outer periphery of the DE flange is not the same as the vibration on the bearing housing.

It is very well known that the special accelerometer used for subsequent envelope analysis should be mounted directly on the bearing housings to reliably diagnose faults in rolling element bearings in SCIMs as discussed in Section 4.2.1.1.

The overall levels of velocity, acceleration and enveloped acceleration recorded by the CM sub-contractor and presented to the author's client are shown in Table 8.4. The oil company has an alarm and shut-down policy for SCIMs with rolling element bearings at 7 mm/s and 11 mm/s r.m.s. respectively when driving the mechanical load.

The enveloped acceleration spectrum presented by the sub-contractor to the client is presented in Figure 8.19, which shows that there was a component at 12.5 Hz and its harmonics. It was proposed by the sub-contractor that this was due to a cage fault in the DE bearing.

The component at 12.5 Hz in Figure 8.19 is not a cage *FTF* frequency component.

8.2.2.1 Prediction of the DE Bearing Defect Frequencies

It was unknown when, or if, this motor had been previously overhauled or if new bearings were fitted at any time, therefore the following points which were presented in the flow chart of Section 4.3 need to be considered.

(i) Although the nameplate specifies an N324 C3 steel cage bearing, two major bearing manufacturers no longer provide this standard steel cage cylindrical roller element bearing as an *off-the-shelf stock* item. They now only supply, as a stock item, an extra capacity bearing replacement, namely an N324 E M C3 to replace an N324 C3. Similarly, for other cylindrical roller element bearings.

(ii) Note that the steel cage version of an N324 C3 has 12 rollers and the N324E M C3 has 13 rollers; therefore the bearing defect frequencies are different.

Table 8.5 Bearing defect frequencies

Bearing type: N324 C3 cylindrical roller element bearing			
Contact angle: 0°	Number of rollers: 12	Motor speed: 1799 r/min	
Bearing frequencies			
BPFO	BPFI	FTF	2 × BSF
145.8 Hz	214 Hz	12.2 Hz	152.6 Hz

Table 8.6 Bearing defect frequencies

Bearing type: N324E M C3 cylindrical roller element bearing			
E: Extra capacity	M: Brass cage	C3: Clearance	
Contact angle: 0°	Number of rollers: 13	Motor speed: 1799 r/min	
Bearing frequencies			
BPFO	BPFI	FTF	2 × BSF
156.3 Hz	233.5 Hz	12 Hz	145.6 Hz

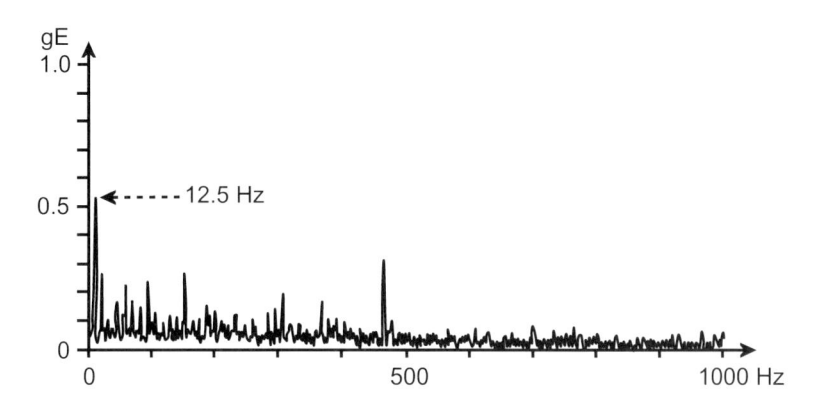

Figure 8.19 Envelope spectrum (gE) versus frequency (Hz), DEH on the flange.

The bearing defect frequencies for both bearings are presented in Tables 8.5 and 8.6.

The bearing defect frequencies predicted by the author for an N324 C3 and an N324E MC3 (it was unknown which bearing was in the motor) are presented in Tables 8.5 and 8.6 for a no-load speed of 1799 r/min and the cage frequency component *FTF* is 12 Hz and 12.2 Hz respectively.

The sub-contractor tested the motor on load on the offshore installation but did not state the operating current or speed, but it would certainly have been less than 1799 r/min (no-load uncoupled speed) and the cage frequency *FTF* would be 11.84 Hz and 12 Hz for an N324 C3 and N324E M C3 respectively if the motor was on full load. This oversight is typical of a mechanical vibration technician not understanding the basics of induction motor operation.

8.2.3 Vibration Measurements and VSA Applied to the SCIM before Strip-Down and Inspection of the DE Bearing

8.2.3.1 Test Conditions
(1) The motor was mounted on a solid steel base-plate in the motor repair workshop as shown in Figure 8.17.
(2) The motor was supplied at the rated nameplate voltage (440 V) and, at the time of testing, the supply frequency was $f_1 = 60.47$ Hz. The supply frequency was subject to variation because it was supplied from an independent generator.
(3) The no-load input line currents were balanced to within ±1.5 amperes of the average line current, which was 56 amperes.
(4) Because the sub-contractor had predicted a cage fault in the DE bearing the focus was on taking vibration measurements at the DE at the same positions as those used by the sub-contractor as shown in Figure 8.18.
(5) Vibration measurements were also recorded directly on the actual bearing housings at the DE and NDE (the fan cowl was removed) because this was possible when the motor was run uncoupled in the repair workshop.
(6) Overall vibration velocity levels (r.m.s.) using a frequency span of 5.3 Hz to 1000 Hz were recorded, based on using the British Standard, BS EN 60034-14, Feb. 2004 [8.1] as a general guide. The maximum permissible velocity (r.m.s.) in either the V, H or axial positions on the bearing housing is 2.3 mm/s ± 10% for this motor, when rigidly mounted on a solid steel base-plate during an uncoupled run.

8.2.3.2 Overall R.M.S Velocities Prior to Motor Strip-Down
The overall r.m.s. velocities measured by the author are presented in Table 8.7 before strip down and removal of the bearings for inspection. Note the differences in velocity between the readings from accelerometers mounted directly on the bearing housings and those from accelerometers mounted on the positions used by the sub-contractor.

For example, at the DEV position on the bearing housing the velocity was 3.2 mm/s r.m.s. but only 1.4 mm/s r.m.s. in the vertical position on the outer periphery of the DE end frame.

The overall r.m.s. velocities presented in Table 8.7 are acceptable for a SCIM that has been operating on an offshore oil production platform, with the highest being 3.2 mm/s at the DEV position on the bearing housing of this SCIM. Therefore, based on these velocities during an uncoupled run the motor was fit for purpose. The DEV bearing velocity is 2.3 times greater than the DEV velocity on the DE flange. No historical records of repairs or PM inspections were available.

8.2 Industrial Case History – Envelope Analysis

Table 8.7 Overall r.m.s. velocities before motor strip down, span 5.3–1000 Hz

Accelerometers mounted at positions shown in Figures 8.17 and 8.18 during the uncoupled run		
DEV	DEH	DEA
Outer periphery of DE flange	Outer periphery of DE flange	See Figure 8.18
1.4 mm/s	2.4 mm/s	1.0 mm/s.

Accelerometers mounted directly on the DE and NDE bearing housings during the uncoupled run					
DEV	DEH	DEA	NDEV	NDEV	NDEA
3.2 mm/s	1.8 mm/s	1.5 mm/s	2.4 mm/s	2.6 mm/s	2.5 mm/s

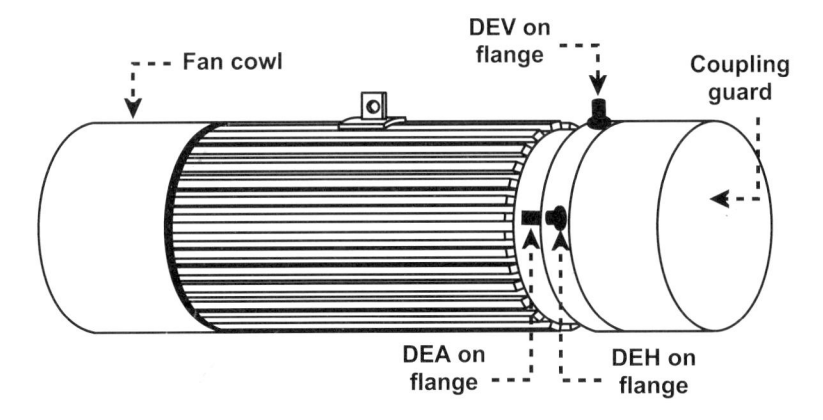

Figure 8.20 Illustration of motor to show where the vibration was measured that produced the spectra shown in Figures 8.21 to 8.24.

8.2.3.3 Vibration Spectrum Analysis – Uncoupled Run before Strip-Down of the Motor

Figure 8.20 shows the accelerometer positions and Figures 8.21 to 8.24 show the spectra from the DE positions on the outer periphery of the DE flange. Recall that these are the positions where the vibration CM sub-contractor took the measurements on the motor when operating on the offshore platform. The vibration spectra are normal during the uncoupled run before strip-down of the motor and there is no evidence of any cage defect frequency at 12 Hz.

However, the vibration zoom spectrum of the DEH vibration on the flange indicates that there are components at 11.6 Hz and 12.5 Hz as shown in Figure 8.22, but these are not due to a cage defect in either an N324 C3 or an N324 EM C3 bearing.

To recap, Figure 8.20 shows the positions that vibration measurements were taken to produce the spectra shown in Figures 8.21 to 8.24.

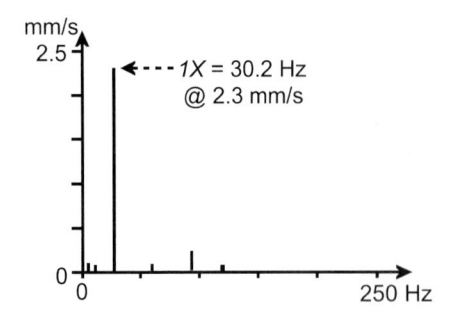

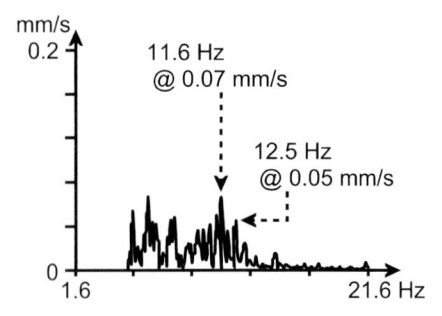

Figure 8.21 DEH spectrum from the DE flange, 0.078 Hz/line.

Figure 8.22 DEH zoom spectrum from the DE flange, 0.078 Hz/line.

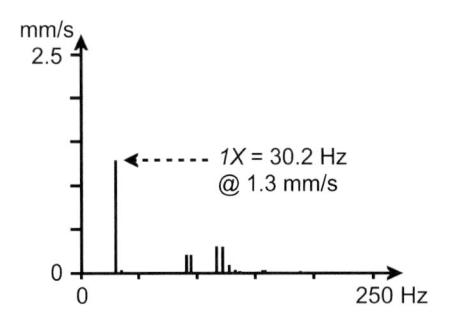

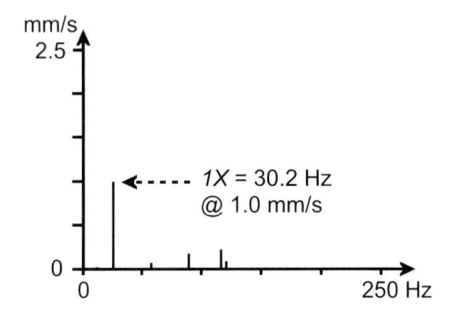

Figure 8.23 DEV spectrum from the DE flange, 0.078 Hz/line.

Figure 8.24 DEA spectrum from the DE flange, 0.078 Hz/line.

The components shown in Figure 8.22 can be the effect of skidding of a cylindrical roller element bearing during an uncoupled run of a SCIM, as was proven in the case history in Section 7.2.

Interestingly, the CM sub-contractor stated that the cage frequency was at 12.5 Hz when the motor was tested on load when driving the pump on the oil production platform.

This was not a cage defect frequency, but note that the DE bearing could have been skidding, particularly if it was an extra capacity rated N324 EM C3 bearing, which is over rated for the motor and would not be in accordance with the OEM's original design specification.

The author predicted that there was no cage defect in the DE bearing, after which the motor was dismantled and the bearings were inspected.

8.2.3.4 Preliminary Inspection of the DE Bearing

The actual bearings in the motor confirmed after motor strip-down, were as follows:

DE: N324 E M C3 (i.e. an extra capacity, brass cage, with C3 clearance).
NDE: 6316 C3.

Figure 8.25 Photograph of the DE bearing which indicates that no cage fault existed; note that the brass cage is very robust.

Clearly this motor had been previously repaired, and an extra capacity bearing was fitted at the DE, which is what the author anticipated would be the case.

This bearing should not have been fitted because the OEM's design for this SCIM did not require an extra capacity bearing.
 The key factor is that an E rated bearing requires more dynamic loading to prevent excessive skidding.

8.2.4 Conclusions

The repair shop reported that there was no cage defect in the DE bearing, and the general condition of the bearing was that it was only slightly worn due to the normal wear process.

Figure 8.25 shows a photograph of the DE bearing and it is in a normal condition and to quote the repair shop:

'The DE bearing, if properly maintained would
 have continued to perform with no problems'.

This verified the author's prediction, via VSA, that there was not a cage defect and confirmed the CM sub-contractor's prediction that a cage fault existing in the DE bearing was a false diagnosis due to the use of vibration envelope analysis in a situation when the accelerometers could not be fitted directly on the bearing housing.

Figures 8.26 and 8.27 present photographs of a new N324 E M C3 bearing and it is obvious that the brass cage is very robust indeed.

A cage fault in this brass cage, of an extra capacity rated bearing would be a rare event.

8.3 Bearing Failures in 800 kW/1072 H.P. SCIMs Driving Sulphate Removal Pumps (SRP) and a FAT on Repaired Motor

8.3.1 Catastrophic Bearing Failure and Broken Shaft

Abstract – There are *two case histories* in Section 8.3. This first case history reports on a catastrophic failure of an NDE bearing and a consequential broken shaft, which

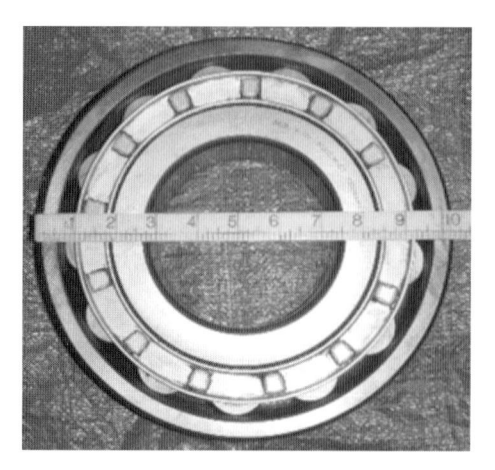

Figure 8.26 Photograph of a new N324E M C3 cylindrical roller bearing – the ruler scaled in inches indicates the overall diameter.

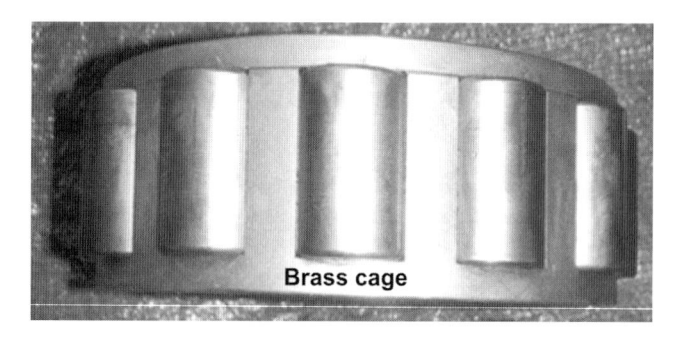

Figure 8.27 Photograph of an N324E M C3 to illustrates the rollers, cage and robustness of this bearing.

occurred in one (A) of five SCIMs driving sulphate removal pumps (SRP) on a large offshore oil production (approximately 200,000 barrels of oil per day) platform.

There were also subsequent failures of DE and NDE 6316 C3 bearings in identical motors. The routine vibration condition monitoring was carried out once per month by a CM vendor who did not have experts on the design and operation of SCIMs. The author was not involved in vibration monitoring and analysis of any of the motors prior to bearing failures. He was initially hired by the oil company to investigate this catastrophic failure.

It was difficult to obtain reliable operational records and measured vibration data from the date of installation of these new motors in 2007 up to the time of a bearing failure. Maintenance crews changed every three weeks, therefore it was difficult to obtain concise and unambiguous records. An RCFA of the failed bearing was impossible because the bearing was seized due to the very high temperature, which was more than 900 °C at the inner raceway, as predicted from an independent metallurgical examination. Thus, any initial evidence of bearing degradation was destroyed.

The reason for presenting the inspection of this failure is to demonstrate the catastrophic damage that can be caused by a failed bearing, and as a lead in to the case histories that follow on identical motors.

The SCIM details are as follows:

- 3-phase SCIM, 6.6kV, 800 kW/1072 H.P., 76 A, 60 Hz, 3580 r/min, 0.9 p.f., eff = 96%
- Star connected, frame E450/1400, IP56, Class F, B temperature rise
- Number of rotor bars: 50, single cage, rotor bars not skewed, airgap length 4 mm
- Bearings: DE and NDE 6316 C3, re-greasing quantity 33 g
- Nominal bearing L_{10} life: 30,000 hours, lubrication details: use Shell Albida R2 grease
- OEM's recommendation: bearings should be greased every 1200 running hours using the quantity of grease given above.

Note that the pump manufacturer supplied the complete drive train and, as is normally the case, various electrical machine manufacturers were invited to tender for the supply of the motors. The oil company contracted the pump manufacturer to supply the motor as part of the complete package.

The L_{10} life for the bearings was 30,000 hours, which is 3.4 years, but these motors do not run continuously because the running regime alternates their usage. The author's opinion was that for the high speed and rating of these motors it would have been wise to have used sleeve bearings.

On the day prior to the failure the vibration was measured on the outer periphery of the end frames at the DE and NDE by a condition monitoring contractor (not the author of this book) and the highest overall r.m.s. velocities were 15 mm/s (0.6 inches/s) and 12 mm/s (0.47 inches/s) in the horizontal and vertical directions respectively; therefore the motor should have been stopped. However, because of essential operational requirements the motor could not be immediately shut down, and the bearing failure and broken shaft occurred the next day.

This is an example where production and possible loss of income take priority over the operational condition of the motor even though there was clearly a vibration problem. Note that the author could not obtain any recorded vibration spectra leading up to the failure because only overall vibration velocities were measured.

8.3.1.1 Inspection

Figures 8.28 and 8.29 show photographs of the motor and broken shaft respectively, which were received by the OEM, and Figure 8.30 indicates it was a classical torque and twist break due to seizure of the bearing. The tip of the broken shaft was within the inner raceway of the NDE bearing. Figure 8.31 presents a photograph of the NDE bearing which was seized.

Figure 8.28 Overall view of NDE.

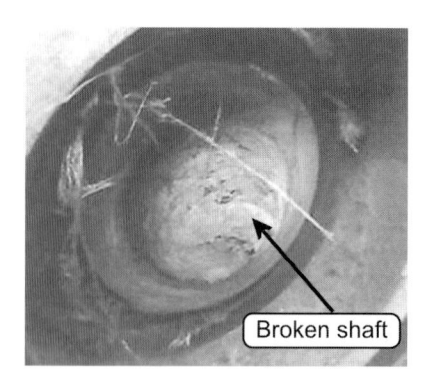

Figure 8.29 Broken shaft end.

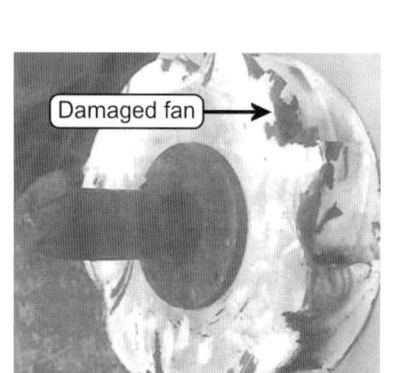

Figure 8.30 Severely damaged fan due to broken shaft.

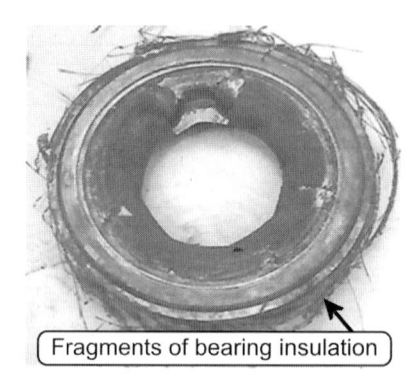

Figure 8.31 Catastrophic failure of NDE bearing; fragments of bearing insulation are on the surface of the outer race.

8.3.2 RCFA of a Faulty 6316 C3 Bearing and FAT of the Repaired Motor (B) with New Bearings

An identical motor (B) to the one described in Section 8.3.1 was returned to the OEM based upon a loud *clunking noise* from the DE, which was reported by maintenance personnel on the offshore platform. Permanent accelerometers were not fitted on the motor and a third-party vendor carried out vibration measurements on a monthly basis using temporary accelerometers mounted via magnets.

A repeat catastrophic bearing failure, as reported in Section 8.3.1, was not acceptable to the oil company and evidence was required to ascertain the cause(s) of the problem in this identical motor. The author was hired to carry out an investigation, which was in three stages:

(i) Inspect the motor received by the OEM and, if possible, run the motor on the OEM's test bed and take vibration measurements before any strip-down.

Figure 8.32 Section of DE bearing which shows a lack of effective grease.

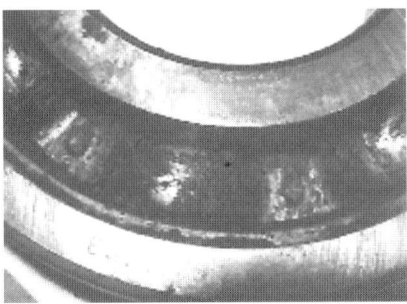

Figure 8.33 Severe spalling (pitting) of the inner raceway.

(ii) Remove the bearings from the DE and NDE to determine the condition of the bearings and grease.
(iii) Carry out a vibration Factory Acceptance Test (FAT) after the fitting of new bearings.

8.3.2.1 Attempt to Run the Motor and Bearing Inspection

The motor was started on a low voltage and after approximately 30 s of running at 200–300 r/min it was obvious that the loud clunking noise was coming from the DE bearing, and therefore the test was stopped to prevent any further damage.

Figure 8.32 shows a section of the DE bearing and there was a distinct lack of effective grease and when it was turned it was very *rough* indeed. The bearing was carefully dismantled, and Figure 8.33 shows the inner raceway. As expected, there was severe fatigue spalling (pitting) at approximately ball pitch intervals. This is indicative of false brinelling due to these five drive trains being adjacent to each other. A maximum of four drives is required at any one time, therefore one is stationary.

8.3.2.2 No-Load Factory Acceptance Test (FAT) – Vibration Measurements

New 6316 M C3 bearings were fitted to the DE and NDE of the motor. The motor was mounted on a solid steel base-plate in the OEM's factory test house and supplied with 6600 V at 60 Hz, and the no-load current in each phase was 10.5 ±2% A.

- Vibration readings were taken in accordance with BS EN 60034-14 [8.1].
- The overall allowable maximum r.m.s. velocity is 2.3 mm/s (0.09 inches/s) ±10% for this speed and shaft height of motor with a frequency span of 10 Hz to 1000 Hz.
- The supply frequency to the motor during the FAT was accurately measured via current signature analysis [8.4]. The frequency was not constant because the motor was supplied from an independent generator whose output frequency varied between 59.83 Hz and 60.1 Hz (see Figure 8.34 for a sample current spectrum). That is the reason for the different frequency values of the vibration component at the fundamental rotational speed frequency ($1X$) in the vibration spectra.

The accelerometers were mounted as shown in Figures 8.35 and 8.36, and the overall r.m.s. velocities on the DE and NDE bearing housings are presented in

Table 8.8 Overall velocities r.m.s. ±10%, span 10–1000 Hz

	DEV	DEH	DEA
DE bearing housing	1.2 mm/s	0.38 mm/s	0.4 mm/s
NDE bearing housing	NDEV 0.65 mm/s	NDEH 0.56 mm/s	NDEA 0.4 mm/s

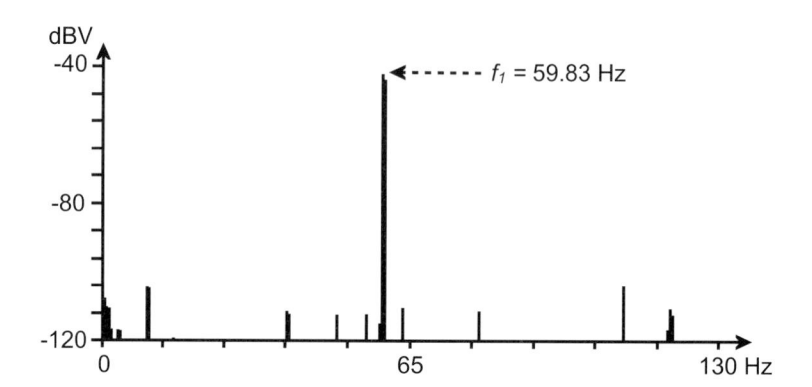

Figure 8.34 Current spectrum; span of 0–130 Hz.

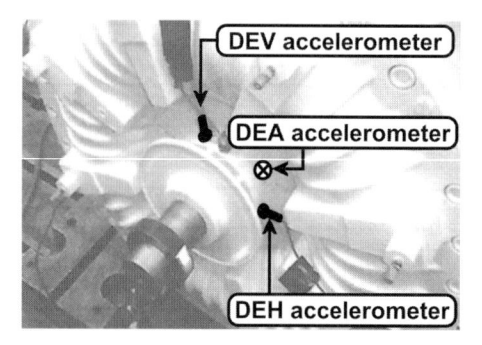

Figure 8.35 Accelerometer positions directly on the DE bearing housing.

Figure 8.36 Accelerometer positions directly on the NDE bearing housing.

Table 8.8. They are normal and well below the upper limit of 2.3 mm/s r.m.s. as specified in BS EN 60034-14, Feb. 2004 [8.1].

The vibration spectra are presented in Figures 8.37 to 8.42, and Table 8.9 gives the predicted bearing defect frequencies.

8.3.3 Conclusions

(i) The vibration spectra are all normal as measured on the DE and NDE bearing housings and, as expected, they are dominated by the *1X* frequency component,

8.3 Bearing Failures in 800 kW/1072 H.P. SCIMs

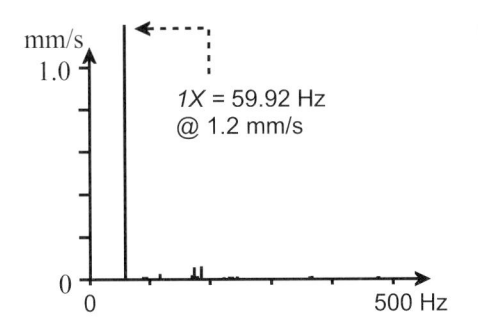

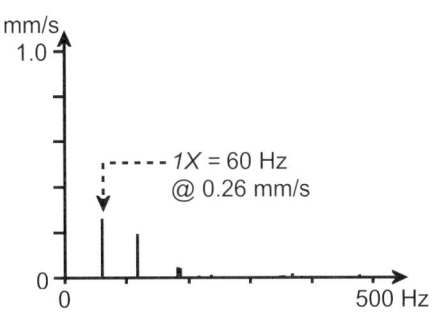

Figure 8.37 DEV vibration spectrum. (1X = 59.92 Hz @ 1.2 mm/s)

Figure 8.38 DEH vibration spectrum. (1X = 60 Hz @ 0.26 mm/s)

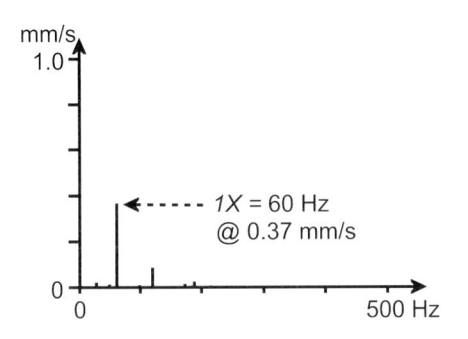

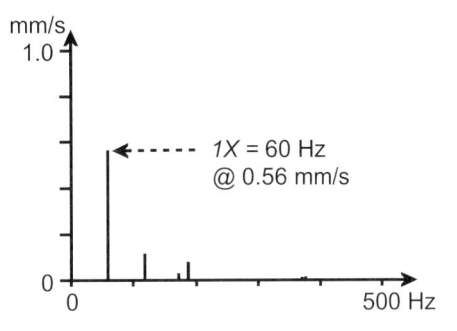

Figure 8.39 DEA vibration spectrum. (1X = 60 Hz @ 0.37 mm/s)

Figure 8.40 NDEV vibration spectrum. (1X = 60 Hz @ 0.56 mm/s)

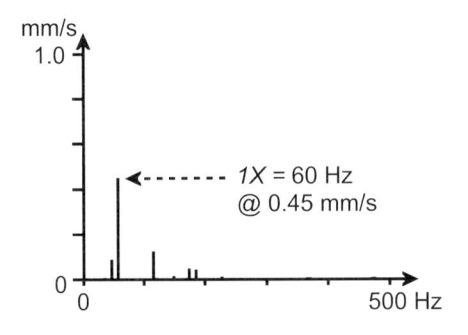

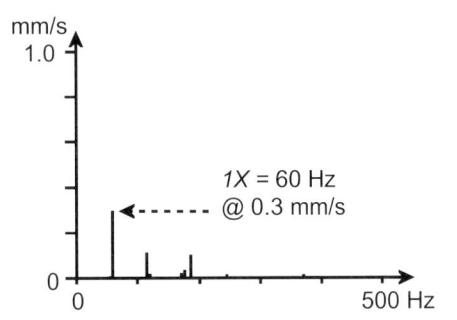

Figure 8.41 NDEH vibration spectrum. (1X = 60 Hz @ 0.45 mm/s)

Figure 8.42 NDEA vibration spectrum. (1X = 60 Hz @ 0.3 mm/s)

which is caused by the centrifugal force from the inherent mechanical unbalance, which was ISO balance grade G1.0 [8.3], in the rotor.

(ii) There were no bearing defect frequencies present, as should be the case with brand new bearings.

(iii) These results provided a base-line of the overall r.m.s. velocities and vibration spectra for a refurbished SRP motor, which was fitted with new bearings.

(iv) The motor was returned to the oil company.

Table 8.9 Bearing defect frequencies

Bearing type: 6316 M C3 deep groove ball bearing	
M: Brass cage	C3: Clearance
No. of balls: 8	Motor speed approximately: 3599 r/min
Bearing frequencies	

BPFO	BPFI	FTF	$2 \times BSF$
185 Hz	295 Hz	23 Hz	249 Hz

Figure 8.43 Photograph of the motor.

8.4 Industrial Case History – False Brinelling and FAT of an 800 kW/1072 H.P. SCIM and Attenuation of Vibration between the Bearing Housing and the Outer Periphery of the End Frame

8.4.1 Summary

Abstract – Sections 8.31 and 8.3.2 presented information on a catastrophic NDE bearing failure and broken shaft in one of five motors that drive sulphate removal pumps on an offshore oil and gas production platform (see Figure 8.43).

Figure 8.33 showed a severely pitted (spalled) DE bearing in an identical motor (B). It was proposed that this may have been initiated by false brinelling.

Motor details:

- 3-phase SCIM, 6.6kV, 800 kW/1072 H.P., 76 A, 60 Hz, 3580 r/min, 0.9 p.f., eff 96%
- Star connected, frame E450/1400, IP56, Class F, B temp rise

8.4 Industrial Case History – False Brinelling and FAT of an 800 kW/1072 H.P. SCIM

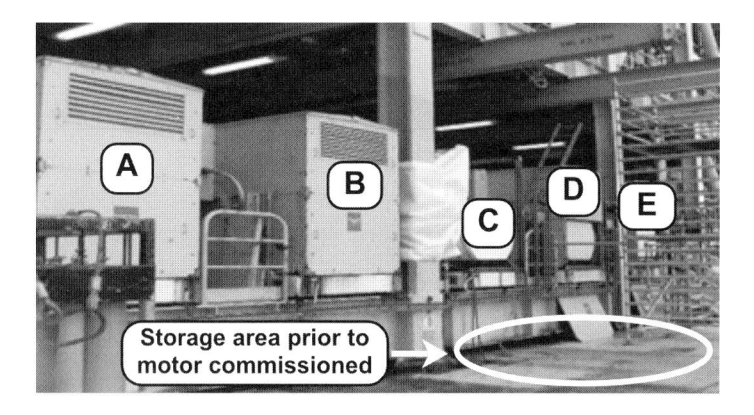

Figure 8.44 Motors mounted on a steel fabricated structure on the offshore oil production platform.

- Bearings: DE and NDE 6316 C3, re-greasing quantity 33 g
- Lubrication details: Shell Albida R2 grease
- OEM's recommendation: bearings should normally be re-greased every 1200 running hours using the correct quantity of grease.

A photograph of the layout of the motors driving the sulphate removal pumps is shown in Figure 8.44. Access to the motors is obtained by removing the external panels.

This case history reports on an identical SCIM, which was not rotated for five months after it had been returned to the offshore oil production platform after refurbishment and fitting of new bearings.

It had been stored for different periods of time (unknown) in various locations, such as in a dedicated storage area that was three decks (levels) above the SRP drive trains, and then on the deck directly below the drive trains as shown (ellipse) in Figure 8.44, before it was installed in its drive train slot.

This motor was returned to the repair shop as a test case to establish if false brinelling was present in the bearings as part of the investigation to ascertain the root cause of the bearing failures in these motors.

Stationary motors were also stored on the deck immediately below (see Figure 8.44) the drive trains and were subjected to vibration from various sources and likewise when installed in a drive train slot because the running SRP motors are mounted on a communal, steel fabricated base.

8.4.1.1 Inspection of Bearings

The DE and NDE bearings were dissected, and Figures 8.45 to 8.49 present photographs of the nominally new bearings from the repaired motor which had not been turned on the offshore installation while being stored in different locations:

(i) Evidence of initial false brinelling, as shown in Figure 8.45.
(ii) Evidence of light fretting corrosion on the DE bearing, as shown in Figures 8.46 and 8.47.

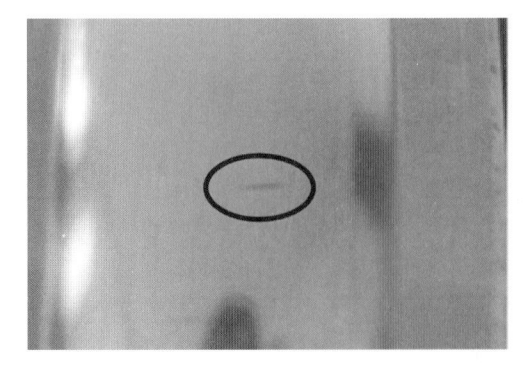

Figure 8.45 DE 6316 C3 inner ring raceway; evidence of false brinelling.

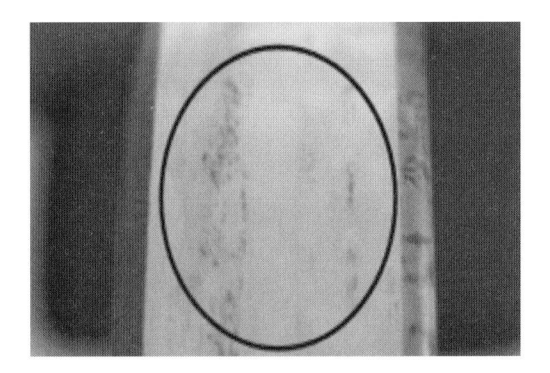

Figure 8.46 DE 6316 C3 outer ring outside diameter; light fretting corrosion.

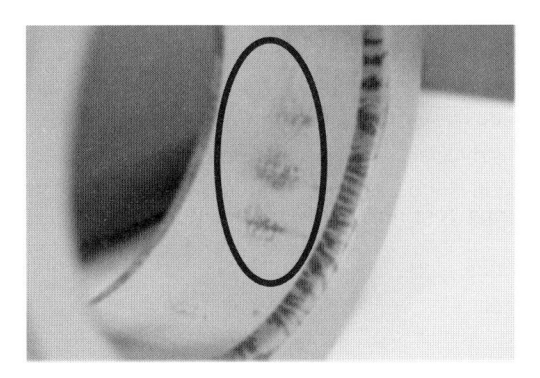

Figure 8.47 DE 6316 C3 inner ring bore – light fretting corrosion on bore surface.

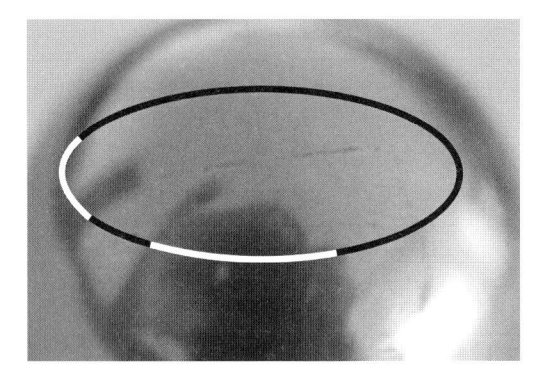

8.4 Industrial Case History – False Brinelling and FAT of an 800 kW/1072 H.P. SCIM

Figure 8.48 DE 6316 C3 multiple light surface abrasions/wear markings on the spherical surface.

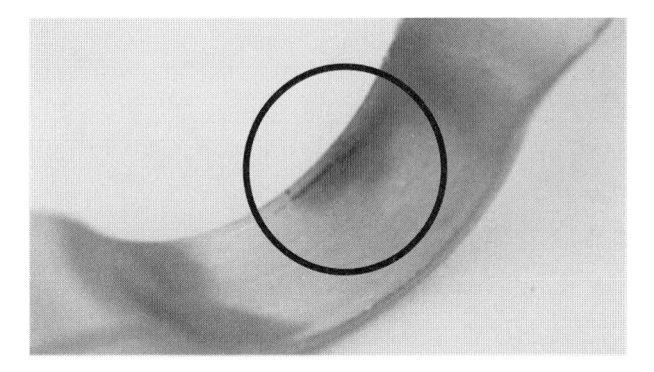

Figure 8.49 DE 6316 C3 steel cage pocket – multiple surface abrasions/wear in pocket cavity.

(iii) Multiple light surface abrasions/wear on the balls, as shown in Figure 8.48.
(iv) Multiple surface abrasions/wear in the steel cage pocket, as shown in Figure 8.49.

8.4.2 Vibration Factory Acceptance Test (FAT) of Repaired Motor

New bearings were fitted to the motor and a vibration FAT was carried out in a motor repair shop. The motor on the test bed is shown in Figure 8.50. The fan cowl was unbolted from the end frame to allow vibration measurements on the NDE bearing housing.

The overall r.m.s. velocities are presented in Table 8.10, and they are well below (maximum 1.5 mm/s r.m.s.) the maximum allowable of 2.3 mm/s (0.09 inches/s) r.m.s. as specified in BS 60034-14-2004 [8.1].

Table 8.10 Overall r.m.s. velocities on bearing housings and end frames

(i) No-load uncoupled run – vibration measurements carried out as per BS 60034-14, Feb. 2004 [8.1].
(ii) Maximum allowable overall r.m.s. velocity = 2.3 mm/s (0.09 inches/s) ±10%; frequency span 10–1000 Hz
(iii) Rated volts applied: 6600 V; measured frequency: 60.05 Hz
(iv) Measured no-load phase currents: R: 10.5, Y: 10.7 and B: 10.2 A

Vibration after two hours of uncoupled run
See Figures 8.51 to 8.54 for positions of accelerometers

Position DEV_B bearing housing 1.2 mm/s	Position DEH_B bearing housing 1.5 mm/s	Position DEA_{EF} end frame 0.7 mm/s
Position DEV_{EF} end frame 0.8 mm/s	Position DEH_{EF} end frame 0.67 mm/s	
Position $NDEV_B$ bearing housing 1.2 mm/s	Position $NDEH_B$ bearing housing 0.9 mm/s	
Position $NDEV_{EF}$ end frame 1.6 mm/s	Position $NDEH_{EF}$ end frame 1.5 mm/s	Position $NDEA_{EF}$ end frame 0.6 mm/s

Figure 8.50 FAT vibration measurements – photograph of the motor with its NDE fan cowl separated from the end frame to allow access to measure the vibration directly on the NDE bearing housing.

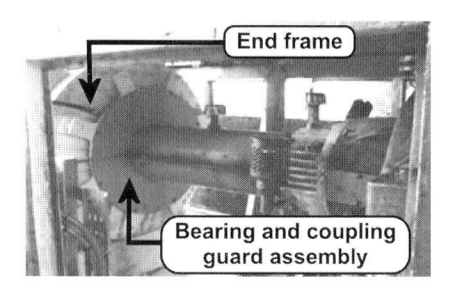

8.4 Industrial Case History – False Brinelling and FAT of an 800 kW/1072 H.P. SCIM

Figure 8.51 Motor in-situ: the DE bearing housing is not accessible.

Figure 8.52 Motor in-situ: photograph of NDE – the NDE bearing housing is not accessible.

Figure 8.53 Accelerometers on the DE bearing housing.

Figure 8.54 Accelerometers on the outer periphery of the DE end frame.

8.4.3 Attenuation of Bearing Defect Frequencies

Because the bearing housings on the SCIM cannot be accessed on the offshore oil and gas production platform, as shown in Figures 8.51 and 8.52, an additional and key objective of the FAT vibration measurements was as follows:

- To compare the vibration directly on the bearing housings with that measured on the outer periphery of the end frame.

The magnitudes of the bearing defect frequencies of vibration generated within the bearing can be attenuated when the vibration is measured on the outer periphery of the end frame or at the end of the stator frame.

This SCIM is a Zone 1, flame-proof EXd design and the construction is very robust, particularly the end frames, which will have a high mechanical stiffness in both the radial and axial directions, as shown in Figures 8.53 to 8.56.

From Table 8.10, the effect of attenuation between the overall r.m.s. velocities measured on the DE bearing housing, compared to the overall r.m.s. velocities values on the DE end frame, is as follows:

Table 8.11 Bearing Defect Frequencies

Bearing type: 6316 M C3 deep groove ball bearing	
M: Brass cage	C3: Clearance
No. of balls: 8	Motor speed approximately: 3599 r/min

Bearing frequencies			
BPFO	*BPFI*	*FTF*	$2 \times BSF$
185 Hz	295 Hz	23 Hz	249 Hz

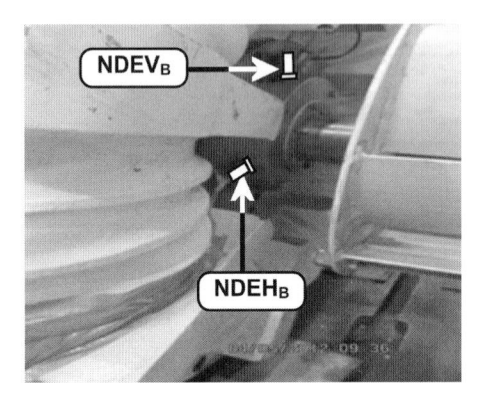

Figure 8.55 Accelerometers on the NDE bearing housing.

Figure 8.56 Accelerometer on the outer periphery of the NDE end frame.

Attenuation Factors **$DEH_B : DEH_{EF} = 1.5 : 0.67$; i.e. an attenuation by a factor of 2:2.**

The horizontal r.m.s. velocity at the DEH_{EF} position on the periphery of the end frame was attenuated by a factor of 2.2 compared to the horizontal velocity at the DEH_B position as measured directly on the DE bearing housing.

Likewise for the NDE velocity in the vertical direction:

$NDEV_B : NDEV_{EF} = 1.2 : 0.8, 1.5:1$; i.e. an attenuation by a factor of 1.5.

For this motor, mounting the accelerometers on the end frames gives a misleading result in comparison to the actual vibration on the bearing housing.

It is accepted that vibration standards, such as BS 60034–14-2004 [8.1] and NEMA MG 1-2006 [8.2], allow vibration measurements on the end frames to assess the vibration level during a FAT when the bearing housings are not accessible.

However, these standards do not include any reference to the diagnoses of bearing defects identified via the bearing defect frequencies from rolling element bearings.

The bearing defect frequencies are presented in Table 8.11 using the no-load, uncoupled speed as the reference speed.

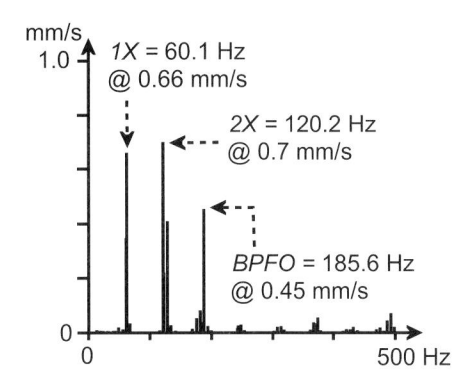

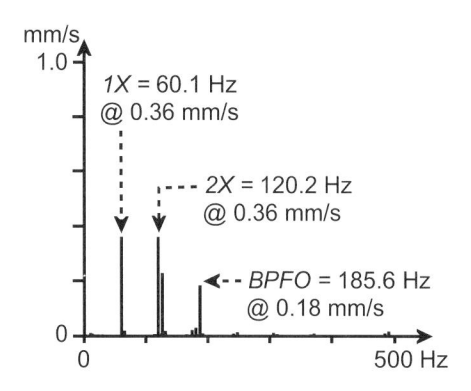

8.4 Industrial Case History – False Brinelling and FAT of an 800 kW/1072 H.P. SCIM

Figure 8.57 DEH$_B$ horizontal bearing housing, vibration spectrum.

Figure 8.58 DEH$_{EF}$ horizontal outer periphery of end frame.

Figures 8.57 and 8.58 present the vibration spectra for the DE horizontal positions directly on the DE bearing housing and on the outer periphery of the DE end frame respectively.

The *1X*, *2X* and *BPFO* components are present in both spectra, but their magnitudes on the outer periphery of the end frames are attenuated due to the mechanical stiffness and frequency response of the DE end frame between the bearing housing and the outer periphery of the end frame. The vibration spectra verified the following attenuation factors for this motor caused by the mechanical stiffness and frequency response of the end frames:

1X is attenuated by a factor of 1.8.
2X is attenuated by a factor of 2.
BPFO is attenuated by a factor of 2.5.

8.4.4 Conclusions and Final Outcome

(i) Measuring the vibration on the outer periphery of the end frames of this motor gives false readings with respect to the actual vibration on the bearing housings.

(ii) Permanent accelerometers were not fitted to these motors and the oil company would not retrofit them because they are required to be intrinsically safe and *re-certification from the OEM* of the motor is required for any retrofits to a Zone 1, flame-proof motor.

(iii) The costs to retrofit permanent IS accelerometers were prohibitive, particularly the installation of cables and relevant instrumentation to suit a Zone 1.0 hazardous area.

(iv) These motors are essential for production and cannot simply be taken out of service for retrofits. The only option is to attach temporary accelerometers on the end frames via strong magnets.

(v) To assess the velocities of bearing defect frequencies from the bearing via measurements on the end frames requires correction factors to be applied.

(vi) More rolling element bearing failures occurred and the outcome was that the deep groove ball bearings (6316 C3) were unsuitable for the operational conditions of these motors and therefore not fit for purpose.
(vii) The author recommended a revised motor design and that sleeve bearings should be fitted. This was carried out over a period of two years during major maintenance outages.
(viii) No failures have been reported since the sleeve bearings were installed.

References

8.1 BS 60034-14-2004, *Rotating Electrical Machines, Part 14 Mechanical Vibration of Certain Machines with Shaft Heights of 56 mm and Higher – Measurement, Evaluation and Limits of Vibration Severity*, 2004.
8.2 NEMA MG 1-2006, *Part 7: Mechanical Vibration-Measurement, Evaluation and Limits*, 2006.
8.3 ISO 1940-1-2003, *Rotor Balancing*, 2003.
8.4 W. T. Thomson and I. Culbert, *Current Signature Analysis for Condition Monitoring of Cage Induction Motors*, IEEE-Press Wiley, ISBN: 978-1-119-02959-5, 2017.

9 Industrial Case Histories on Vibration Measurements and Analysis Applied to Large Induction Motors with Sleeve Bearings

9.1 Introduction and Basic Operation of a Sleeve Bearing

Abstract – This chapter is focused on the presentation of industrial case histories on vibration measurements and analysis from fluid film bearings, which are referred to as sleeve or journal bearings in large induction motors. It is not about reproducing already-published knowledge on the design, construction and operating principles of fluid film bearings because that is a pointless exercise. The reader wants new information rather than already-published knowledge. There are numerous publications on fluid film bearings and a sample of references is cited; see [9.1] to [9.10]. A thorough tutorial paper titled Fluid Film Bearing Fundamentals and Failure Analyses by F. Y. Zeidan and B. S. Herbage can be accessed via the Web [9.1].

This tutorial (21 pages) was presented at the 20th Turbo Machinery Symposium in Houston, Texas, in 1991, and it can provide all the fundamental knowledge that the reader will require to underpin the original industrial case histories presented in this chapter. This paper [9.1] covers the following:

(1) fundamentals on hydrodynamic lubrication and bearing load
(2) common fluid journal bearings
(3) measurement of bearing clearance
(4) bearing materials and construction
(5) bearing dynamics and basic concepts
(6) bearing failure analysis.

As a lead into the industrial case histories which follow, it is appropriate to explain the basic principle of operation of a fluid film bearing. Figures 9.1 and 9.2 show a sleeve or journal bearing housing and bearing shells respectively [9.12].

In a fluid film bearing, a thin film of oil supports the rotating shaft and during steady-state operation there is no contact between the rotating shaft (journal) and the fixed white metal sleeve or Babbitt.[1]

---
[1] Note that Babbitt metal is also named white metal and is a soft, white non-ferrous alloy used to provide a bearing surface. Its composition helps to reduce friction, which makes it a good material to use in a sleeve bearing.

Figure 9.1 Typical sleeve or journal bearing housing assembly in an induction motor.

Figure 9.2 Bearing shells removed from a sleeve bearing in an induction motor.

Figure 9.3 is a simple diagrammatic illustration of the oil film wedge in a fluid film bearing. A rotating shaft in a fluid film bearing produces a high-pressure wedge of oil, for example, at 10 bars or 145 pounds/square inch (p.s.i.), which lifts the shaft off the bearing Babbitt and prevents metal-to-metal contact, and in theory the bearing should have an infinite life-time. This oil film pressure profile naturally moves the shaft into an eccentric operating position and there must be sufficient clearance between the journal and the bearing sleeve to prevent a rub. For detailed information on bearing clearances please refer to the references; however, a general rule-of-thumb guideline that is often used for diametral journal bearing clearance is 0.0015 inches/1.5 mils-per-inch of journal diameter. Considerable caution must be exercised because it depends on many factors, particularly the application, in which it is being used; see references [9.11] and [9.12].

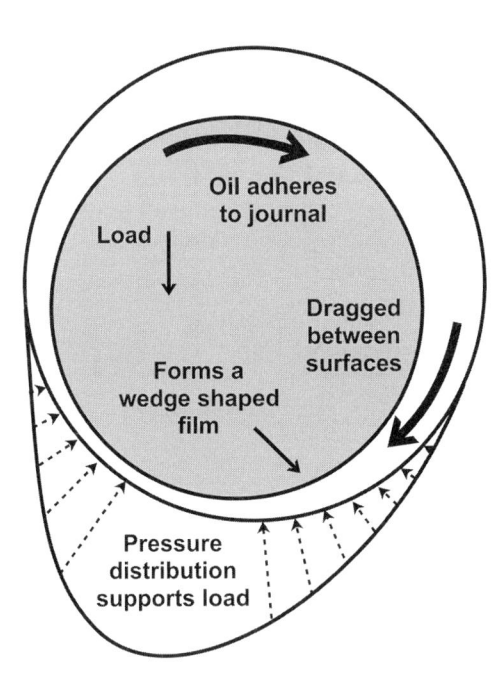

Figure 9.3 Illustration of the fluid film profile in a sleeve bearing.

The oil film thickness in the operating load zone of a sleeve bearing is different for each SCIM and depends on factors such as:

(i) bearing design
(ii) bearing clearance
(iii) shaft diameter
(iv) operating speed
(v) power rating.

Oil film thickness has a very wide range and reference [9.5] quoted the following:

Typically, minimum oil film thickness in the load zone during operation ranges from 1.0 μm to 300 μm but values of 5 μm to 75 μm are more common in midsized equipment.

Reference [9.5] also quoted that: 'normally the minimum oil thickness is also the operating clearance of the bearing'.

9.1.1 Illustrations of the Practical Installation of an Oil-Fed Pressure Sleeve Bearing

Many publications do not present the actual practical installation and operation of oil-fed pressure bearings that are used in large induction motors. The following knowledge in Sections 9.1.1 to 9.1.2.4 was provided by Mr Alistair Carr, formerly head of test of high voltage machines and chief trouble-shooter of on-site problems in large SCIMs operating in industry, during his employment for 42 years with Parsons Peebles in Scotland.

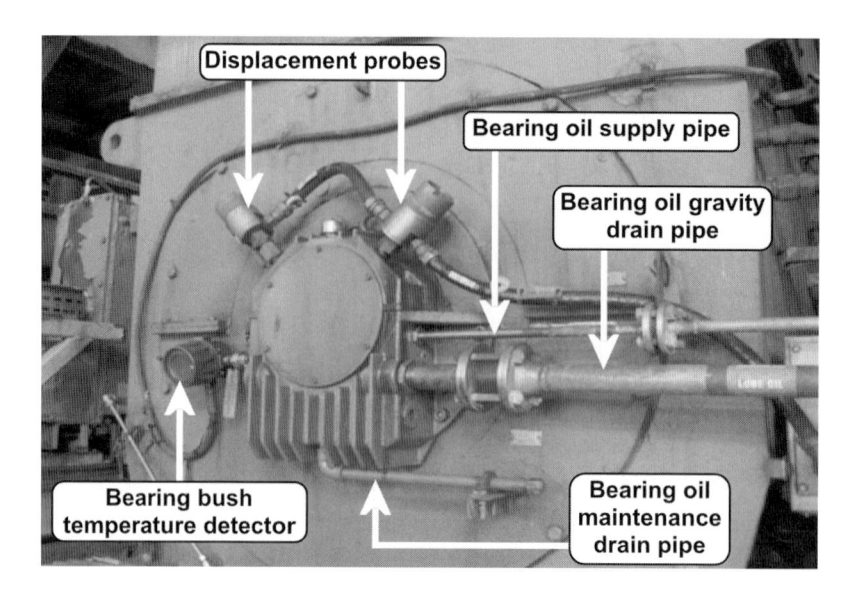

Figure 9.3a SCIM pressure-fed sleeve bearing assembly.

Mr Carr has vast technical knowledge and practical experience; he is now employed by EM Diagnostics Ltd as a senior electrical machines engineer. The author of this book is the Director of EM Diagnostics Ltd, and Mr Carr has given permission to the author to present the following knowledge.

Figure 9.3a shows the NDE sleeve bearing assembly of a SCIM driving a condensate export pump on an offshore oil and gas production platform.

This bearing is oil-pressure fed and has a gravity drain. It is also fitted with a bearing bush temperature detector and, in the top cap, proximity probes for vibration measurement.

The bearing is fitted with a loose oil ring to provide short-term lubrication in the event of oil-supply failure. There is also provision for draining the bearing oil, during maintenance and inspection, utilizing the drain pipe fitted into the bottom of the sump.

The bearing supplier or motor OEM will specify the oil quantity and viscosity and may also give examples of suitable oil types from main suppliers. The oil lubrication system is a closed circulatory system that consists of a tank, main and standby delivery pumps, control valves and filtration. This system may be common to other items of plant in the drive string including pumps, gearboxes and compressors.

9.1.1.1 Oil Feed

Oil is fed to the bearing through the small stainless-steel pipe line, typically 12 mm to 20 mm bore (1/2 inch to 3/4 inch). The flanges that connect the feed line to the bearing stub pipe have an orifice plate clamped between them.

The oil flow rate for the bearing illustrated will be of the order of 5 litres/minute (1.3 US gallons per minute). This flow rate is achieved by taking the specified oil

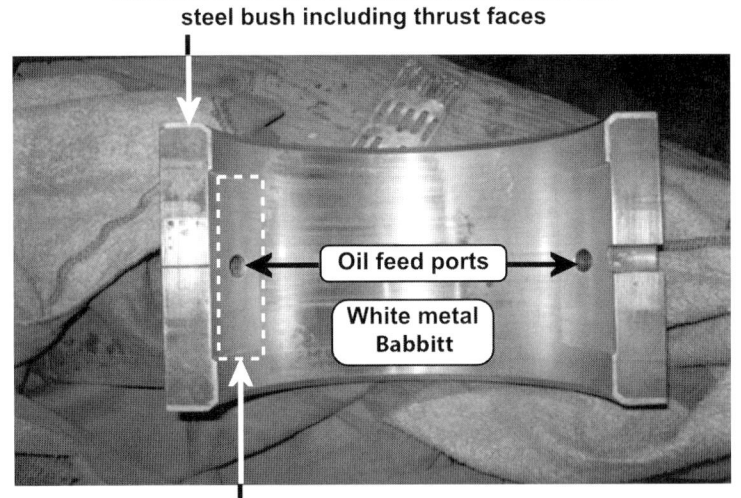

Figure 9.3b Bearing bottom bush.

(Photo annotations: Interface between the white metal babbitt and steel bush including thrust faces; Oil feed ports; White metal Babbitt; Dotted rectangle: area where the white metal is relieved to allow oil to enter the bush)

system feed pressure and sizing the orifice accordingly. A typical oil-feed pressure would be 25 p.s.i. (1.724 bar). It is essential that this specified oil-supply pressure is maintained to protect the bearing from wear and possible failure. This pressure is usually controlled by a pressure reducing valve at the oil system and there is a gauge in the motor supply line for monitoring purposes.

The oil-feed ports are clearly visible in the bearing's bottom bush, as shown in Figure 9.3b, and the area where the white metal is relieved to allow the oil to enter the bush can also be seen. This photograph also illustrates the interface between the white metal Babbitt and the steel bush, including the thrust faces. Wear is clearly visible in this white metal Babbitt but, despite this, the bearing bush will still be serviceable.

The thrust faces on the bush illustrated are designed only for limited and short-duration axial load. This may occur during starting for example.

9.1.1.2 Oil Drainage

Oil drains from the bearing and returns to the main oil tank by gravity only. This larger oil drain line is typically 38 mm to 75 mm (1.5 to 3 inches) bore. This line must have adequate fall to ensure that the oil can flow freely out of the bearing housing.

The oil level in the bearing housing is determined by the height of this drain pipe and the weir that is located inside the pipe at the bearing housing end.

It is not unusual, if the fall on the drain line is marginal, that flooding of the bearing housing will occur. This can happen if the oil flow rate is on the high side or if the oil is cold or at least well below the design operating temperature. For the gravity drain or oil return to be effective, the bearing housing must be ventilated.

Looking at Figure 9.3a the breather can be seen at the 12 o'clock position at the top of the housing. The motor internal fans produce a negative pressure in the motor enclosure adjacent to the bearing housing and the vent ensures that this negative pressure is not transferred into the bearing via the seals.

9.1.1.3 Bearing Housing Connections

Figure 9.3a shows that there are various access points into the bearing housing, some are utilized, and some are plugged. On most bearings the ports provided are, starting from the top: oil feed, bearing bush temperature detector, oil drain/ level indicator and sump oil cooler. Jacking oil ports can also be provided when specified. These ports are available on both sides of the housing and allow the motor manufacturer flexibility regarding the location of pipework and the temperature detector cabling. Generally, either side can be used for the port application. Note that the oil-level sight glass is fitted in the opposite port to the oil drain line.

9.1.1.4 Oil Viscosity

The viscosity of the oil refers to its resistance to flow, or thickness. Lubricating oil will fall somewhere between kerosene and treacle regarding thickness. The viscosity grade number is the midpoint of the viscosity range, in centistokes, when the oil is at 40 °C/ 104 °F.

For example, a VG 46 oil will have kinematic viscosity limits between approximately 41/0.064 and 51/0.079 centistokes per square inch per second when at 40 °C. In most cases, sleeve bearings for SCIMs will be specified for use with ISO VG32 or ISO VG46 grade oil.

9.1.2 Brief Overview of Operational Problems with Sleeve Bearings

External malfunctions are the predominant reasons why sleeve bearings fail and the following sub-sections justify this statement. A sleeve bearing that is in good order and is correctly lubricated will last almost indefinitely. The very high resilience and reliability of sleeve bearings has been demonstrated via numerous routine inspections by Mr A. Carr.

The reason is that although the motor bearings were found to have been *wiped* [2] at some time, these bearings were still performing well within acceptance criteria with respect to temperature and vibration and there was no indication that the Babbitt metal was damaged. With lubrication restored, the design of the sleeve bearing is sufficiently forgiving to allow it to continue to function.

---

[2] The term *wiped* is used to describe the situation when movement of white metal has occurred due to localized excess surface heat from short-term loss of lubrication, which may occur during an abnormal shutdown.

Figure 9.4 Photograph of a self-contained ring sleeve lubricated bearing to show the oil scoop ring.

(i) **External malfunctions are the predominant reasons why sleeve bearings fail.**

9.1.2.1 Inadequate Lubrication

In most cases the failure of a sleeve bearing will be associated with the prolonged loss of suitable lubricant.

(1) **Pressure-Fed Bearings – Typical Causes of Failures**
- Blocked oil filters resulting in a severely restricted oil supply.
- Oil flow rate set well below the bearing supplier's recommended tolerance.
- Water in the oil, often due to the failure of the lubrication system oil cooler.
- Contaminated oil or a mix of lubricants which are not compatible. In some instances, the motor lubrication systems are shared with other items of plant in the drive string.

(2) **Self-Contained Ring Lubricated Bearings – Typical Causes of Failure**
- Loss of sump oil, thus the oil ring cannot lift sufficient oil to lubricate the bearing. This is often the result of poor seals allowing oil to be drawn into the motor enclosure. The coupling assembly can also draw the oil past the bearing seals.
- Low oil level. Oil-level sight glasses become stained and opaque with time and can give a false indication of the oil level in the bearing sump.
- Sticking oil rings/scoops (see Figure 9.4). This is most likely to occur after maintenance work. The ring may be oval due to poor handling or the joint can become stepped. The rotation of the oil ring can also be restricted by an incorrectly fitted bearing bush temperature probe.

9.1.2.2 Axial Load

Most SCIM sleeve bearings are not designed to withstand continuous axial load. The sleeve bearing has axial clearance which allows the motor's rotor to find its magnetic centre. The magnetic centre is the electro-magnetic axial alignment of the rotor and stator steel cores when the motor is energized. Axial movement of 5 mm (200 mils) either side of the magnetic centre is a typical value.

The sleeve bearing bushes have white metal end faces (see Figure 9.3b), which are designed to accept and limit rotor axial movement when the motor is started uncoupled. These faces are not designed for prolonged or high loading.

Excess or continuous axial load will result in the metal faces overheating and breaking up. This will lead to failure of the bearing particularly when the motor shaft shoulder finally contacts the steel portion of the bearing bush.

9.1.2.3 Vibration and Alignment

Prolonged severe vibration will result in deterioration of a sleeve bearing. The shaft movement will cause disruption of the oil film and lead to momentary metal-to-metal contact during rotation. In extreme cases, the white metal Babbitt can separate from the steel bearing shell.

Damage to bearing shells can be caused by misalignment of the bearing shells with respect to the motor shaft journals or when misalignment of the coupled load is outwith the limits of the coupling flexible spacer.

Therefore, the load will not be evenly distributed along the bearing shell; this causes excessive loading, for example at the end of the shell, and can lead to localised loss of oil film and resulting hot spots.

9.1.2.4 Effect on Sleeve Bearings of Starting SCIMS

Other than the very small film of oil (see Section 9.1) there is metal-to-metal contact between the journal and the white metal Babbitt, when the SCIM is stationary. Residual oil will sit in the wedge and as soon as the shaft starts to rotate oil is drawn in. Flood lubrication ensures that there is ample oil available, and ring lubrication will ensure ample oil as soon as the shaft turns. The result of each start causes negligible wear or deterioration, but the combination of this and wear from small foreign particles in the oil are the natural wear factors. Standby motors should be barred over periodically, to ensure that there is some oil residue around the journal and bearing sleeve.

9.2 Introductory Case History – Vibration Factory Acceptance Test of a New 6800 kW/9115 H.P. SCIM

Abstract – The objectives of this introductory industrial case history are as follows:

(i) To explain how vibration measurements are carried out on a large, 6800 kW/ 9115 H.P. SCIM with sleeve (or journal) bearings.

9.2 Vibration Factory Acceptance Test of a New 6800 kW/9115 H.P. SCIM

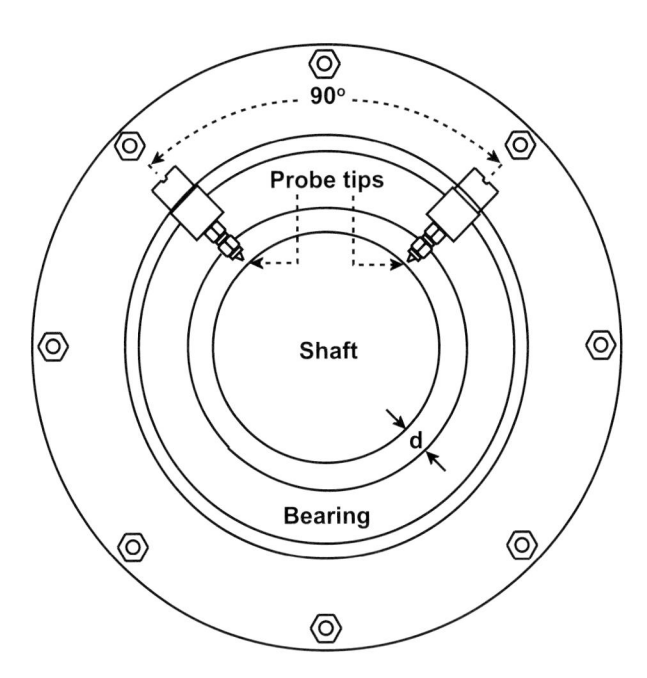

Figure 9.5 Diagrammatic Illustration of eddy current displacement or proximity probes to measure shaft displacement.

(ii) To present results from vibration Factory Acceptance Tests (FATs) on a new 6800 kW/9115 H.P. SCIM in accordance with the British vibration standard (BS 60034-14-2004 [9.13]), and as per the client's requirements.

(iii) To demonstrate the distinct difference between the measured shaft and bearing housing displacements from a sleeve bearing in this large SCIM.

9.2.1 Measurement of Shaft Displacement

The crucial parameter to measure in a sleeve (or journal) bearing is the relative displacement (d) between the rotating shaft (journal) and the fixed, white metal Babbitt.

This is achieved by using eddy current displacement probes, and a diagrammatic representation of the basic installation of these probes to measure the displacement of the rotating shaft (journal) in a sleeve bearing is shown in Figure 9.5.

The photograph shown in Figure 9.5a illustrates how external access to fit and remove the probe is achieved. The probe tip is typically positioned at a distance (d) of 50 mils/0.05 inches/1.27 mm from the rotating shaft or journal surface.

The basic principles of operation of an eddy current displacement probe are as follows:

- The coil in the probe's tip is supplied from a high-frequency oscillator and the current in the coil produces an alternating magnetic field, as illustrated in Figure 9.6.

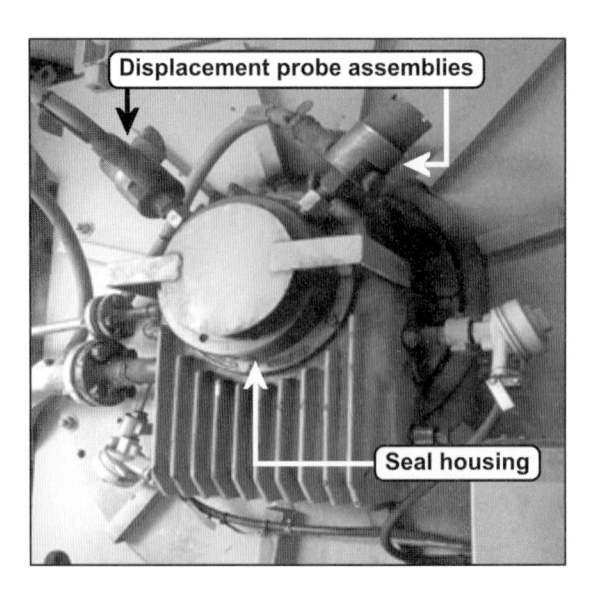

Figure 9.5a Photograph of a sleeve bearing on a SCIM with displacement probes fitted.

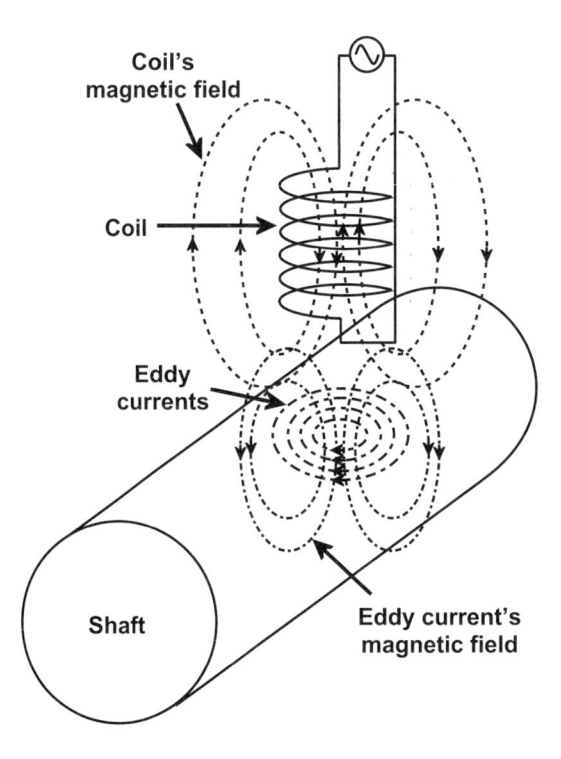

Figure 9.6 Illustration of the main flux from the probe's coil and the magnetic field created by induced eddy currents.

- This high-frequency magnetic field couples with the conductive target, which in this case is the rotating shaft (journal), and eddy currents are induced in the steel shaft. Eddy currents are closed loops of current, which move perpendicular to the flux produced by the coil but effectively parallel to the shaft.
- The magnetic flux produced by the eddy currents opposes the main flux from the coil. A change in the distance (d) between the shaft and the fixed probe tip in Figure 9.5 results in a change in the resultant flux and therefore a change in inductance and impedance.
- This changing coil impedance is directly proportional to the distance (d) between the fixed coil in the probe's tip and the rotating shaft.

Two displacement probes, as illustrated in Figure 9.5a, are mounted 90° apart so that polar plots, orbits and the shaft centre line position data can be produced. The displacement probes can be fitted through the centre of the bearing cap housing rather than via the seal housing, and the advantage of this is that the surface is truly concentric with the bearing, thus eliminating any possible error due to machining.

9.2.1.1 Comments on Measurement of Shaft Displacement and Bearing Housing Vibration

In Chapters 4 to 8 the focus was on the diagnosis of faults in roller element bearings via the measurement and analysis of vibration as measured on the bearing housings. During a FAT of an induction motor that has sleeve bearings, the shaft displacement and bearing housing vibration are both measured. It is emphasized that the shaft displacement measurement as described in Section 9.2.1 is distinctly different from the displacement derived from an accelerometer on the bearing housing of a sleeve bearing by integrating (twice) the acceleration signal. Note that the following applies:

A direct comparison between the shaft displacement of the rotor and the displacement on the bearing housing of a sleeve bearing in a SCIM is not valid because they are *two distinctly different measurements.*

9.2.2 Motor Nameplate Data and Vibration Test Specification

The motor's nameplate data are presented in Table 9.1, which refers to its full-load rated output, and a photograph of the motor under test is shown in Figure 9.7.

The vibration tests were carried out in accordance with BS 60034-14-2004 [9.13], which applies to electric machines running uncoupled at rated voltage and frequency on a solid steel base-plate (rigidly mounted), as does NEMA MG 1, Part 7 [9.14]. These standards do not apply when the motor is coupled to a mechanical load because vibration FATs must refer to a common reference so that all OEMs have a standard test condition. The limits set in BS 60034-14, for this 13.8 kV, 6800 kW/9115 H.P., 2-pole, 50 Hz, SCIM during an uncoupled run are as follows:

Table 9.1 Motor nameplate data

3-phase SCIM	6800 kW	333 A	60 Hz
3567 r.p.m.	0.89 p.f.	eff 96%	Frame EL710/2340JS
Enclosure CACW	Rating MCR S1	Starting current, 5 FLC	Starting torque 0.6 FLT

Figure 9.7 Photograph of the 13.8 kV, 60 Hz, 2-pole, 6800 kW/9115 H.P., SCIM under test.

(1) The maximum allowable shaft displacement is **65 μm** peak–peak.
(2) The maximum allowable vibration levels on the bearing housings are:
    Velocity: 2.3 mm/s,
    Acceleration: 3.6 m/s^2,
    Displacement: 35 μm,
    ±10% to allow for different tolerances in vibration instruments used by OEMs.

9.2.3 No-Load FAT Results from Vibration Measurements on the Bearing Housing

Figure 9.8 shows a photograph of the NDE sleeve bearing with the accelerometers on the housing and the positions of the displacement probes.

Table 9.2 presents the overall r.m.s. velocity, displacement and acceleration on the bearing housings, when the motor was run uncoupled on a rigid base-plate supplied at rated voltage and frequency. The no-load current was 64 amperes.

From Table 9.2, all the measured r.m.s. velocity (max. 1.7 mm/s), acceleration (max. 1.6 m/s^2) and displacement (max. 5.5 μm) levels on the bearing housings are below the allowable limit levels in the British Standard BS 60034-14-2004 [9.13].

Table 9.2 Overall r.m.s. vibration levels ±10%; span: 10–1000 Hz; BS 60034-14 2004 limits:

$v = 2.3$ mm/s r.m.s., $a = 3.6$ m/s^2 r.m.s., $d = 35$ µm r.m.s.

Motor DE bearing housing

Vertical	**Horizontal**	**Axial**
Velocity: 1.7 mm/s	Velocity: 1.0 mm/s	Velocity: 1.0 mm/s
Acceleration: 1.4 m/s^2	Acceleration: 1.2 m/s^2	Acceleration.: 1.0 m/s^2
Displacement: 3.5 µm	Displacement: 1.67 µm	Displacement.: 2.5 µm

Motor NDE bearing housing

Vertical	**Horizontal**	**Axial**
Velocity: 1.3 mm/s	Velocity: 1.0 mm/s	Velocity: 1.7 mm/s
Acceleration: 1.2 m/s^2	Acceleration: 1.6 m/s^2	Acceleration: 1.2 m/s^2
Displacement: 3.4 µm	Displacement: 0.9 µm	Displacement: 5.5 µm

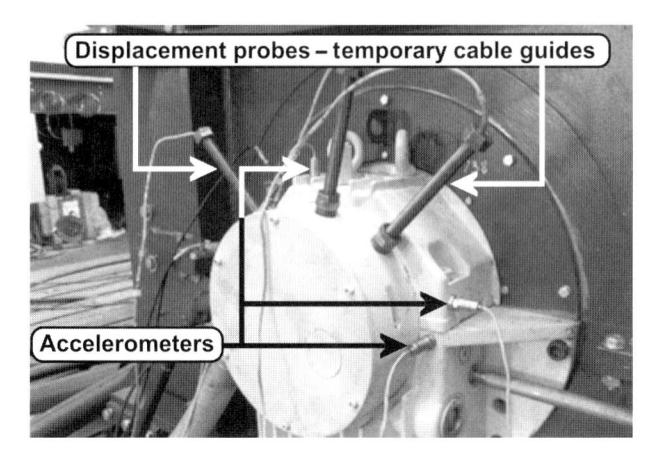

Figure 9.8 Photograph of accelerometers on the NDE bearing housing and position of shaft displacement probes.

The displacement levels in the vertical and horizontal positions on the bearing housings are negligible (maximum of 3.5 µm) but they do not indicate the actual displacement of the rotating shaft (bearing journal), which must be measured by displacement probes.

Recall that there is no physical contact between the rotating journal and the white metal Babbitt and bearing shell housing.

9.2.3.1 Sample of Vibration Spectra on the Bearing Housings – No-Load Run

During an uncoupled run the rotor of an induction motor is rotating at very close to synchronous speed, N_s.

For a 2-pole SCIM supplied at 60 Hz, the synchronous speed, $N_s = 3600$ r/min, therefore the *1X* frequency component is virtually at 60 Hz, because the rotor speed in a sleeve bearing motor will typically differ by only 1 to 2 r/min from the synchronous speed (3600 r/min).

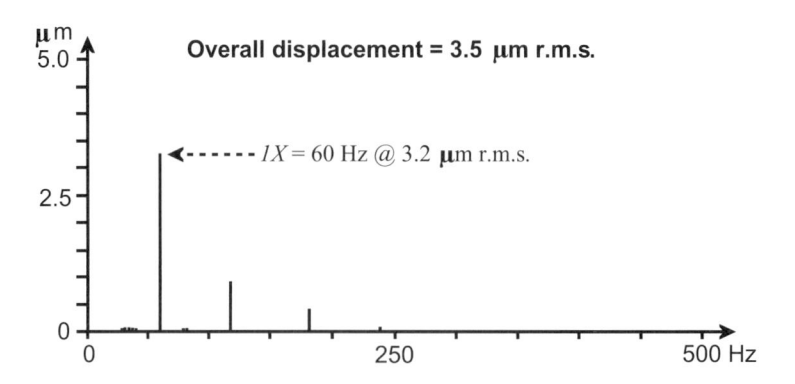

Figure 9.9 DE vertical bearing housing, r.m.s. displacement spectrum.

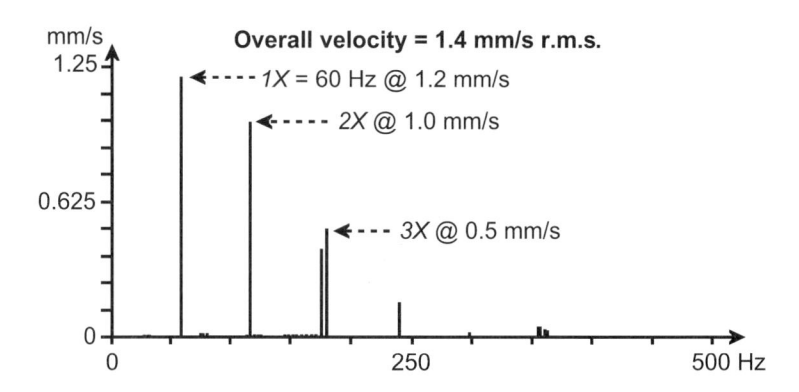

Figure 9.10 DE vertical bearing housing, r.m.s. velocity spectrum.

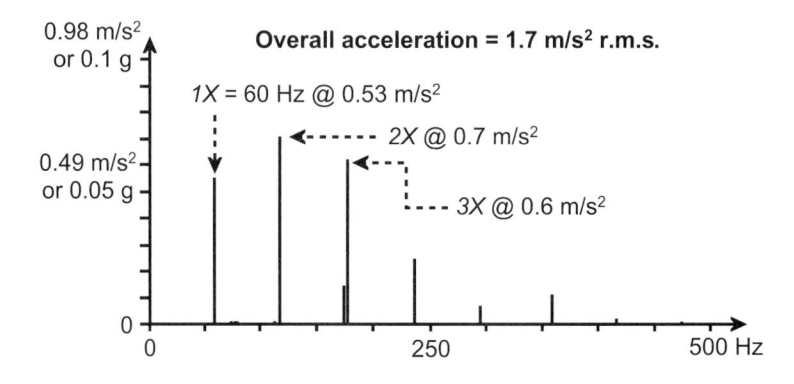

Figure 9.11 DE vertical bearing housing, r.m.s. acceleration spectrum.

For example, the no-load rotor speed was in fact, N_r = 3598 r/min and $1X$ = 59.97 Hz. The vibration spectra were taken between 10 Hz and 500 Hz using 12,800 spectral lines, giving a line resolution of 0.0398 Hz/line, which equates to 4.6 r/min, therefore the $1X$ component is displayed as being at 60 Hz.

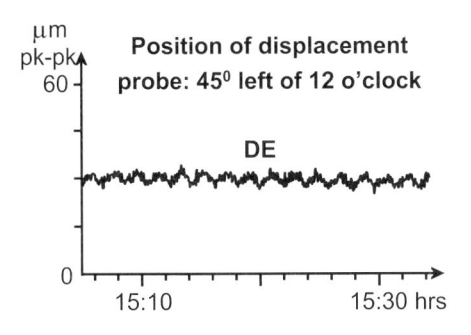

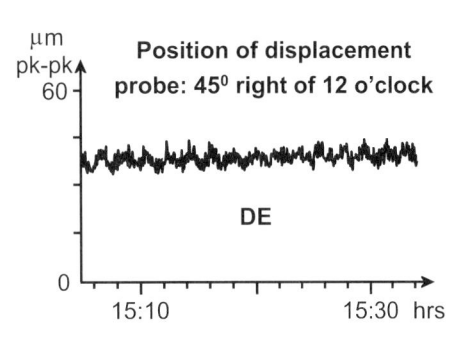

9.2 Vibration Factory Acceptance Test of a New 6800 kW/9115 H.P. SCIM

Figure 9.12 DE peak–peak shaft displacement during a 27-minute recording.

Figure 9.13 DE peak–peak shaft displacement during a 27-minute recording.

Figures 9.9 to 9.11 show the displacement, velocity and acceleration r.m.s. spectra respectively at the vertical position on the DE bearing housing.

The vibration spectra are perfectly normal. Note that the classical differences in frequency profile between the displacement, velocity and acceleration are present. Obviously the displacement spectrum is dominated by the $1X$ component (Figure 9.11) whereas, considering the acceleration spectrum, the higher frequencies at $2X$ and $3X$ are greater (Figure 9.10) in magnitude than the $1X$ component.

9.2.4 No-Load FAT Results from Shaft Displacement Probe Measurements

Plots of the shaft displacements at the DE and NDE are presented in Figures 9.12 to 9.15. The displacements are tabulated in Table 9.3 which also shows the position of the probes.

Main observations from the no-load uncoupled vibration FAT results are as follows:

(i) The highest displacement at the DE position at 45° to the right of 12 o'clock, was 38 μm peak–peak, which is only 58% of the maximum allowed of 65 μm peak–peak in BS 60034-14-2004 [9.13].

(ii) The American Petroleum Institute's (API) 541 vibration standard (item 6.3.3.12) [9.15] has a limit of 38 μm peak–peak (1.5 mils peak–peak). This new 6800 kW, 2-pole SCIM manufactured by Peebles in Scotland had a maximum displacement of 38 μm peak–peak, and would have satisfied the API shaft displacement requirement.

(iii) The displacement measured on the DE bearing housing was 3.5 μm r.m.s. but, as shown in the displacement spectrum (Figure 9.9), the $1X$ component was 3.2 μm r.m.s., therefore the peak–peak displacement of the $1X$ component on the DE bearing housing was 9 μm peak–peak. This is only 24% of the actual shaft displacement at the DE.

Table 9.3 Displacements peak–peak (in µm)

Position of probe at DE and NDE, 45° left of 12.00 hours	Position of probe at DE and NDE, 45° right of 12.00 hours
DE displacement probe: 28 µm peak–peak	DE displacement probe: 38 µm peak–peak
NDE displacement probe: 20 µm peak–peak	NDE displacement probe: 15 µm peak–peak

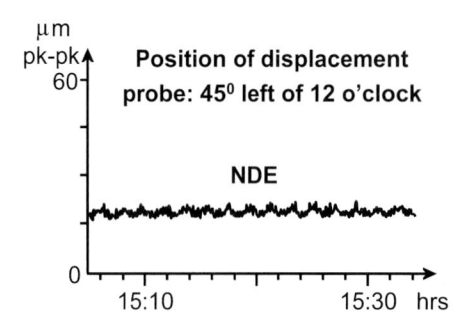

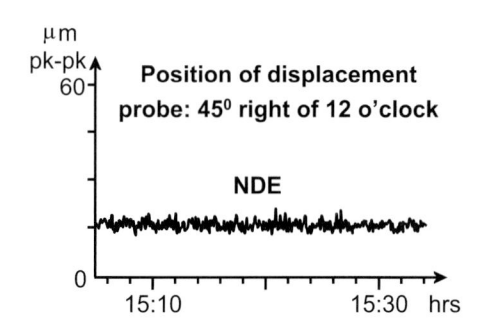

Figure 9.14 NDE peak–peak shaft displacement during a 27-minute recording.

Figure 9.15 NDE peak–peak shaft displacement during a 27-minute recording.

(iv) This demonstrates that the displacement values are vastly different and the peak–peak displacement on the bearing housing does not indicate the peak–peak shaft displacement of the shaft (journal) in the sleeve bearing.

9.2.5 Full-Load Shaft Displacement Results

Table 9.4 presents the shaft (journal) displacements at the end of a four-hour, full-load heat run when the motor was delivering 6800 kW/9115 H.P. at 3566 rev/min and the motor was supplied at rated volts (13.8 kV) and frequency (60 Hz) with an input full-load current of 332 amperes.

At the DE the displacements of the shaft at the 45° right and left of 12 o'clock were up to 70 µm peak–peak during the first 20 minutes of the full-load heat run. The bearing temperatures at the DE and NDE stabilised to 45.6 °C and 49 °C respectively, and after all the motor parts reached their stable and full-load operating temperatures, these displacements dropped to 56 µm and 38 µ peak–peak respectively, as shown in Figures 9.16 to 9.19.

9.2.5.1 Conclusions
(i) Recall that the vibration standard (BS 60034-14-2004 [9.13]) only applies to a vibration FAT during a no-load uncoupled run.

9.2 Vibration Factory Acceptance Test of a New 6800 kW/9115 H.P. SCIM

Table 9.4 Displacements peak–peak (in μm) after a four-hour full-load run

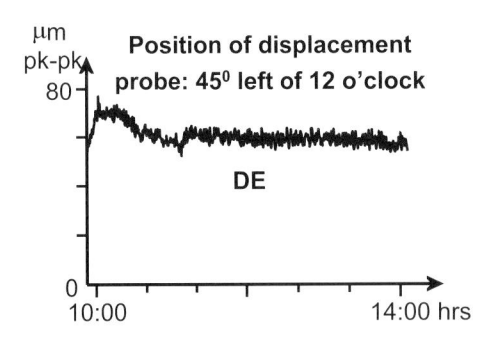

Position of probe at DE and NDE, 45° left of 12.00 hours	Position of probe at DE and NDE, 45° right of 12.00 hours
DE displacement probe: 56 μm peak–peak	DE displacement probe: 38 μm peak–peak
NDE displacement probe: 35 μm peak–peak	NDE displacement probe: 47 μm peak–peak

Figure 9.16 DE peak–peak shaft displacement during a two-hour recording.

Figure 9.17 DE peak–peak shaft displacement during a two-hour recording.

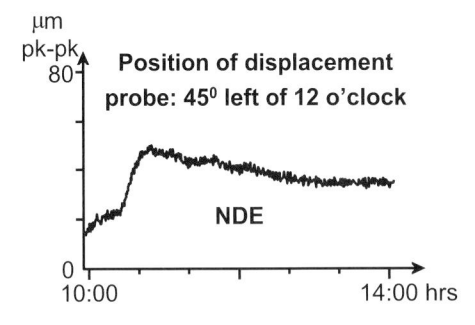

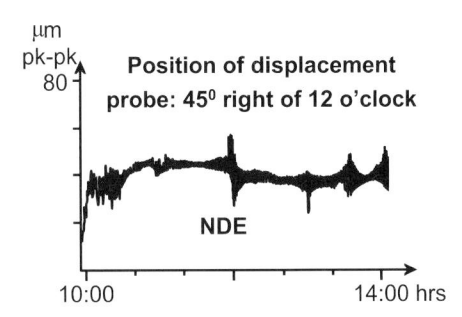

Figure 9.18 NDE peak–peak shaft displacement during a two-hour recording.

Figure 9.19 NDE peak–peak shaft displacement during a two-hour recording.

(ii) However, the peak–peak displacements after the four-hour full-load heat run were still below the upper limit of 65 μm peak–peak, and this was acceptable to the client who did not require bearing housing vibration measurements to be taken during the full-load heat run for the following reasons:
- When this motor is installed on the offshore oil and gas production platform, it will be driving a sea-water injection pump (SWIP), which is quite different from the load machine in the OEM's factory.

- The transmitted vibration to the SCIM's bearing housings from the SWIP and from adjacent rotating plant on the platform will be quite different to that transmitted from the OEM's load test facility, and thus comparisons would not be valid.

9.3 Industrial Case History – Analysis of Shaft Displacement and Subsequent Strip-Down Inspections Diagnosed Faults in a Journal Bearing of a 6250 kW/8380 H.P. Slip Ring Induction Motor

9.3.1 Background

Abstract – The author was contacted by a major oil company to determine the root cause of a recurrent vibration problem in a 6250 kW/8380 H.P., variable speed, Slip-Ring Induction Motor (SRIM) that drives a centrifugal compressor. Prior to the client contacting the author the motor had had its stator and rotor rewound as part of a major overhaul.

After the client (not the repair company) re-commissioned the motor, it could not be run continuously because it tripped out on high shaft displacement at the pre-set level of 120 µm peak–peak at the DE. This occurred during uncoupled and coupled runs, and the run times, before trip-out occurred, were typically between 5 minutes and 10 minutes.

A key feature of the shaft displacement problem was that after the motor tripped, the shaft displacement (peak–peak) at the DE continued to increase during the initial run-down period up to a level of 180 µm peak–peak at around half the synchronous speed and thereafter the shaft displacement did decay to zero. Note that this was not the result of a rotor resonance problem because the high vibration did not occur during the controlled run-up.

This was an unusual phenomenon, and the client, the repair company that rewound the stator and rotor and a third-party vibration CM company could not establish the root cause of the problem. This phenomenon should not theoretically occur because the fundamental force on the shaft is due to centrifugal force (*C.F.*) caused by mechanical imbalance in the rotor:

$$C.F. \propto mw^2 r(N_r), \qquad (9.1)$$

$C.F.$ = centrifugal force in Newtons (N)
m = unbalanced mass of the cylindrical rotor, kg
w = angular rotational speed = $\frac{2\pi N_r}{60}$ radian/s
N_r = rotor speed in r/min
r = radii of unbalanced mass(es); (rotor has more than one plane in comparison to a simple rotating disc).

The mechanical imbalance m and radius r are normally constant, therefore the *C.F.* should drop in proportion to the square of the speed and therefore the shaft displacement should decrease after trip-out.

9.3 Analysis of Shaft Displacement and Subsequent Strip-Down Inspections

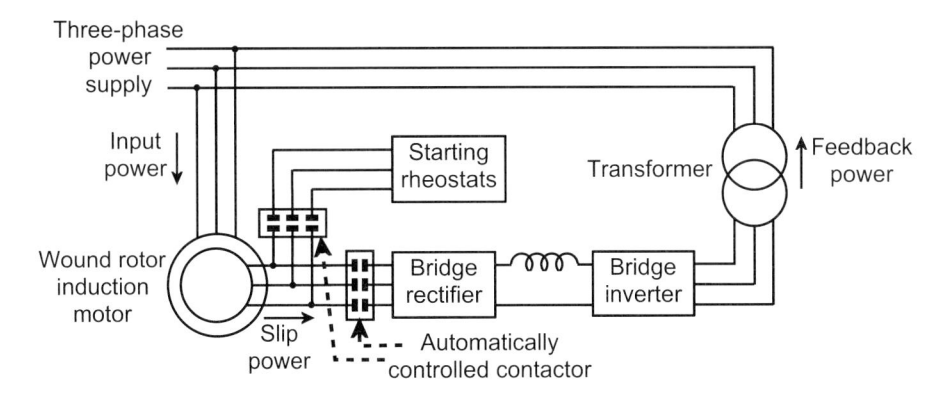

Figure 9.20 Schematic of a static Kramer slip energy recovery, speed control system.

The client reported that this shaft displacement problem had existed (for six months) from the time that the repaired motor was re-commissioned. Fortunately, a duplicate drive train was available to deliver the gas process.

Various parties had been involved for six months in trying to determine the root cause of the problem with no success, despite numerous on-site test runs and vibration measurements and analysis.

It was therefore obvious to the author that the motor had to be removed for controlled tests at the operating speed range of the motor during uncoupled and load runs in a factory test bed. The client agreed, and the motor was sent to an independent OEM of large induction motors but not to the repair company that had carried out the major refurbishment of the motor.

9.3.2 On-Site Operation and Motor Nameplate Data

This SRIM was manufactured in 1992 and is started via external rotor resistance to limit the starting current. After it reaches operational speed, resistors are automatically replaced by a static Kramer variable-speed electronic control system.

A basic schematic diagram of the closed-loop system is shown in Figure 9.20. The voltage and current in the rotor are at slip frequency (sf_1) and frequency conversion to the supply frequency (f_1) is necessary so that the power loss in the rotor circuit can be returned to the 3-phase supply to the motor.

In this slip energy recovery system, the slip frequency voltage in the rotor circuit is rectified in the bridge rectifier and smoothed before being returned to the mains supply via a 3-phase bridge inverter. Speed control is achieved by varying the firing angle of the controlled inverter and thus the torque-speed curve for each firing angle.

In effect, a variable back voltage is inserted into the rotor circuit; this can change the torque-speed characteristic as a function of the firing angle in the bridge inverter.

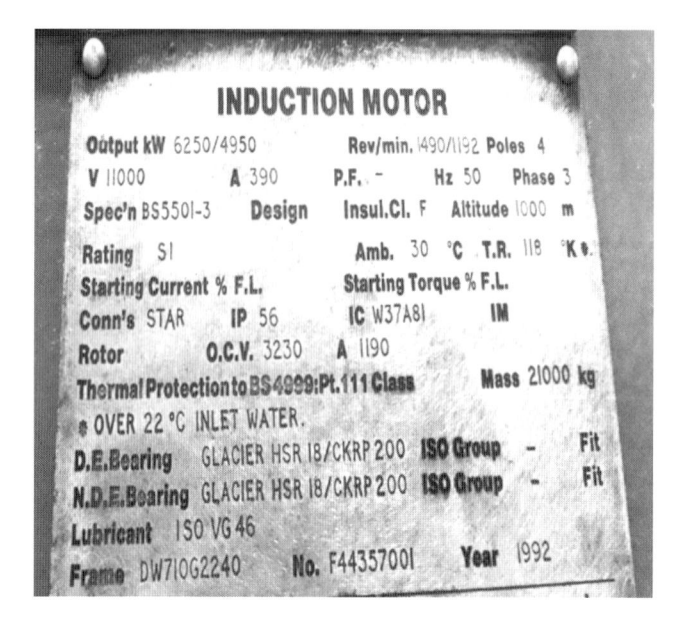

Figure 9.21 A photograph of the SRIM motor's nameplate.

Power is recovered rather than wasting power by varying the rotor resistance using external rotor resistors to obtain speed control.

In the drive string (see Figure 9.22) there was a very large coupling, of 1800 kg, which, in the author's opinion, also acts as a flywheel to reduce any torsional oscillations that may occur in this variable speed drive. This could not be confirmed by either the original OEM or the client for the following reasons:

(1) The original OEM had been brought over by another company and original records were not available.
(2) The motor was 22 years old when the motor vibration problem occurred, and the personnel employed by the client when the motor was originally specified and commissioned had retired.

Records and fundamental knowledge were therefore lost, and this is one of the real practical difficulties that can be experienced in industry when a vibration problem occurs in a large 22-year-old SRIM. Photographs of the motor and the motor's nameplate are shown in Figures 9.21 and 9.22 respectively.

9.3.3 Deliverables and Vibration Test Results

A definitive test specification was presented to the client by the author and the following main deliverables were agreed:

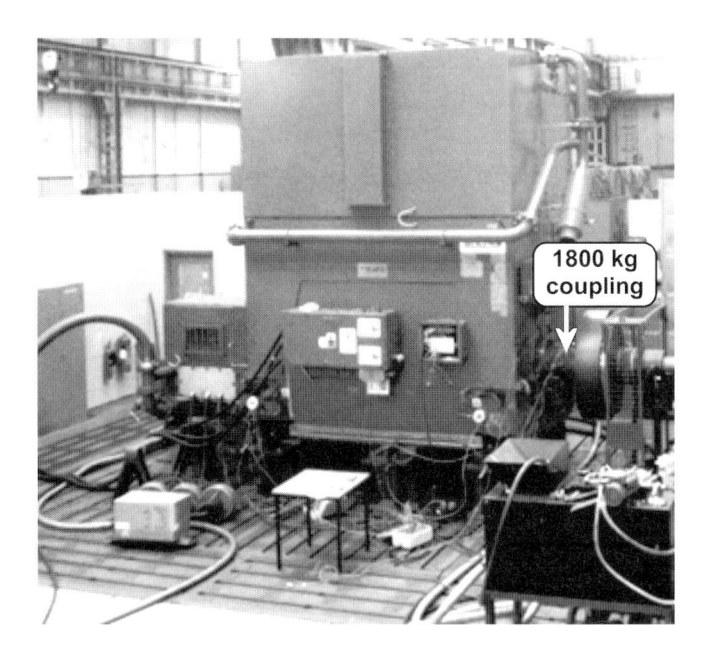

9.3 Analysis of Shaft Displacement and Subsequent Strip-Down Inspections

Figure 9.22 Photograph of 11 kV, SRIM, 6250/4950 kW, 1490/1120 r/min.

Figure 9.23a Displacement probes at the DE.

(1) The first test would be a no-load uncoupled run of the received SRIM.
(2) The client instructed the independent OEM to discontinue the no-load run when the shaft displacement reached 100 μm peak–peak, to avoid any damage to the bearings.
(3) The main goal of the first test run was to confirm that the vibration phenomenon which had occurred during an uncoupled run on-site was repeated on the test bed.

The client informed the author and the independent OEM that it would be impossible to remove the 900 kg half coupling hub without damaging the shaft and therefore no attempt should be made to do so.

Figure 9.23b Displacement probes at the NDE.

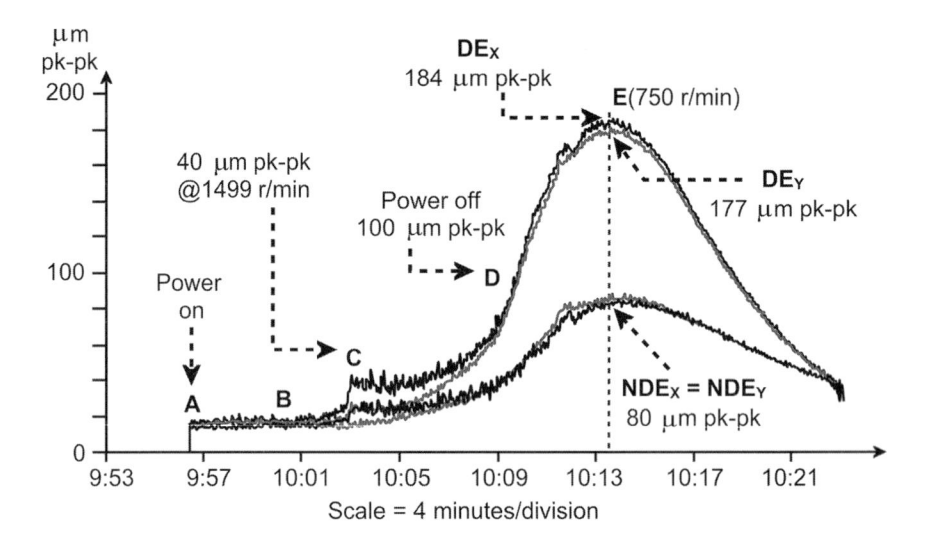

Figure 9.24 Shaft displacements during the no-load uncoupled run versus time.

9.3.3.1 Vibration Results from the First No-Load Uncoupled Run

The motor was started on reduced voltage, which was increased in small steps until the voltage was 11 kV at 50 Hz, and a no-load speed of 1499 r/min was reached. Figures 9.23a and b present photographs of the positions of the displacement probes to measure the shaft displacement at the DE and NDE as already described in the introductory case history presented in Section 9.2. The shaft displacement plots at the DE and NDE during the test are presented in Figure 9.24.

(a) **Interpretation of Figure 9.24**

(i) The slow roll run out of the shaft's peak–peak displacement is shown in section A to B of Figure 9.24, and is acceptable at 15 μm peak–peak, because BS 60034-14-2004 allows 16 μm peak–peak for electric motors of this speed and shaft height.

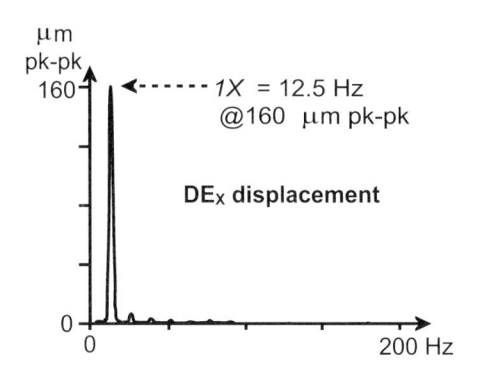

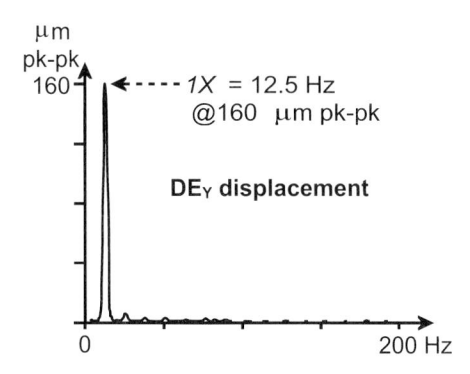

9.3 Analysis of Shaft Displacement and Subsequent Strip-Down Inspections

Figure 9.25 Spectrum DE$_X$ shaft displacement versus frequency – see position E on plot in Figure 9.24.

Figure 9.26 Spectrum DE$_Y$ shaft displacement versus frequency – see position E on plot in Figure 9.24.

(ii) The maximum shaft displacement from the four probes was 40 μm peak–peak at the DE$_X$ position when the motor reached its no-load speed of 1499 r/min at position C.

(iii) After 6.5 minutes at 1499 r/min the DE shaft displacement had risen to 65 μm peak–peak, which is the maximum allowed as per BS 60034-14.

(iv) After a total of eight minutes running at 1499 r/min the DE shaft displacement had reached 100 μm peak–peak at position D, and the client instructed the independent OEM to switch off the motor. This was an increase of 150% during a period of only eight minutes and corresponds to a rate of rise of 7.5 μm peak–peak/min.

(v) The shaft displacements were still recording data after the switch off, but during the deceleration period between D to E on Figure 9.24, the DE$_X$ and DE$_Y$ shaft displacements increased to a maximum of 184 μm and 177 μm peak–peak at position E respectively, at a speed of 750 r/min. This was not due to rotor resonance at 750 r/min because it did not occur during the controlled run-up.

(vi) The phenomenon recorded at the client's refinery plant was therefore replicated, which confirmed the vibration problem was inherent in the motor.

(b) **Uncoupled Run – Shaft Displacement Spectra**

The vibration spectra from the displacement probes which correspond to the displacements at position E (750 r/min) on Figure 9.24 are presented in Figures 9.25 and 9.26. As expected the spectra are dominated by the *1X* rotational speed frequency component.

(c) **Uncoupled Run – DE Bearing Housing Overall R.M.S. Velocities**

Figure 9.27 shows the r.m.s. velocities on the DE bearing housing in the vertical, horizontal and axial positions during the run-up and run-down period.

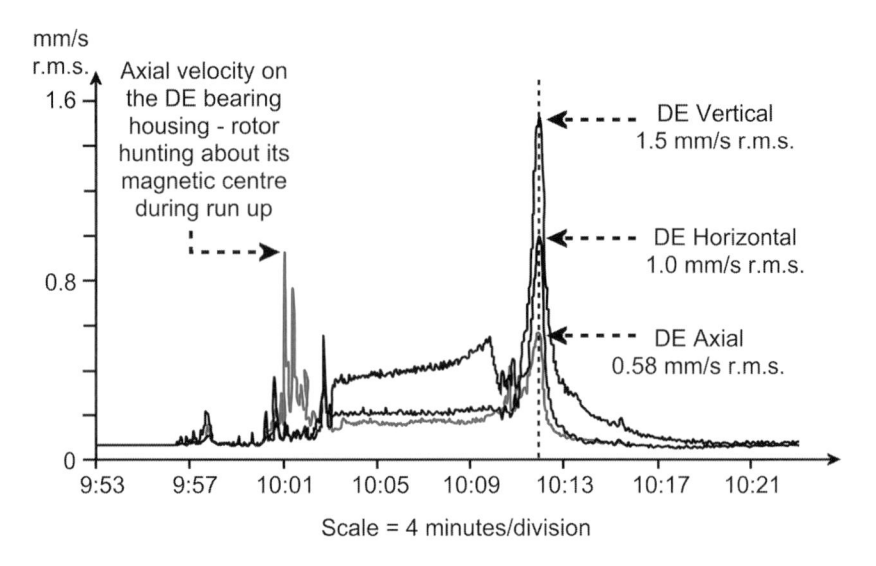

Figure 9.27 Overall r.m.s. velocities on the DE bearing housing during an uncoupled run from start-up versus time.

The fundamental reason for presenting the bearing housing velocities is to demonstrate that the levels are perfectly acceptable at a maximum of 1.5 mm/s r.m.s. in the vertical position on the DE bearing housing *but* if shaft displacements were not being measured the serious shaft displacement problem would not have been identified.

9.3.4 Ingress Protection (IP) Rope Seal Malfunction – Source of High Shaft Displacement

It was Alistair Carr (senior electrical engineer), an employee of the author's company, who predicted that a malfunctioning IP (ingress protection) rope seal, which is used to prevent contaminants damaging the Babbitt, was the source of the high shaft displacement at the DE during the run-down.

Photographs of the DE bearing's IP seal assembly and the actual rope seal and channel are shown in Figures 9.28 and 9.29 respectively.

The reasons for suspecting that the rope seal was the possible source of the shaft vibration problem were as follows:

- The bearing seals on this SRIM consist of floating labyrinths to contain the oil and an IP seal that also serves to retain enclosure pressure.
- The IP seal consists of fibrous material (6 mm^2), as shown in Figure 9.29, which is impregnated with polytetrafluoroethylene (PTFE) or tallow to provide a lubricant.
- Heating can be very localised and is confined to the inboard seal housing and the motor shaft at the point of contact. As a result, this cannot be readily detected thermally and would not be seen by the bearing bush temperature detectors.

9.3 Analysis of Shaft Displacement and Subsequent Strip-Down Inspections

Figure 9.28 IP seal assembly.

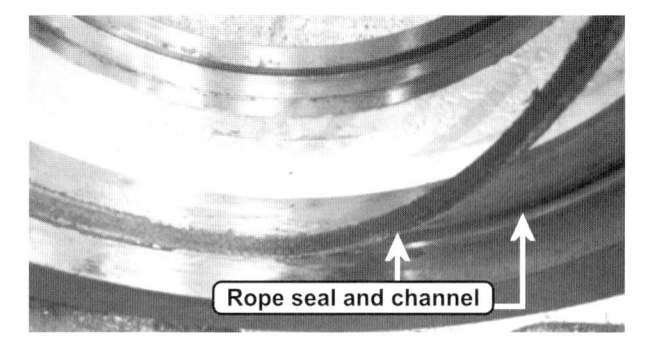

Figure 9.29 Rope seal and channel in the IP assembly.

- The frictional heating generated by the seal is cumulative. As the seal area heats up the seal tightens and, in addition, the lubricant dries out. Generally, the tight area will only be over a segment of the seal and not the full circumference. As the IP seal binds on the shaft the shaft tends to climb in the bearing.
- The shaft movement relative to the white metal Babbitt increases *but* the bearing housing absolute vibration remains comparatively low as this is only influenced by what is transmitted through the shaft seal or the oil film.
- Once the IP seals have been *over heated* the material's original characteristics cannot be recovered as the lubricant has dried out and the material will no longer be evenly distributed in the locating groove.
- Note that after careful fitting of a correctly specified IP seal material it may still be necessary to carry out a seal bedding procedure, and this may not have been properly done by the motor repair company. There must be a balance where the seal surface sits sufficiently close to the shaft to serve its purpose, without any contact friction, which could generate excess heat and subsequent binding.
- Seal bedding is achieved when the motor is run up to speed, in stages, from a variable frequency supply. Shaft vibration must be closely monitored and if seen to start to climb at a given speed, the motor speed is reduced, or the motor stopped, until the seal area cools.

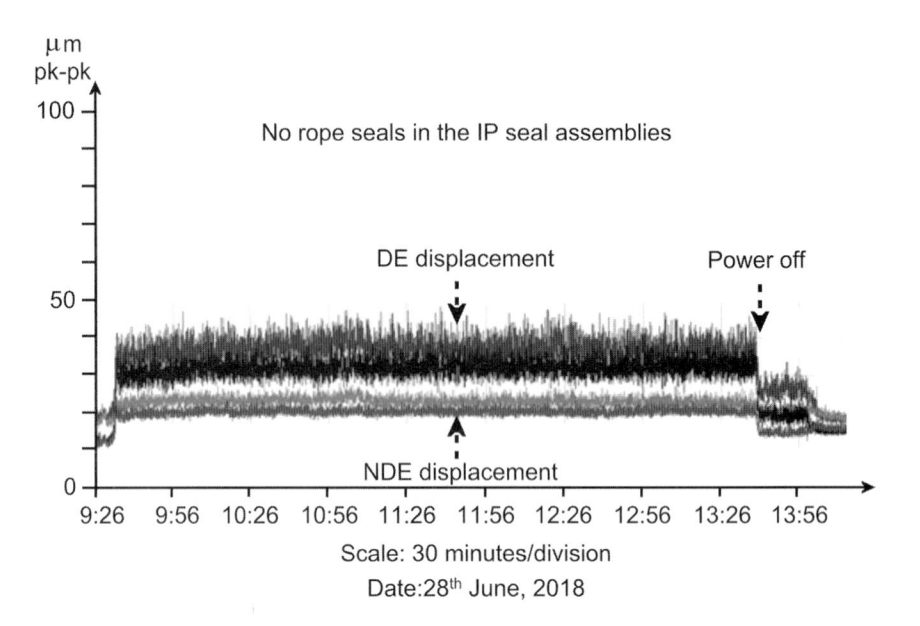

Figure 9.30 Shaft displacements (unfiltered) at the DE and NDE during a four-hour uncoupled run.

9.3.5 Shaft Displacement with the IP Rope Seals Removed from the DE and NDE

The tests were being carried out on a factory test bed; therefore the IP rope seals from the DE and NDE could be removed and the vibration measurements repeated during a no-load run.

This procedure was used to verify whether the IP rope seals were the source of the problem. Figure 9.30 presents the shaft displacements at the DE and NDE during a no-load run of four hours from start-up.

The shaft displacements were not greater than 40 μm peak–peak throughout the complete run period, which therefore confirmed that the rope seals were not fit for purpose.

Figure 9.31 presents the shaft displacement spectrum for the DEx position and this confirms that the displacement is normal with the $1X$ frequency component at 25 Hz being only 22 μm peak–peak compared to 160 μm peak–peak when the operationally unfit for purpose rope seal was in the DE bearing.

9.3.6 Shaft Displacements with New IP Rope Seals at the DE and NDE – Full-Load Heat Run

9.3.7 Conclusions

(i) New IP seals were fitted into the DE and NDE bearing assemblies and the seals were subsequently bedded-in.

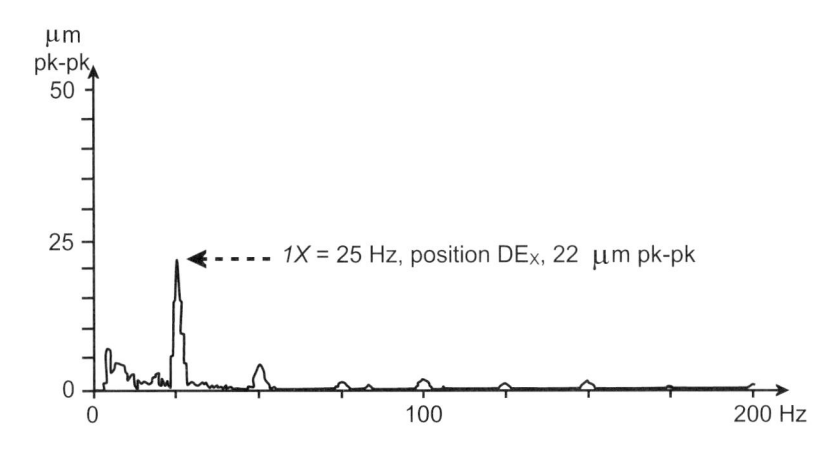

9.3 Analysis of Shaft Displacement and Subsequent Strip-Down Inspections

Figure 9.31 Spectrum of DE_x shaft displacement versus frequency.

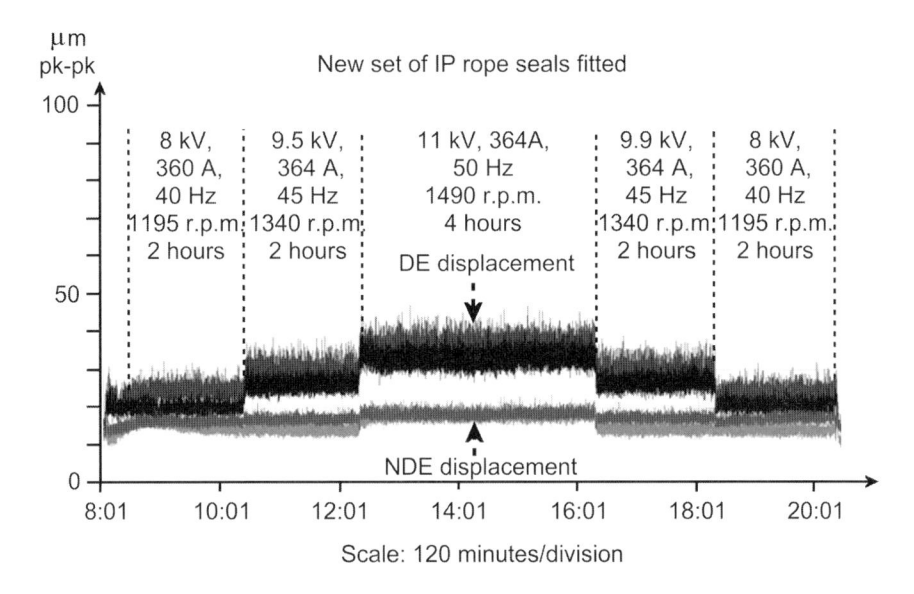

Figure 9.32 Shaft displacements versus time during a 12-hour load run with different supply frequencies but with an approximately constant V/f ratio to keep the flux constant.

(ii) Because the SRIM had an operational speed range between 1192 r/min and 1490 r/min, the shaft displacements were measured at full-load current covering this speed range.

(iii) Figure 9.32 presents the shaft displacements during a 12-hour load run and the V/f ratio was kept constant at the different operating speeds to ensure the main airgap flux was constant.

Table 9.5 Motor nameplate data

3-phase SCIM	6800 kW	326 A	60 Hz
3570 r.p.m.	0.89 p.f.	eff 97.2%	Frame 710/2-2
Enclosure CACW	Rating MCR S1	$I_s = 5$ FLC	$T_s = 0.7$ FLT
Insulation class F, 900 °C by temperature rise by resistance			

(iv) The results of Figure 9.32 show that the maximum shaft displacements (DE_X and DE_Y) were at the DE, but they were perfectly acceptable at 40 µm peak–peak. The new seals were fit for purpose and the motor was successfully returned to service.

9.4 Industrial Case History – Excessive Shaft Displacement during First 100 Minutes of a Five-Hour Heat Run of a Re-Furbished 2-Pole 6800 kW/9115 H.P. SCIM – Function of Temperature Change and Rotor Design

9.4.1 Summary

Abstract – The author of this book diagnosed broken rotor bars using Motor Current Signature Analysis (MCSA) in a 13.8 kV, 6800 kW, SCIM.

The rotor cage winding was re-barred, and the motor was Factory Acceptance Tested (FAT) for vibration levels during a no-load uncoupled run and a full-load heat run. It was found during the heat run that shaft displacement was abnormally high (130 µm peak–peak) for 30 minutes during the first 40 minutes after switch on but it settled down to a normal level of 30 µm peak–peak during the last 2.5 hours of a five-hour full-load heat run.

The results will verify that this phenomenon was a function of the rotor design and rate of change of temperature. The motor's nameplate data are presented in Table 9.5, which refers to its full-load rated output, and a photograph of the motor under test is shown in Figure 9.33.

9.4.2 Shaft Displacements at the NDE during the Full-Load Heat Run

The motor was supplied at rated volts and frequency for the full-load heat run at the rated input current of 326 amperes. Figures 9.34 and 9.35 present the shaft displacements at the NDE at 45° left and 45° right of the 12 o'clock position respectively.

9.4.2.1 Interpretation of Shaft Displacements

With reference to Figure 9.34, the key observations are as follows:

(i) During the first 10 minutes on full load, after start-up, the vibration levels (shaft displacement) were low at, for example, a maximum of 30 µm peak–peak at

9.4 Industrial Case History – Excessive Shaft Displacement

Figure 9.33 Motor under test.

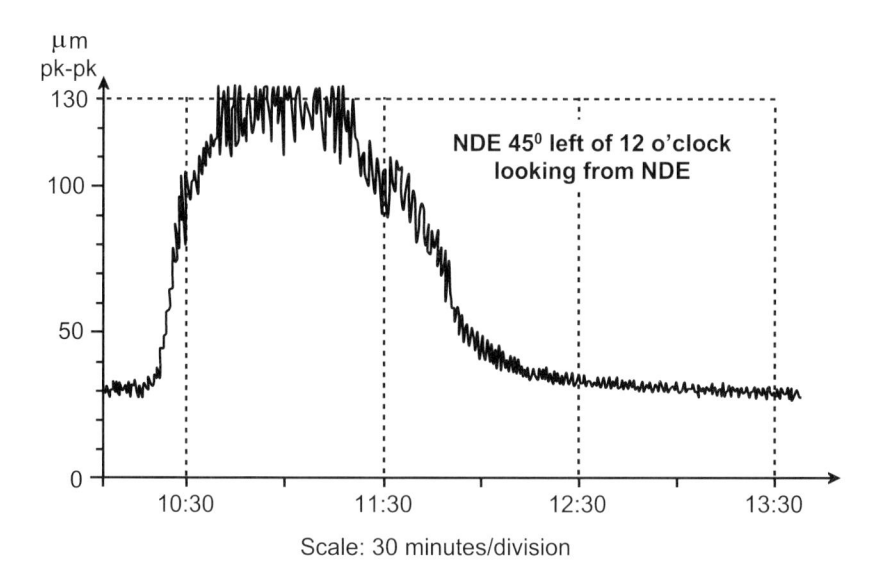

Figure 9.34 NDE shaft displacement, 45° left of 12 o'clock versus time of full-load run.

the NDE, but there was a dramatic rise up to 130 μm peak–peak at the NDE which lasted for 30 minutes. This magnitude of shaft displacement was initially not acceptable to the client.

(ii) The refurbished rotor was balanced to grade 1.0 before the tests. The shaft displacement trip level was previously set at 125 μm peak–peak when the motor was operating on the offshore oil production platform.

(iii) After 100 minutes on full load, the vibration at the NDE was still 100 μm peak–peak (alarm level) but it then dramatically dropped down to 30 μm peak–peak

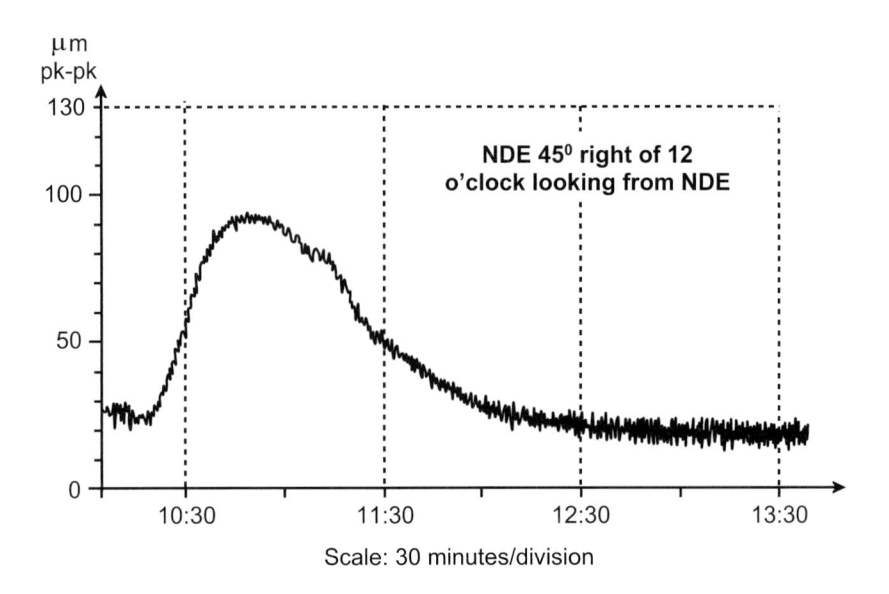

Figure 9.35 NDE shaft displacement, 45° right of 12 o'clock versus time of full-load run.

and remained at that level for the remainder of the full-load run – a further 2.5 hours.

(iv) A second full-load heat run was carried out two days after the first heat run so that the motor had cooled down to the ambient temperature of 20 °C, and the vibration measurements were repeated and the same phenomenon occurred.

9.4.2.2 Shaft Displacement as a Function of Stator Winding Temperature

Figure 9.36 presents the stator winding temperature and shaft displacement at the NDE (45° left of 12 o'clock) versus time during the full-load heat run.

The temperatures of the bearings were perfectly normal throughout the full-load heat run because at start-up the NDE and DE bearing temperatures were both 14 °C and they steadily increased to 55 °C after five hours, which in effect rules out a bearing fault.

The plots in Figure 9.36 indicate the temperature of the rotor and stator assemblies were a function of time between start-up until the temperatures of all parts of the motor stabilised during the full-load heat run.

The rotor temperature is never measured in a SCIM, but the temperatures of different parts of the stator windings are measured via RTDs (resistance temperature detectors) and that was the reference used to plot temperature versus time so that cross correlation can be made with shaft vibration versus time.

The average temperature of the nine RTDs in the stator winding was 27 °C at the start of the full-load heat run. Figure 9.36 indicates the following:

- The shaft displacement increased dramatically during the highest rate of rise of stator winding temperature.

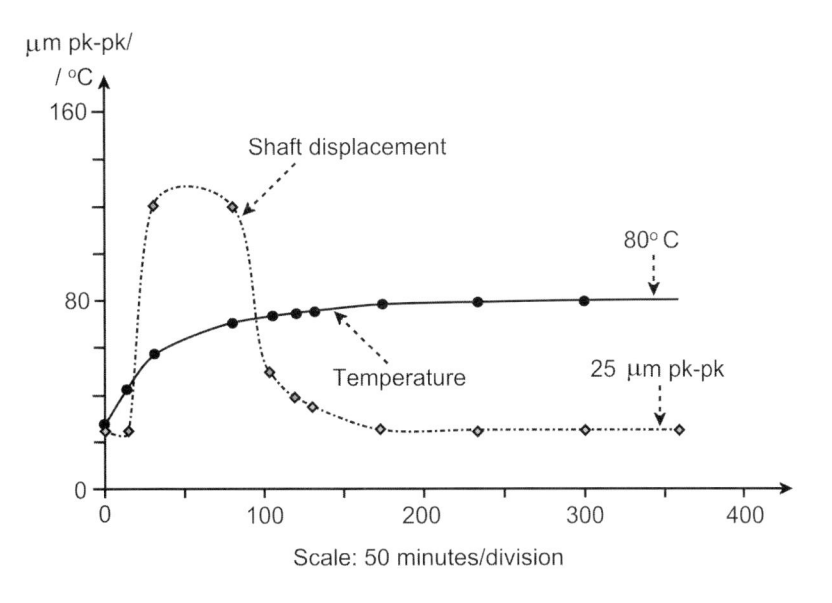

9.4 Industrial Case History – Excessive Shaft Displacement

Figure 9.36 Vibration and stator winding temperature versus time of full load heat run.

Figure 9.37 Squirrel cage rotor before replacement of the cage winding.

- When the temperature reached its steady-state value of 80 °C the vibration dramatically dropped to 25 μm peak–peak. The vibration and stator winding temperatures remained constant thereafter, as shown in Figure 9.36.
- This result confirmed a correlation between the rate of change of stator winding temperature and shaft displacement.

9.4.3 Conclusions

The rotor with broken rotor bars prior to it being re-barred is shown in Figure 9.37.

The rotor core is a heat shrink fit onto the shaft, and of particular significance is the fact that there are no radial cooling ducts in this rotor from a 6800 kW SCIM.

The likely reasons for the vibration phenomenon were as follows:

(i) During the tests the temperatures of the copper cage winding, laminated steel core and solid steel shaft will all be different and, in particular, the copper bars and end ring will heat up much more quickly than the rotor core, and the former will expand external to the rotor core ends.

(ii) There will, of course, be a re-distribution of the rotor's mechanical imbalance due to temperature differentials between start-up from cold and the motor reaching its steady-state temperature after four to five hours. The rotor's mechanical imbalance then takes up a steady-state value, after all parts of the rotor reach their full-load temperatures and the vibration returns to normal.

(iii) It was subsequently found that this was a characteristic of the rotor design, and carrying out a trim balance of the rotor did not reduce the high shaft displacement (130 μm peak–peak) during a period of 30 minutes, as shown in Figure 9.36.

(iv) The client required the motor to be returned for operational service as soon as possible, and therefore the only option was to set the vibration trip level on the offshore platform to 135 μm peak–peak to prevent the motor tripping out during this early running period after start-up.

(v) The client approved this increase in the vibration trip level and. despite this higher level of displacement, after start-up, the motor has run without incident since its return to service in 2010.

References

9.1 F. Y. Zeidan and B. S. Herbage, Fluid Film Bearing Fundamentals and Failure Analyses, *Proceedings 20th Turbo Machinery Symposium,* Texas A&M University, College Station, Houston, TX, 1991, pp. 161–86. Can be accessed via: https://pdfs.semanticscholar.org/676e/8d91559fc3a6275f2924b436767d957c73e3.pdf.

9.2 W. R. Finley and M. M. Hodowanec, Sleeve vs. Anti-Friction Bearings: Selection of the Optimal Bearing for Induction Motors, *Record of Conference Papers, IEEE Incorporated Industry Applications Society, Forty-Eighth Annual Conference, 2001 Petroleum and Chemical Industry Technical Conference* (Cat. No.01CH37265), 2001, pp. 305–17.

9.3 M. E. Leader, Understanding Journal Bearings, *Proceedings 25th Turbo Machinery Symposium,* Texas A&M University, College Station, Houston, TX, 2001.

9.4 H. P. Bloch and F. K. Geitner, *Machinery Failure Analysis and Trouble Shooting: Practical Machinery Management for Process Plants,* fourth edition, Elsevier, 2012.

9.5 R. Scott, Journal Bearings and their Lubrication, *Machinery Lubrication Magazine,* July 2005.

9.6 G. J. Willis, *Lubrication Fundamentals,* Marcel Dekker Inc., ISBN 0-8247-696-7, 1980.

9.7 M. M. Khonsani and R. E. Booser, *Applied Tribology – Bearing Design and Lubrication*, John Wiley & Sons Inc., ISBN 0-471-28302-9, 2001.

9.8 A. Muszynska, *Rotor Dynamics*, CRC Press/Taylor and Francis, FL, USA, 2005.

9.9 J. M. Vance, *Rotordynamics of Turbo Machinery*, John Wiley & Sons, ISBN 0-471-80258-1, 1988.

9.10 B. J. Hamrock, S. R. Schmid and B. O. Jackson, *Fundamentals of Fluid Film Lubrication*, Marcel Decker, New York, 2004.

9.11 C. Yung, Sleeve bearing clearance depends on many factors, www.plantengineering.com/single-article/sleeve-bearing-clearance-depends-on-many-factors.html.

9.12 Type E Slide Bearings – RENK AG, http://www.renk.biz/download.php, Slide_Bearings_Type_E_Series_EF.pdf.

9.13 BS 600 34-14-2004, *Rotating Electrical Machines, Part 14 Mechanical Vibration of Certain Machines with Shaft Heights of 56 mm and Higher – Measurement, Evaluation and Limits of Vibration Severity*, 2004.

9.14 NEMA MG 1-2006, *Part 7: Mechanical Vibration – Measurement, Evaluation and Limits*.

9.15 American Petroleum Institute (API), 541 Vibration Standard, item 6.3.3.12.

10 Industrial Case Histories on Magnetic Forces and Vibration from Induction Motors

10.1 Electromagnetic Forces and Vibration in Induction Motors

10.1.1 Twice Supply Frequency Vibration

It is well known – see references [10.1] to [10.9] – that the fundamental electromagnetic force and consequential radial vibration in an induction motor is at **twice the supply frequency = $2f_1$**.

The mathematical derivation is given in Appendix 10A and it is also shown that vibration harmonics of $2f_1$ can occur at $4f_1$, $6f_1$, $8f_1$, etc. Electromagnetically induced stator core vibration therefore occurs at $2f_1$ and its harmonics and this is verified via vibration measurements presented in case histories in this chapter, for example see Figure 10.1.

It is also well known – see references [10.4] and [10.5] – that the electromagnetic force at $2f_1$ has a mode wave number equal to twice the number of pole-pairs and this is illustrated in Figure 10.1a [10.5], and to quote:

- Figure 10.1a shows the developed airgap of an induction motor with 2-poles of the revolving flux wave.
- The magnetic flux exerts a radial pull across the airgap, proportional at each point to the flux density squared ($B^2 \propto \Phi_2$).
- The force wave is therefore a fully displaced sine wave of double the frequency of the magnetic flux wave as shown in Figure 10.1a [10.5].

A brand new SCIM will therefore inherently produce stator core vibration at $2f_1$, but an increase in magnetic asymmetry will change the characteristics of the vibration measured on the stator core back and will normally increase its magnitude. However, this increase will only occur at specific positions on the core back as shown in the case history in Section 10.4, later in this chapter.

The causes of an increase in magnetic asymmetry include, for example:

(i) Thermal bowing of the rotor, which causes a change in dynamic airgap eccentricity.
(ii) Soft foot that may cause frame or stator core distortion in certain designs of SCIMs.
(iii) Single phasing or an increase in supply voltage unbalance.

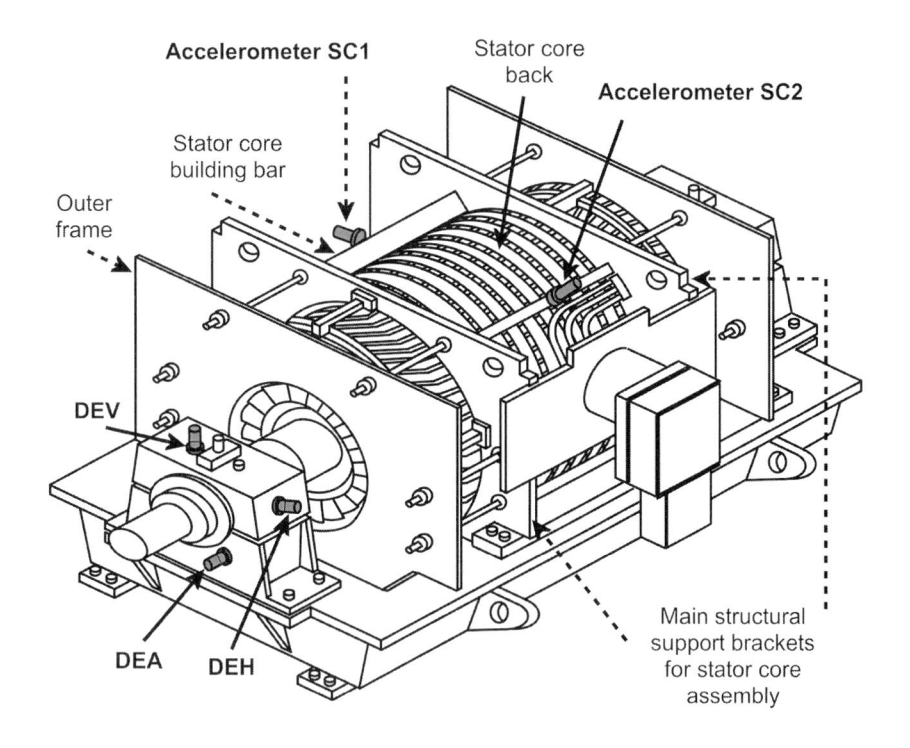

Figure 10.1 Schematic diagram of stator core and winding and positions of vibration measurements.

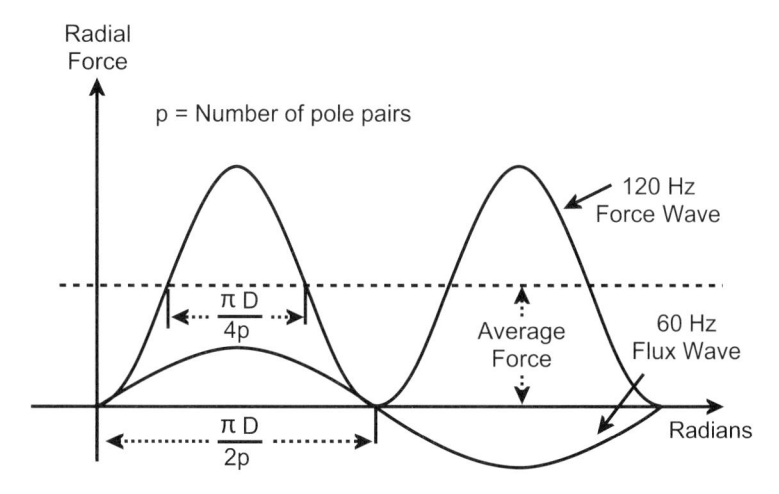

Figure 10.1a Illustration of the radial magnetic flux and force wave; see reference [10.5], *Transactions AIEE*, 73 (IIA), 1954, pp.118–25.

Different designs and ratings of SCIMs will have different stator cores and frame dimensions and different flux densities, therefore the magnitude of vibration on the stator core at $2f_1$ is different for each design of motor.

Unless the client specifically requests such measurements, the vibration on the actual stator core is not normally measured during Factory Acceptance Testing (FAT) of large induction motors (e.g. 1000 H.P./746 kW and above). There are no vibration standards or guidelines that cover acceptable levels of velocity or acceleration **directly on the stator core** back.

Acceptable vibration criteria on the bearing housings are published in vibration standards by NEMA, IEEE, API and BS, which are in fact different; please refer to references [10.10] to [10.13] for full details.

Twice supply frequency vibration at $2f_1$ when it is measured on the bearing housings does not reflect the actual stator core back vibration at that frequency. The former will normally be less than that which exists on the stator core back as demonstrated in the first industrial case history (Section 10.2).

In 2014, The American Petroleum Institute published the following criterion:

The API 541 standard recommends [10.11] that the vibration on the outer frame, in line with support brackets [10.14] for the stator core should not be greater than twice the permitted velocity on the bearing housings.

10.1.2 Vibration Components Caused by Rotor Slotting

It has been known for many years that rotor slotting results in additional radial electromagnetic forces between the rotor and stator of a SCIM and the frequencies of the consequential vibration are given by Equation 10.1; see references [10.3] to [10.5] and [10.15] to [10.20]:

$$f_{rv} = f_1 \left\{ \frac{R}{p}(1-s) \pm n_{rv} \right\}. \tag{10.1}$$

This equation can be expressed in the form given by Equation 10.2:

$$f_{rv} = \left\{ \frac{RN_r}{60} \right\} \pm f_1 n_{rv}, \tag{10.2}$$

where:

f_{rv} = classical and inherent rotor slot vibration frequency components
f_1 = supply frequency
R = number of rotor slots
p = pole-pairs
s = operating slip in p.u., where $s = \frac{(N_s - N_r)}{N_s}$
n_{rv} = integers 0, 2, 4, 6
N_r = rotor speed in r/min
N_S = synchronous speed in r/min.

Example: consider a 4-pole, 60 Hz, 1782 /min SCIM with 56 rotor slots. Substituting these values in Equation 10.2, and with $n_{rv} = 0$, f_{rv} can be calculated as

$$f_{rv} = \left\{ \frac{RN_r}{60} \right\} \pm f_1 n_{rv} = \left\{ \frac{56 \times 1782}{60} \right\} \pm 60 \times 0 = 1633 \text{ Hz},$$

where: $f_{rv(n_{rv}=0)}$ = the principal rotor slot passing vibration component.
With n_{rv} = 2, 4, 6, 8 the rotor slot passing vibration components can also occur at:

n_{rv} = 2, f_{rv} = 1633 ± 2f_1 = 1513 and 1753 Hz
n_{rv} = 4, f_{rv} = 1633 ± 4f_1 = 1393 and 1873 Hz
n_{rv} = 6, f_{rv} = 1633 ± 6f_1 = 1273 and 1993 Hz
n_{rv} = 8, f_{rv} = 1633 ± 8f_1 = 1153 and 2113 Hz.

That is, a series of rotor slot passing frequency components equally spaced apart at twice the supply frequency from each other.

Important points

The electromagnetic forces at these rotor slot passing frequencies are normal and inherent in a SCIM and are not due to a fault; see references [10.3] to [10.5].
Their magnitudes in mm/s or m/s^2 are a function of load current taken by the motor, for example as the load current increases the current in the rotor bars also increases.
The leakage flux from each rotor bar therefore increases and likewise the magnitude of the rotor slot passing flux components and hence the rotor slot vibration components.
Their magnitudes in mm/s or m/s^2 can increase with an increase in airgap eccentricity, but they are not unique frequency characteristics which can be used to truly diagnose an airgap eccentricity problem. The reason is that their magnitudes in mm/s change as a function of the rotor current and load applied to the motor.

10.2 Industrial Case History – Magnetic Forces and Vibration on the Stator Core and Outer Frame of a New 4000 kW/5362 H.P. SCIM

Abstract – This investigation had the following objectives, which were carried out and achieved by the author in 1986:

(i) To measure and analyse the vibration on the stator core back and at strategic positions on the outer box frame of a new 6.6 kV, 4000 kW/5362 H.P., 60 Hz, 2-pole, SCIM.
(ii) To compare the vibration measurements and analysis from item (i) with the measurements and analysis of vibration on the housings of the journal bearings.

This was carried out during FATs of the motor, and accelerometers were mounted directly on the stator core back.

The results of the investigation have not been previously published by the author of this book, but he did carry out the vibration measurements and analysis as part of a research and development project, which was sponsored by BP Exploration, Aberdeen, Scotland. At that time he was a senior lecturer at Robert Gordon University, Aberdeen, Scotland, and leader of the electrical machines monitoring research and development team.

Table 10.1 Motor nameplate data

Type of motor	3-phase SCIM
Voltage	6.6 kV
Power	4000 kW/5362 H.P.
Current (full-load)	404 A
Frequency (nominal)	60 Hz
Full-load speed	3540 r/min (2-pole)
p.f.	0.89
Efficiency	95.4%

In 2014, the American Petroleum Institute's standard API 541 [10.11] was amended to include vibration limits on the frame of an electrical machine with guidelines as to where the vibration measurements should be taken. This was 28 years after the author carried out the tests presented in this case history.

10.2.1 Vibration Measurement and VSA during Factory Acceptance Tests

The motor's nameplate data are presented in Table 10.1.

The motor was supplied at rated voltage and frequency. A schematic diagram of the motor without its external box frame is shown in Figure 10.1.

Additional data

No-load current: 75 A (i.e. 18.6% of $I_{f.l.}$).
Number of stator and rotor slots: 60 and 46 respectively.

The overall r.m.s. velocities were measured on the bearing housings at the DE and NDE in the vertical, horizontal and axial positions. Accelerometers were mounted directly on the stator core building bars, one (SC1) at 36° clockwise and the other (SC2) at 36° anticlockwise with respect to the 12 o'clock position, looking from the NDE – see Figure 10.1.

Table 10.2 presents the overall r.m.s. velocities and accelerations measured on the stator core back and on the bearing housings. Because the motor was Factory Acceptance Tested in the UK, the OEM proposed to the oil company that the vibration standard to be used was BS 60034-14-2004 [10.12] and the client accepted this proposal.

The upper velocity limit for this motor is 2.3 mm/s r.m.s. (±10%) on the bearing housings. From Table 10.2 the r.m.s. velocities are perfectly acceptable, with the highest level being 1.5 mm/s r.m.s.

With respect to the velocity and acceleration levels which were measured ***directly on the stator core back*** there were no vibration standards in 1986 that covered limits on the stator core vibration, and that is still the case at the time of writing this book (from 2016 to 2019).

However, in 2014 the American Petroleum Institute (API) amended its API 541 standard to include vibration measurements on the outer frame of electrical machines but that is not the same as measuring on the stator core back [10.14].

Table 10.2 Overall r.m.s. vibration levels ±10%; span: 10–1000 Hz

---

BS 60034-14 2004 [6.12]
Limits on the bearing housings:
Velocity = 2.3 mm/s r.m.s., acceleration: 3.6 m/s² r.m.s.

Motor DE bearing housing

Vertical	**Horizontal**	**Axial**
Velocity: 1.0 mm/s	Velocity: 1.3 mm/s	Velocity: 1.5 mm/s
Acceleration: 1.5 m/s²	Acceleration: 1.3 m/s²	Acceleration: 1.2 m/s²

Motor NDE bearing housing

Vertical	**Horizontal**	**Axial**
Velocity: 1.2 mm/s	Velocity: 1.0 mm/s	Velocity: 1.2
Acceleration: 1.3 m/s²	Acceleration: 1.0 m/s²	Acceleration: 1.0 m/s²

Stator core back

Position SC1	**Position SC2**
Velocity: 5.2 mm/s	Velocity: 5.2 mm/s
Acceleration: 5 m/s²	Acceleration: 5 m/s²

The key conclusions from the measured overall velocities and accelerations on the bearing housings and stator core back are as follows:

(i) The overall velocities and acceleration levels on the bearing housings are all perfectly acceptable and well within the upper limits in BS 60034-14-2004, as shown in Table 10.2.

(ii) The overall velocity and acceleration levels on the stator core back *are up to 3.5 times greater* than in the radial directions on the bearing housings.

(iii) The bearing pedestals are separate from the end frames in this motor and the vibration transmitted from the stator core to the bearing housings will be attenuated. The attenuation would be less if the bearing housings were an integral part of the end frames.

The reasons for the vibration being much higher (see item (ii) above) on the stator core back of this 2-pole SCIM require an analysis of the time domain waveforms and vibration spectra.

10.2.2 Time Domain and VSA of Vibration on the Stator Core Back

Figures 10.2 and 10.3 present the time domain velocity waveforms on the core back at positions SC1 and SC2, and it is very clear that there is a dominant pattern with a complete time cycle of period, $T = 8.33$ ms, which corresponds to 120 Hz.

The peak velocity was 8 mm/s at position SC1 on the core back, which is 0.3 inches/s peak. The NEMA MG1 standard [10.10] allows up to 0.15 inches/s peak on the bearing housings when mounted on a rigid base, which was the case with this

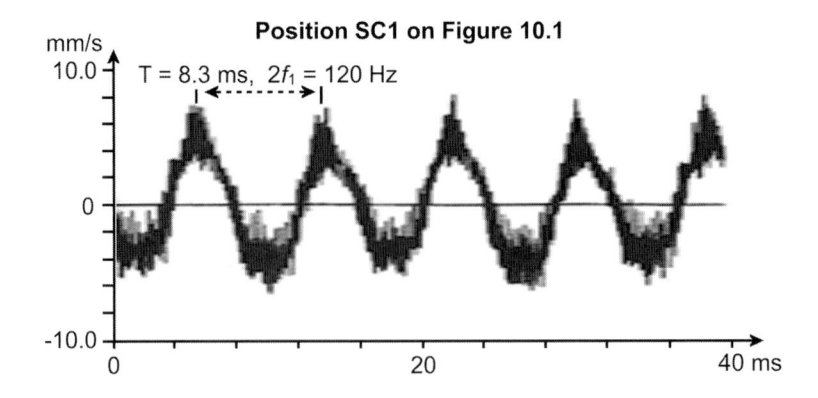

Figure 10.2 Velocity time waveform at position SC1 on the stator core back.

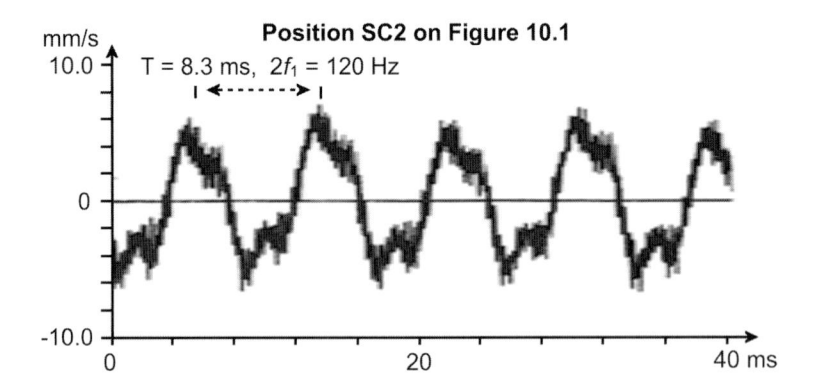

Figure 10.3 Velocity time waveform at position SC2 on the stator core back.

motor, thus the core back vibration was twice the NEMA limit for the bearing housing vibration.

The API 541 [10.11] standard only allows a maximum of 0.1 inches/s peak on the bearing housing, therefore the core back vibration at 0.3 inches/s was three times larger in this motor: an interesting result.

There are also numerous higher frequencies superimposed on the 120 Hz time domain waveform. The velocity spectrum on the stator core back at SC1, as shown in Figure 10.4, is given as a sample using a dB versus frequency display so that all the components can be simultaneously displayed on the same spectrum.

With respect to Figure 10.4 the following nomenclature applies:

A = $1X$ rotational speed frequency = 60 Hz at 92 dB (i.e. 0.76 mm/s r.m.s.)
B = $2f_1$ = 120 Hz at 108 dB (i.e. 4.8 mm/s r.m.s.)
C = $4f_1$ = 240 Hz at 88 dB (i.e. 0.48 mm/s r.m.s.)
D = $6f_1$ Hz
E = $8f_1$ Hz
F = $f_{rv(n_{rv} = 0)}$ = 2760 Hz at 78 dB (i.e. 0.15 mm/s r.m.s.).

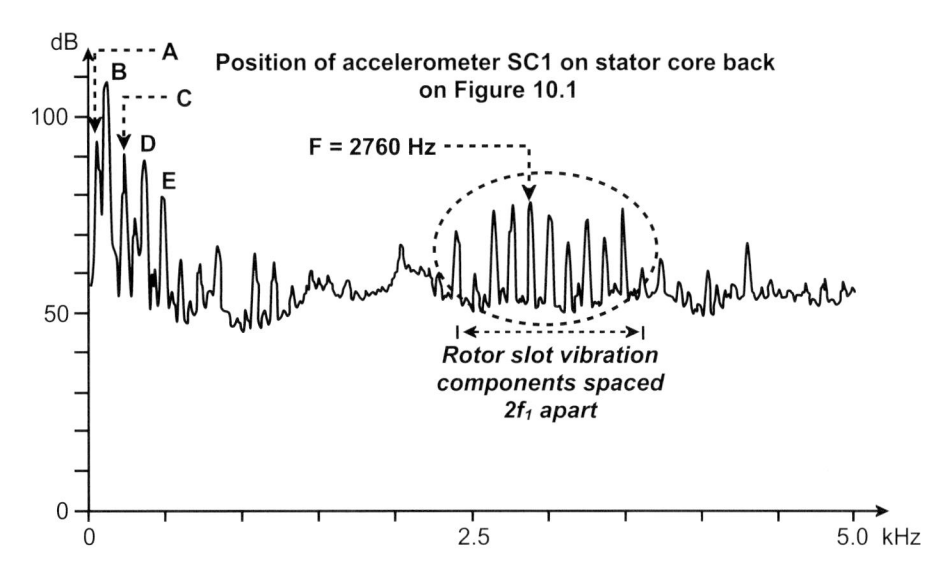

Figure 10.4 Velocity spectrum, position SC1 on the stator core back.

The key observations from the velocity spectrum on the stator core back are as follows:

(i) The spectrum is dominated by the $2f_1$ component at 120 Hz at a velocity of 4.8 mm/s r.m.s. (overall 5.2 mm/s r.m.s.).
(ii) As predicted, there are harmonics of $2f_1$ at 240 Hz, 360 Hz and 480 Hz of magnitudes at 0.24 mm/s, 0.48 mm/s and 0.5 mm/s r.m.s. respectively.
(iii) As predicted, there is a series of rotor slot vibration components spaced at $2f_1$ apart, and from Equation 10.2, and with $n_{rv} = 0$, we have:

$$f_{rv} = \left\{\frac{RN_r}{60}\right\} \pm f_1 n_{rv} = \left\{\frac{46 \times 3600}{60}\right\} \pm 60 \times 0 = 2760 \text{ Hz.}$$

The motor was running uncoupled; therefore the rotor speed was very close to the synchronous speed of 3600 r/min, and for the calculation N_r is taken to be 3600 r/min.

This is often termed the principal rotor slot passing vibration component. There are also additional and inherent rotor slot passing frequencies, which are spaced $2f_1$ apart, as shown in Figure 10.4:

With $n_{rv} = 2, 4, 6, 8$ the rotor slot passing vibration components occur for this motor as

$n_{rv} = 2$, $f_{rv} = 2760 \pm 2f_1 = 2880$ and 2640 Hz
$n_{rv} = 4$, $f_{rv} = 2760 \pm 4f_1 = 3000$ and 2520 Hz
$n_{rv} = 6$, $f_{rv} = 3120 \pm 6f_1 = 3120$ and 2400 Hz
$n_{rv} = 8$, $f_{rv} = 2760 \pm 8f_1 = 3240$ and 2280 Hz.

This proved the previous predictions and was in line with numerous publications by eminent electrical machine designers and academics; see references [10.1] to [10.5].

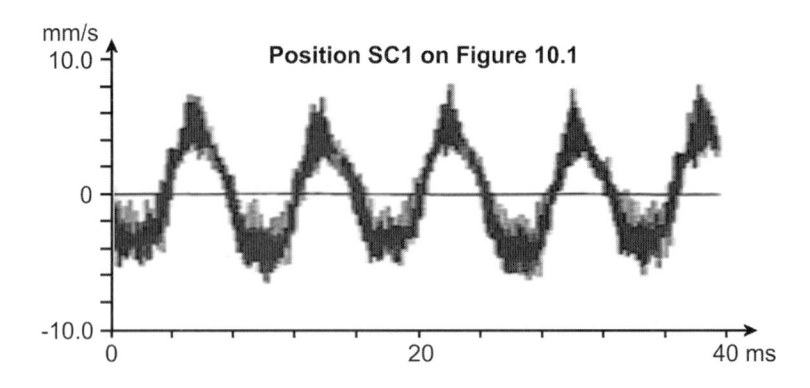

Figure 10.5 Velocity time waveform on stator core back at position SC1.

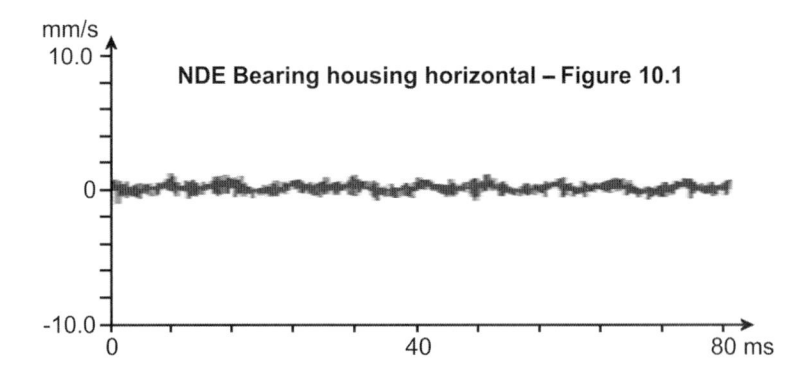

Figure 10.6 Velocity time waveform in the horizontal direction on the NDE bearing housing.

These results were from a brand new 2-pole SCIM that had an airgap eccentricity (static plus dynamic) of 8% of the radial airgap, which is normal and was confirmed by the OEM.

The electromagnetic forces and consequential vibration on the stator core back at these frequencies were inherent in this brand new SCIM and were not the result of a fault.

10.2.3 Bearing Housing Vibration – Comparison with Stator Core Vibration

Figures 10.5 and 10.6 show the time domain velocity waveforms from the NDE bearing housing and the stator core back at SC1 respectively. The bearing housing velocity is only 10% of the stator core back velocity; this is evident by comparing Figures 10.5 and 10.6.

With respect to Figure 10.4 the following nomenclature applies:

B = $2f_1$ = 120 Hz at 108 dB (i.e. 4.8 mm/s r.m.s.)
C = $4f_1$ = 240 Hz at 88 dB (i.e. 0.48 mm/s r.m.s.)

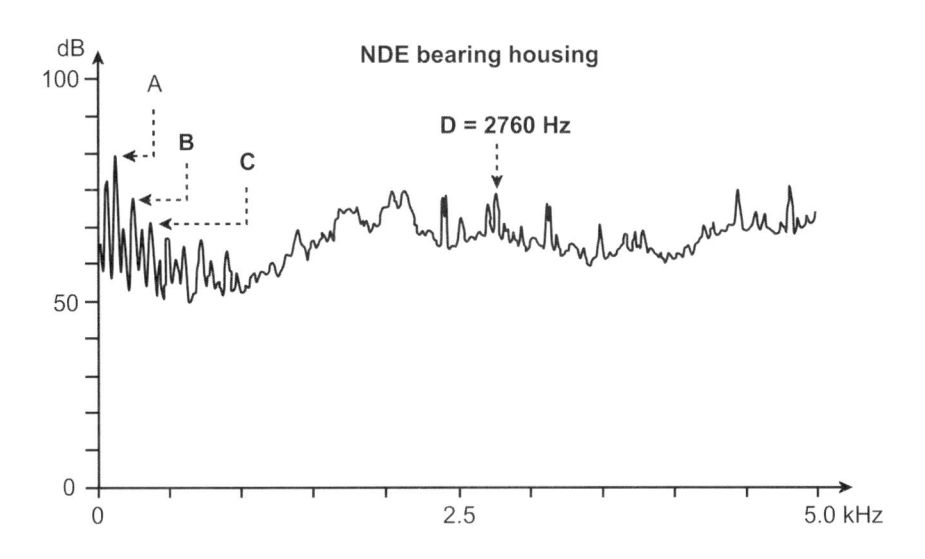

10.2 Industrial Case History – Magnetic Forces and Vibration 217

Figure 10.7 Vibration spectrum, NDE horizontal on the bearing housing.

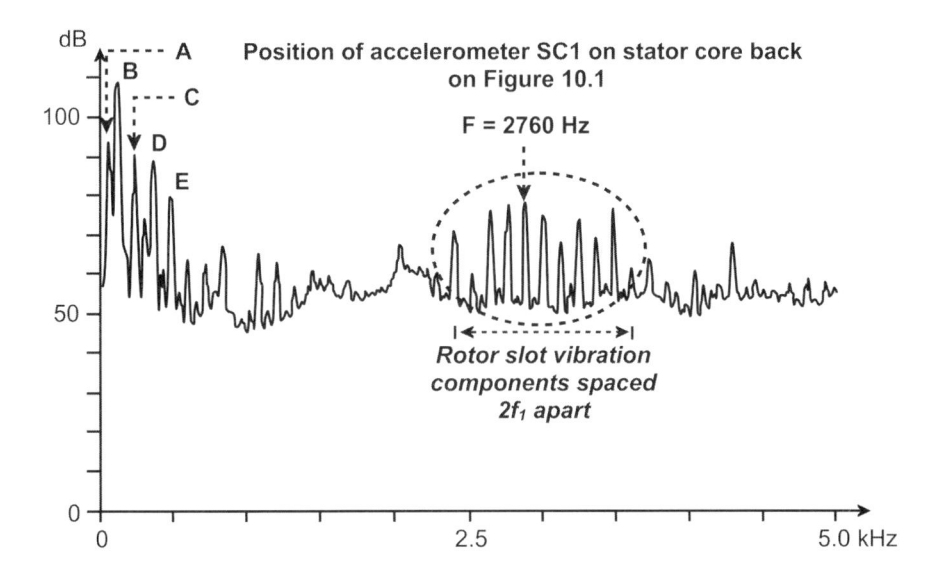

Figure 10.7a Velocity spectrum, position SC1 on the stator core back.

$D = 6f_1$ Hz
$F = f_{rv(n_{rv} = 0)}$ = 2760 Hz at 78 dB (i.e. 0.15 mm/s r.m.s.).

The velocity spectrum at the NDE horizontal position on the bearing housing is shown in Figure 10.7, and Figure 10.7a is a repeat (for a direct comparison between the two spectra) of the spectrum shown in Figure 10.4, which was from the stator core back at position SC1; the key observation is as follows:

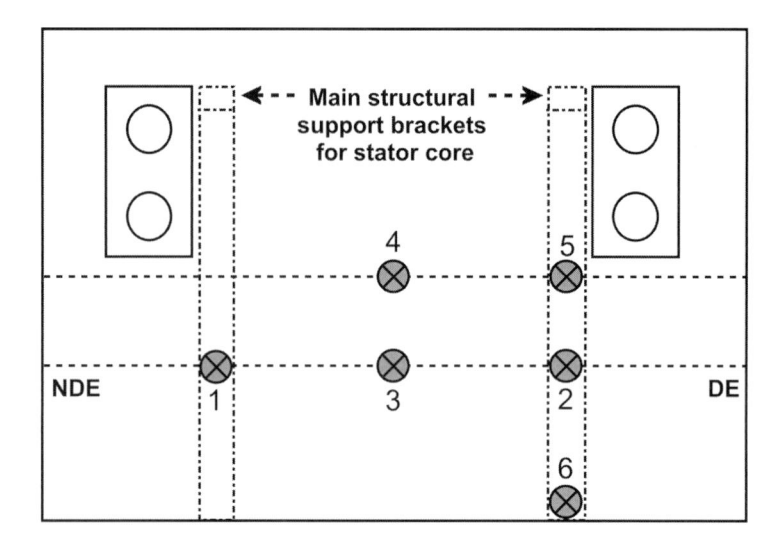

Figure 10.8 Overall r.m.s. velocities and accelerations on the outer frame at the side *without* the terminal box.

(i) The twice supply frequency vibration component at 120 Hz was only 0.48 mm/s r.m.s. on the bearing housing; this was **one tenth** that at the SC1 and SC2 positions (see Figures 10.1 and 10.7a) on the stator core back.

With respect to Figure 10.7a the following nomenclature applies:

A = 1X rotational speed frequency = 60 Hz at 92 dB (i.e. 0.76 mm/s r.m.s.)
B = $2f_1$ = 120 Hz at 108 dB (i.e. 4.8 mm/s r.m.s.)
C = $4f_1$ = 240 Hz at 88 dB (i.e. 0.48 mm/s r.m.s.)
D = $6f_1$ Hz
E = $8f_1$ Hz
F = $f_{rv(n_{rv} = 0)}$ = 2760 Hz at 78 dB (i.e. 0.15 mm/s r.m.s.).

10.2.4 Outer Frame Vibration and Analysis at Full Load

In 1986, the author measured and analysed the vibration on the outer frame of this new motor during a full-load heat run test. The motor was supplied at rated voltage and frequency, and at the full-load current of 404 amperes.

Figures 10.8 and 10.9 show the positions where the vibration was measured and the corresponding overall r.m.s. velocities and acceleration levels.

The velocities and acceleration overall r.m.s. levels at positions '1' to '6' in Figure 10.8 were as follows:

1: 5.5 mm/s and 6 mm/s^2 (0.61 g)
2: 5 mm/s and 6 mm/s (0.61 g)
3: 9 mm/s and 8 mm/s^2 (0.82 g)

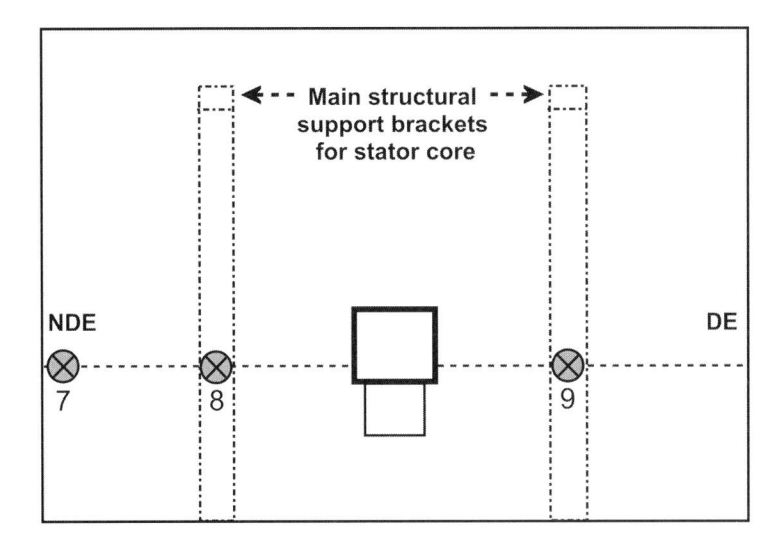

[Diagram: Rectangular outer frame with "Main structural support brackets for stator core" labeled at top with arrows pointing to two vertical dashed columns. Central square block. Three circled-X measurement positions along horizontal centreline: position 7 (far left, labeled NDE), position 8 (left of centre), position 9 (right of centre, with DE labeled far right).]

Figure 10.9 Overall r.m.s. velocities and accelerations on the outer frame at the side with the terminal box.

4: 5 mm/s and 6.5 mm/s^2 (0.66 g)
5: 5 mm/s and 3.6 mm/s^2 (0.37 g)
6: 2 mm/s and 4 mm/s^2 (0.41 g).

The velocity and acceleration overall r.m.s. levels at positions '7' to '9' in Figure 10.9 were as follows:

7: 2 mm/s and 4 mm/s^2 (0.41 g)
8: 5 mm/s and 6 mm/s^2 (0.61 g)
9: 4 mm/s and 5.5 mm/s^2 (0.56 g).

Before concluding on the measurements, recall that these measurements were taken in 1986, which was 28 years before API 541, in 2014, recommended the following limits of vibration on the outer frame of an electrical machine:

The limit of frame vibration at the shaft centre-line level, on a loaded structure, is twice the velocity allowed for the bearing housings.

Figure 10.10, as reported in reference [10.14], illustrates one mode of excitation along with an example of why it is important to measure the vibration on a loaded structural member.

10.2.4.1 Conclusions

The key observations from the velocities and accelerations are presented in Figures 10.8 and 10.9 in comparison to the velocities measured on the stator core back at positions SC1 and SC2 (see Table 10.1).

(1) From Figure 10.8, which is on the side without the terminal box, the overall r.m.s. velocities and accelerations at positions '1' and '2' *were 6 mm/s at each position and 5.5 m/s^2 and 5 m/s^2 respectively.*

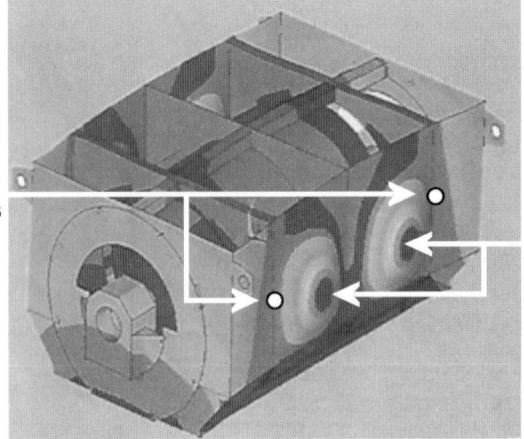

Measure opposite the stator support brackets to assess the stator frame vibration

Panels which are unloaded structural parts – <u>do not</u> measure vibration on these panels to assess the stator frame vibration

Figure 10.10 Stator frame vibration should be measured on a loaded structure as shown in the illustration [10.14].

Note the following:
(a) At positions '1' and '2' there is a *direct transmission path to the stator core via the stator core support brackets*, which are structurally loaded members.
(2) The overall r.m.s. velocities at both positions (SC1 and SC2 – see Section 10.2 and Figure 10.1) on the stator core back were 5.2 mm/s, and at positions '1' and '2' on the outer frame the velocities were 6 mm/s, which is 15% higher (see Figure 10.8).
(3) The measurements at positions '1' and '2' on the stator frame give a reasonably good correlation with the actual vibration measured on the stator core back at positions SC1 and SC2 for this motor.
(4) From Figure 10.9, on the side with the terminal box, the overall r.m.s. velocities at positions '8' and '9' were 5 mm/s and 4 mm/s, which are lower than at positions '1' and '2' on the opposite side.
   Note the following:
   (a) Positions '8' and '9' are on the side that is mechanically stiffer and has a greater mass due to the terminal box being bolted to the panel. This is in comparison to the side panel without the terminal box.
(5) From Figure 10.8, the overall r.m.s. velocity at position '3' was 9 mm/s on the side without the terminal box, which was on the panel mid-way between the stator support brackets. This is 73% higher than the stator core back velocity of 5.2 mm/s at the SC1 and SC2 positions on the core back, which is in fact in-line with position '3' on the outer frame at that side.
(6) This verified that vibration measurements on a panel that does not have a direct structural and stiff transmission path to the stator core do not reflect the actual stator core vibration and normally amplify it.
(7) The vibration measurements verified that vibration measurements should not be taken on unsupported panels of an outer frame of a SCIM.

10.3 Industrial Case History – False Positive of Cage Winding Breaks by a Vibration Condition Monitoring (CM) Sub-Contractor – Identified the True Cause as Normal and Inherent Stator Core Vibration at a Rotor Slot Passing Frequency

10.3.1 History and Summary

Abstract – In 2009, an end user employed a vibration condition monitoring sub-contractor to assess the mechanical condition of a SCIM driving a reciprocating compressor operating in an LNG processing plant. However, the sub-contractor did not have the required theoretical knowledge of the design, construction and operation of cage induction motors to reliably identify the cause of the vibration characteristic, which they measured on the motor's frame.

The vibration that they measured was due to electromagnetic forces that were inherent to its design. The sub-contractor reported that this motor had broken rotor bars. The end user removed the motor from service and sent it to a repair workshop, who reported that the aluminium, die cast rotor cage winding seemed to be normal based on a thorough visual inspection.

The motor was returned to site and 12 months later a different technician from the same sub-contractor reported that the motor had broken rotor bars.

In 2010, the end user called in the author to carry out vibration measurements and analysis and Motor Current Signature Analysis (MCSA) – see references [10.21] to [10.30] – to check the operational integrity of the cage winding.

This case history shows that the sub-contractor, on two instances, erroneously deduced that the vibration component at the principal rotor slot passing frequency (see Section 10.1.2) measured on the motor's frame at the NDE was caused by broken rotor bars. This prediction resulted in unnecessary costs for the end user, because the motor was removed from the LNG plant for strip-down and an independent inspection at a motor repair workshop.

To quote the electrical team leader at the LNG site, the total cost was of the order of £10,000.00 because removal and inspection involved the following:

(i)   Lifting equipment to remove the motor shown in Figure 10.11a.
(ii) Time for personnel to remove the motor.
(iii) Transportation to repair shop.
(iv) Strip-down cost.
(v)   Inspection of rotor.
(vi) Fitting of new bearings which is normal practice and a requirement by the end user when a motor is stripped down.
(vii) Transport back to site.
(viii) Re-installation.

The vendor's diagnosis was false because, as explained in Section 10.1.2, electromagnetic forces caused by rotor slotting produce consequential vibration at the rotor slot vibration frequencies on the motor's stator core and frame. The consequential vibration caused by rotor slotting is **not due** to broken rotor bars; see [10.15] and [10.17].

The first and crucial oversight by the sub-contractor was to not obtain information on the type of cage rotor in this induction motor, because it was an aluminium die cast rotor that had been in operation for 17 years.

It is well known [10.21] that broken rotor bars in aluminium cage rotors are rare events.
They either exist due to flaws in manufacture, such as porosity in the die casting process or when the rotor has been subjected to excessive overheating from, for example, a stall.
There were no reported incidents of severe overloading or a stall.

The relatively high overall vibration of 6.3 mm/s r.m.s. on the outer frame compared to the highest velocity on the DE bearing housing at 1.8 mm/s r.m.s. was due to an electromagnetic force at the principal rotor slot passing frequency (defined in Section 10.1.2).

This caused vibration at that frequency on the outer frame, which has an interference and keyed fit to the stator core back. It will be shown that MCSA verified that there were no $\pm 2sf$ sidebands (or pole-pass frequencies) around the supply component of current [10.21], and the aluminium die cast cage winding was perfectly healthy. As of 2020 this motor was still running normally without any incidents of cage winding defects.

The motor's nameplate data are as follows:

- 3-phase SCIM, 415 V, 132 kW/177 H.P.
- 237 A, 50 Hz, 592 r/min, p.f. = 0.85, eff = 91%
- DE bearing: N324 C3 and NDE bearing: 6316-C3.

The author obtained the number of rotor bars (84) from the OEM, who also confirmed that it was an aluminium die cast cage winding.
The motor was driving a reciprocating cold gas compressor.

10.3.2 Overall Vibration Measurements

This is a 10-pole SCIM supplied at 50 Hz, and the nominal full-load speed (N_r) and slip ($s_{f.l.}$) are 592 r/min and 1.33% respectively at a full-load current of 237 amperes. The operational current was 220 amperes during the vibration and MCSA measurements, therefore the motor was running close to full load. Figures 10.11 and 10.11a show the positions of the accelerometers. As is very often the case, there is an external fan in this design of LV SCIM, which means that neither the NDE bearing housing nor the NDE end frame was accessible.

Placing an accelerometer on a thin fan cowl can be futile because the vibration from the bearing defect frequencies may well not be detectable and, in addition, the fan cowl often acts as a resonating source of vibration and swamps other vibration from within the motor. The closest position to the NDE end frame and bearing housing is on the stator outer frame at the NDE as shown in Figure 10.11a.

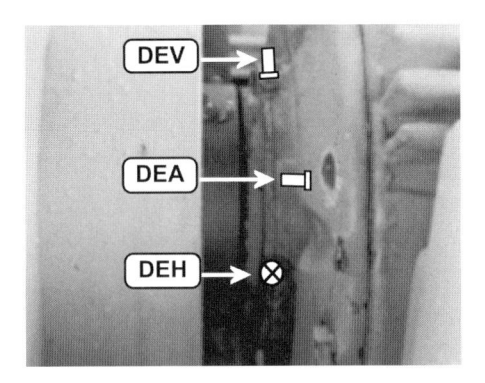

Figure 10.11 Positions of accelerometers on DE bearing housing.

Figure 10.11a Position of accelerometer on the frame at the NDE.

Table 10.3 Overall r.m.s. velocity mm/s ±10%, span 10–1000 Hz

Motor full-load current = 237 A; operating current = 220 A			
Position: bearing housing	Vertical	Horizontal	Axial
DE	1.4 mm/s	1.8 mm/s	1.0 mm/s
NDE F motor outer frame	6.3 mm/s or 0.25 inches/s		

The overall r.m.s. velocities at the DE were normal, but the overall velocity at the NDE F position on the outer frame of the motor was up to 3.5 times higher than from any of the positions on the DE bearing housing; see Table 10.3. Spectrum analysis was required to explain the reason for this result.

10.3.3 Vibration Spectrum Analysis

From Figure 10.12 the *2X* component at 19.76 Hz is evident, which means that the *1X* component is 9.88 Hz and hence the rotor speed is 593 r/min, which gives a

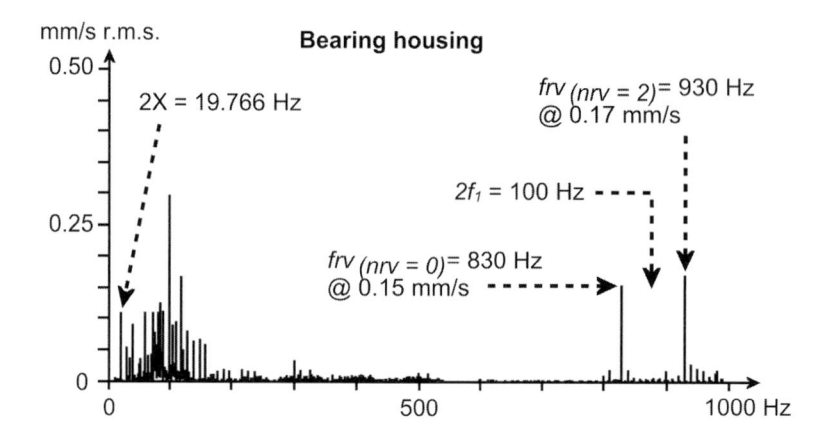

Figure 10.12 Vibration spectrum, bearing housing on DEV, coupled to the compressor, baseband frequency span was 10–1000 Hz, 7.9 mHz/line.

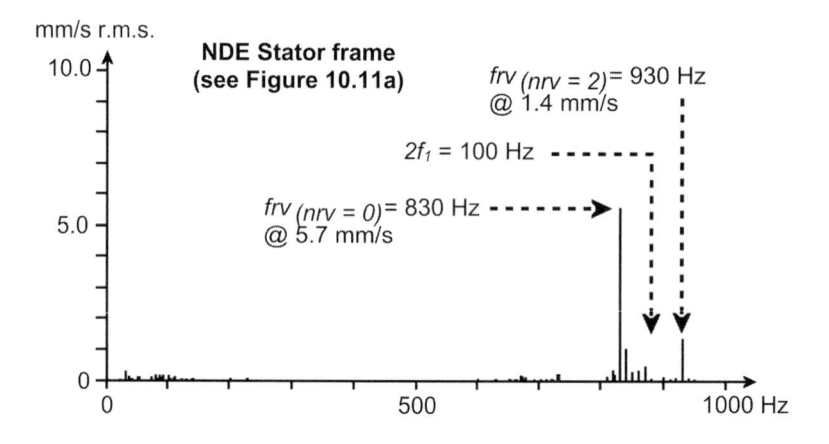

Figure 10.13 Vibration spectrum, NDE F stator frame position, coupled to the compressor, baseband frequency span was 10–1000 Hz, 7.9 mHz/line.

corresponding slip of 0.0117 (1.17%). This is in line with the expected operational slip at a current of 96%$I_{f.l.}$ Particular features of the spectrum are two velocity components at 830 Hz and 930 Hz on the DE bearing housing. They are at a very low velocity level; for example, the 830 Hz component is only 0.15 mm/s.

In comparison, Figure 10.13 shows the velocity of the 830 Hz and 930 Hz components on the frame at the NDE F position on the motor's outer frame are 5.7 mm/s and 1.4 mm/s respectively, compared to 0.15 mm/s and 0.17 mm/s on the DE bearing housing.

The components at 830 Hz and 930 Hz are not harmonic integer multiples of the *1X* component, nor are they at any of the predicted bearing defect frequencies and integer multiples from the DE or NDE bearings as predicted in Table 10.4.

10.3 Industrial Case History – False Positive of Cage Winding Breaks

Table 10.4 Predicted defect frequencies for bearings

N324 C3			
Contact angle: 0°	No. of balls: 12	Full-load speed: 592 r/min	
BPFO: 48 Hz	BPFI: 70.5 Hz	FTF: 4 Hz	2 × BSF: 50.3 Hz
6316 C3			
Contact angle: 0°	No. of balls: 12	Full-load speed: 592 r/min	
BPFO: 48.6 Hz	BPFI: 30.5 Hz	FTF: 4.64 Hz	2 × BSF: 41 Hz

The fundamental source of the components at 830 Hz and 930 Hz on the frame of the motor is caused by inherent and normal components due to vibration rotor slot slotting from electromagnetic forces, as explained in Section 10.1.2. Using Equation 10.2 to predict the rotor slot vibration components gives:

$$f_{rv} = \left\{\frac{RN_r}{60}\right\} \pm f_1 n_{rv},$$

where:

f_{rv} = classical and inherent rotor slot vibration frequency components
f_1 = supply frequency
R = number of rotor slots = 84
n_{rv} = integers 0, 2, 4, 6
N_r = rotor speed in r/min.

Using these values in Equation 10.2, and with $n_{rv} = 0$, f_{rv} can be calculated as

$$f_{rv} = \left\{\frac{RN_r}{60}\right\} \pm f_1 n_{rv} = \left\{\frac{84 \times 593}{60}\right\} \pm 60 \times 0 = 830 \text{ Hz},$$

where: $f_{rv(n_{rv} = 0)}$ = the principal rotor slot passing vibration component.
With n_{rv} = 2, 4, 6, 8 the rotor slot passing vibration components can also occur at:

$n_{rv} = 2$, $f_{rv} = 830 \pm 2f_1 = 730$ and 930 Hz
$n_{rv} = 4$, $f_{rv} = 830 \pm 4f_1 = 630$ and 1030 Hz
$n_{rv} = 6$, $f_{rv} = 830 \pm 6f_1 = 530$ and 1130 Hz
$n_{rv} = 8$, $f_{rv} = 830 \pm 8f_1 = 430$ and 1230 Hz.

The stator frame and core of this particular motor were dynamically responsive to vibration caused by the electromagnetic force at the principal rotor slot passing frequency of 830 Hz in comparison to the relatively low dynamic response to that frequency on the DE bearing housing. In other words, the mechanical impedance at the position on the outer frame of the motor will be less at that high frequency, compared to the mechanical impedance between the DE bearing housing position and

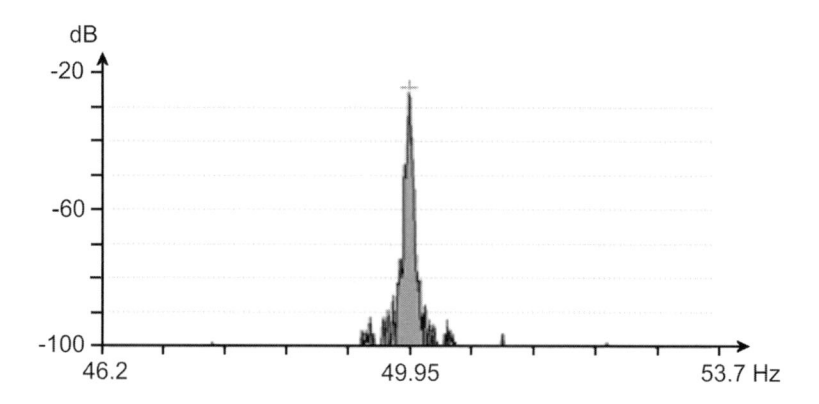

Figure 10.14 Current zoom spectrum base-band span 0–120 Hz, 12,800 spectral lines.

the origin of the rotating electromagnetic force wave that produced the principal rotor slot passing vibration at 830 Hz.

10.3.4 Conclusions

(i) The vibration condition monitoring sub-contractor misunderstood the source of the frame vibration and assumed that because it was a function of the number of rotor bars and was at a relatively high magnitude it must be due to broken rotor bars.

(ii) Figure 10.14 shows the current spectrum produced by the author via MCSA and there are no $\pm 2sf_1$ sidebands (or pole-pass) frequencies around the supply component f_1 and this categorically proves that the aluminium die cast cage winding was perfectly healthy.

(iii) The vibration condition monitoring sub-contractor did not offer MCSA services to check for broken rotor bars.

(iv) For full details on the industrial application of MCSA to diagnose broken rotor bars, the reader is referred to reference [10.21], which contains 50 industrial case histories on MCSA.

10.4 Case Study – Measurement of the Stator Core Vibration Proved that Broken Rotor Bars in a SCIM Modulate the Vibration Rotor Slot Passing Frequencies at Twice the Slip Frequency

10.4.1 Background

Abstract – The author of this book, when he was leader of the electrical machine monitoring research group at Robert Gordon University, Aberdeen, Scotland, initiated a research and development (R&D) project in 1980 on condition monitoring to diagnose faults in 3-phase induction motors. The author's approach to research and

10.4 Case Study – Measurement of the Stator Core Vibration

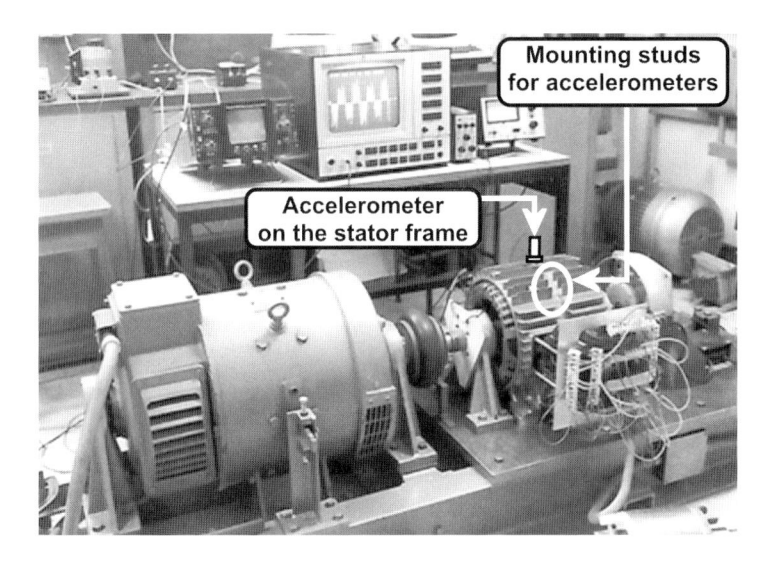

Figure 10.15 Special SCIM test rig for monitoring and analysing vibration and current during normal and fault conditions.

development was to design and build special test rigs [10.21], to introduce different faults in them and to analyse signals such as vibration and current. The latter, known as MCSA – see references [10.21] to [10.30] – is now extensively used by industry, and the author published reference [10.21], which contains 50 industrial case histories on the application of MCSA.

The objective of this study is as follows:

(i) To verify via controlled experiments that broken rotor bars modulate the vibration rotor slot passing frequency components at twice the slip frequency ($2sf_1$).

10.4.2 Measurement of Vibration Rotor Slot Passing Frequencies on the Stator Frame

A special SCIM test rig, shown in Figure 10.15, was designed and constructed in 1981. The outer frame of the motor had mounting studs installed to secure accelerometers on the frame's outer periphery.

The first objective, during normal operation and with no faults introduced into the motor, was to measure the vibration rotor slot passing frequency components on the stator frame, which is in direct contact with the stator core back.

Motor details

- 3-phase, SCIM, 415 V, 11 kW/14.75 H.P., 21 A, 50 Hz, 1430 r/min
- Aluminium die cast rotor with 51 slots.

The motor was operated at full-load current (21 amperes) and speed (1430 r/min) and the predicted rotor slot vibration components from Equation 10.2 are as follows:

$$f_{rv} = \left\{\frac{RN_r}{60}\right\} \pm f_1 n_{rv},$$

where:

f_{rv} = classical and inherent rotor slot vibration frequency components
f_1 = supply frequency
R = number of rotor slots = 51
n_{rv} = integers, 0, 2, 4, 6...
N_r = rotor speed in r/min;

$$f_{rv} = \left\{\frac{RN_r}{60}\right\} \pm f_1 n_{rv} = \left\{\frac{51 \times 1430}{60}\right\} \pm 60 \times 0 = 1215 \text{ Hz},$$

where: $f_{rv(n_{rv} = 0)}$ = the principal rotor slot passing vibration component.
With n_{rv} = 2, 4, 6, 8 the rotor slot passing vibration components can also occur at:

$n_{rv} = 2$, $f_{rv} = 1215 \pm 2f_1 = 1115$ and 1315 Hz
$n_{rv} = 4$, $f_{rv} = 1215 \pm 4f_1 = 1015$ and 1415 Hz
$n_{rv} = 6$, $f_{rv} = 1215 \pm 6f_1 = 915$ and 1515 Hz
$n_{rv} = 8$, $f_{rv} = 1215 \pm 8f_1 = 815$ and 1615 Hz.

A brand new 51 slot skewed cage rotor was used for the tests, and the airgap eccentricity (combination of static and dynamic) was only 5% of the design value for the radial airgap length of 0.6 mm (0.024 inches).

Figures 10.16 and 10.17 show the dB versus frequency acceleration spectra at positions 12 o'clock and 3 o'clock on the outer frame (see Figure 10.15). The stator core is in direct contact with the outer frame and there were no faults in the motor. The key observation from Figures 10.16 and 10.17 is as follows:

- The vibration rotor slot passing frequency components, as predicted from Equation 10.2, are present in both spectra at 100 Hz apart.

There were differences between the acceleration magnitudes at different positions on the stator frame. For example, the magnitudes of the principal rotor slot passing frequency (1215 Hz) at full load, were 76 dB and 70 dB at positions 12 o'clock and 3 o'clock respectively (i.e. a difference factor of 2).

This is because of the complex vibratory response of a stator core assembly and the magnitudes of vibration components are a function of position around the frame.

With respect to Figure 10.16 the following nomenclature applies:

A = $2f_1$ = 100 Hz at 80 dB
B = 1015 Hz at 69 dB
C = 1115 Hz at 76 dB
D = 1215 Hz at 76 dB
E = 1315 Hz at 72 dB
F = 1415 Hz at 86 dB
G = 1515 Hz at 72 dB.

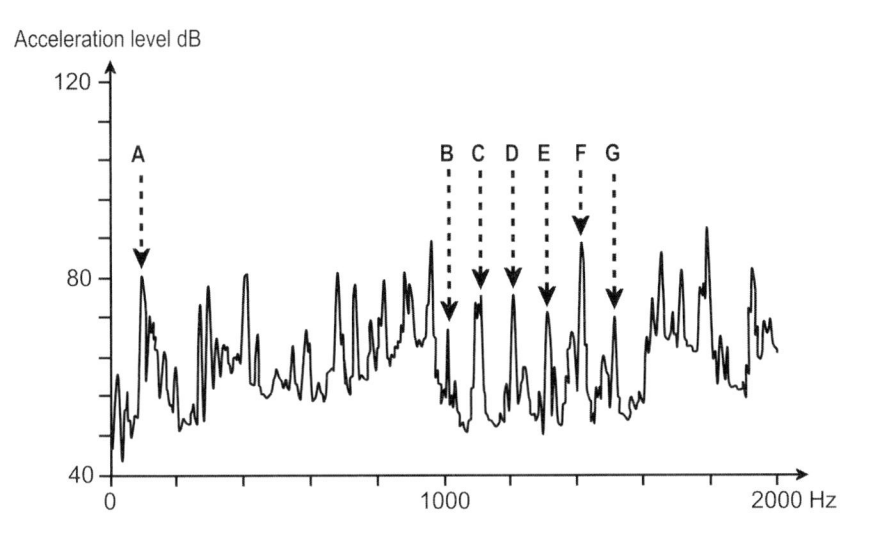

10.4 Case Study – Measurement of the Stator Core Vibration

Figure 10.16 Acceleration spectrum, dB versus frequency, at 12 o'clock on the stator frame as shown in Figure 10.15.

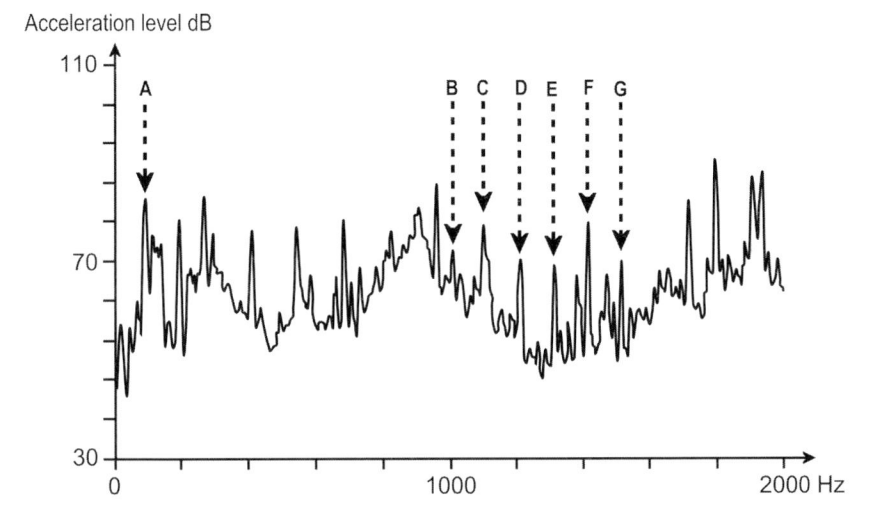

Figure 10.17 Acceleration spectrum, dB versus frequency, at 3 o'clock on the stator frame as shown in Figure 10.15.

With respect to Figure 10.17 the following nomenclature applies:

A = $2f_1$ = 100 Hz at 86 dB
B = 1015 Hz at 72 dB
C = 1115 Hz at 76 dB
D = 1215 Hz at 70 dB

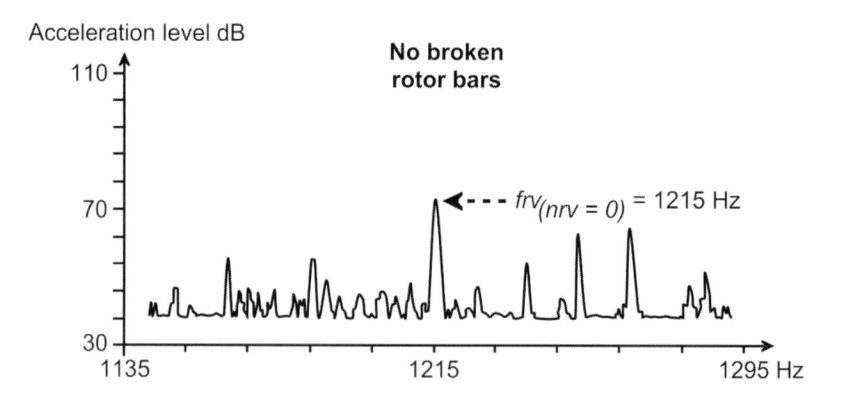

Figure 10.18 No broken rotor bars, acceleration zoom spectrum, dB versus frequency at 12 o'clock on the outer frame – see Figure 10.15.

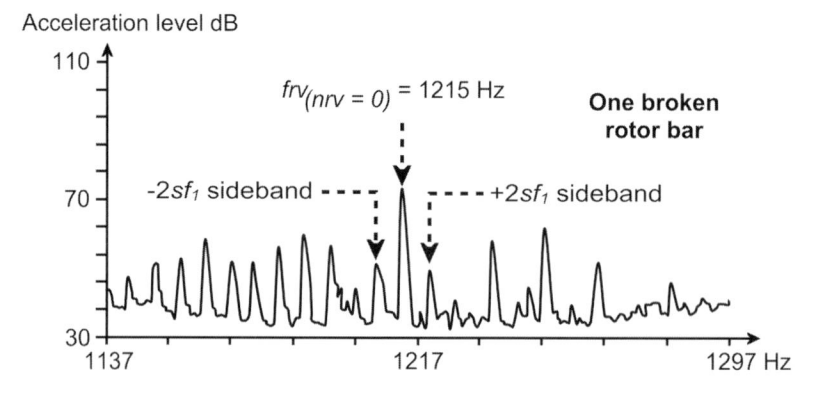

Figure 10.19 One broken rotor bar, acceleration zoom spectrum, dB versus frequency at 12 o'clock on the outer frame – see Figure 10.15.

E = 1315 Hz at 68 dB
F = 1415 Hz at 78 dB
G = 1515 Hz at 70 dB.

10.4.3 VSA of the Stator Frame Vibration with Broken Rotor Bars

A zoom acceleration spectrum of the frame vibration at 12 o'clock from a perfectly healthy 51 slot aluminium cage rotor is shown in Figure 10.18, and Figure 10.19 shows the zoom acceleration spectrum at the same position with one broken rotor bar introduced into the same cage rotor.

At the full-load speed of 1430 r/min, the operational slip is given by:

$$s = \frac{N_s - N_r}{N_s} = \frac{1500 - 1430}{1500} = 0.0467 \, \text{pu}$$

and $2sf_1 = 4.67$ Hz.

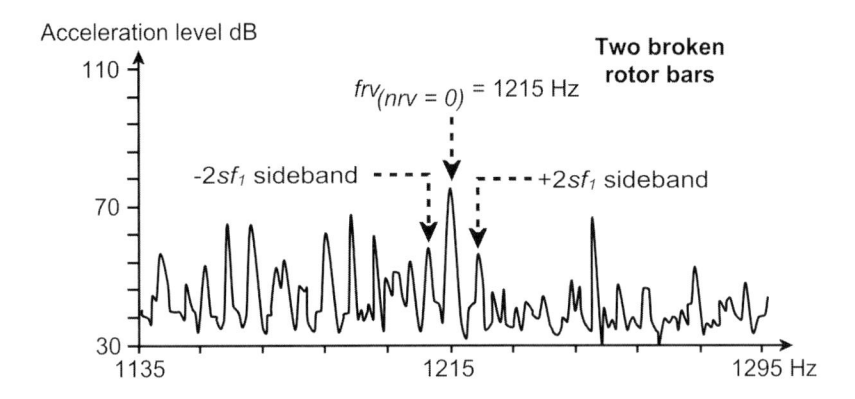

Figure 10.20 Two broken rotor bars, acceleration zoom spectrum, dB versus frequency at 12 o'clock on the outer frame – see Figure 10.15.

It is very well known that broken rotor bars in a squirrel-cage rotor cause the following – see references [10.21] to [10.30]:

(i) Modulation of the rotor's speed and torque at twice the slip frequency ($2sf_1$).
(ii) Amplitude modulation of the airgap flux and supply frequency current at twice the slip frequency ($2sf_1$), which is the correct terminology. These sidebands are sometimes referred to as pole-pass frequencies (but only in the USA) by vibration condition monitoring companies but induction motor experts do not use that term [10.21] and use the correct terminology.
(iii) Equation 10.3, which gives the vibration rotor slot passing frequencies, is a function of the rotor speed, N_r; therefore the rotor slotting vibration components are also modulated at twice the slip frequency ($2sf_1$).

Figure 10.20 shows the acceleration spectrum with two broken rotor bars. The principle rotor slot vibration component is still at 74 dB but the lower ($-2sf_1$) and upper ($+2sf_1$) sidebands around $f_{rv(n_{rv}\ =\ 0)}$ = 1215 Hz are 16 dB and 18 dB down on $f_{rv\ (n_{rv}\ =\ 0)}$. The lower sideband relative to $f_{rv(n_{rv}\ =\ 0)}$ has increased by 6 dB, which equates to a factor increase of 2.

10.4.4 Conclusions on Motor Current Signature Analysis (MCSA) versus Vibration Analysis to Diagnose Broken Rotor Bars in SCIMs

The previous sections in this case study were a very small part of a major R&D project which commenced in 1982 on condition monitoring for SCIMs. For the avoidance of doubt, due cognizance of the following must be borne in mind:

(i) Since 1982, MCSA has become the established diagnostic technique to diagnose broken rotor bars in SCIMs; see references [10.21] to [10.30].
(ii) Dedicated MCSA instruments have been in the market place for 30 years, but that is not the case with the application of VSA to detect broken bars.

(iii) Vibration from a SCIM is a second-order effect compared to MCSA (a first-order effect) that detects the $\pm 2sf_1$ sidebands around the supply current component. The condition of the cage winding can now be estimated via MCSA; this is not the case via VSA.

(iv) To diagnose broken rotor bars requires VSA of the frame vibration to identify the $\pm 2sf_1$ sidebands around the vibration rotor slot passing frequencies. However, there is a disadvantage with this analysis because the number of rotor slots is required, but that information is not on the nameplate and must be obtained from the OEM.

(v) There are now many industrial case histories on the diagnosis of broken rotor bars in SCIMs [10.21] using MCSA, but that is certainly not the case with vibration monitoring to detect broken rotor bars.

(vi) However, with, for example, 440 V and 660 V SCIMs that have starters, it is normally impossible, due to safe operation interlocks, to open the starter and clip a current transformer (CT) around the secondary side of an existing CT while the motor is running. The starter door cannot be left open for an MCSA cable to exit for the MCSA measurement. Also, the supply cable to the motor is normally always a 3-core cable and an external clip-on ammeter around one phase cable is not an option [10.21]. Therefore, in this situation permanent MCSA CTs would need to be installed at a PM outage or, alternatively, vibration analysis could be applied.

10.5 Case Study – Vibration Monitoring to Identify Changes in the Stator Frame Vibration of a SCIM due to Supply Voltage Unbalance

10.5.1 Introduction

Abstract – This study reports on research work carried out in 1983 as part of the R&D project that was described in Section 10.3.1. A special SCIM test rig designed and constructed in 1981 is shown in Figure 10.21. The outer frame of the motor had mounting studs installed to secure accelerometers on the frame's outer periphery. The nameplate data of the test motor shown in Figure 10.21 are as follows:

3-phase, SCIM, 415 V, 11 kW/14.75 H.P., 21 A, 50 Hz, 1430 r/min.

The objectives of this experimental study were as follows:

(i) To verify that the magnitude of the acceleration and velocity of the twice supply frequency vibration component ($2f_1$) is a function of position on the outer circumference of the stator frame.

(ii) To measure the magnitude of the acceleration and velocity of the twice supply frequency vibration component ($2f_1$) at no load and full-load.

(iii) To verify that the magnitude of the velocity and acceleration of the twice supply frequency component around the stator frame is highly sensitive to an increase in voltage unbalance and that there are optimum positions on the frame that must be selected for measuring the velocity to identify that change.

10.5 Vibration Monitoring to Identify Changes in the Stator Frame Vibration

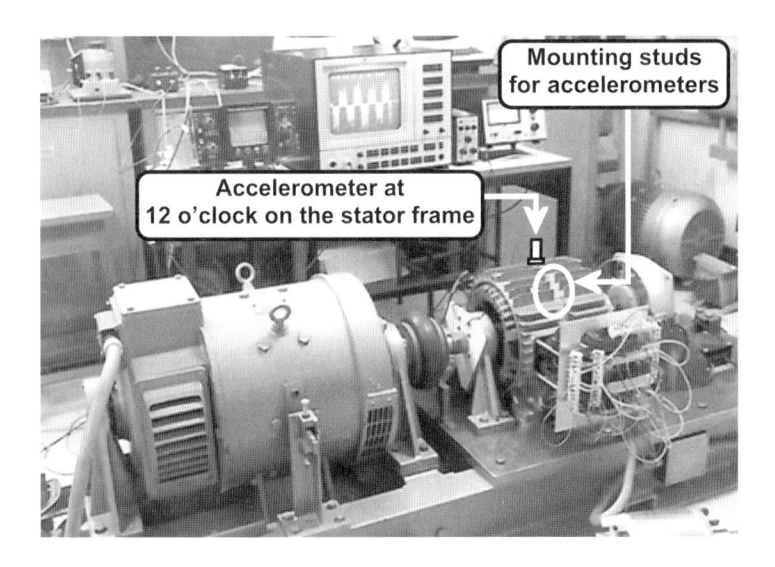

Figure 10.21 Special SCIM test rig for monitoring and analysing vibration and current during normal and fault conditions.

10.5.2 Measurement of the Velocity and Acceleration of the Twice Supply Frequency Component on the Stator Frame as a Function of Position during No-Load and Full-Load Operation

It has already been shown in Section 10.1.1 that the fundamental vibration due to electromagnetic forces in a SCIM is at:

twice the supply frequency = $2f_1$.

Note that the *stator core and outer aluminium frame are in direct contact* in this 11 kW/14.75 H.P. SCIM because the core is pressed into the frame under a very high pressure. The outer frame and the stator core act as a solid body to the magnetic force and consequential vibration at 100 Hz ($2f_1$). Owing to the foot mounting it was impossible to measure the vibration around the complete circumference of the outer frame as is clear in Figure 10.21, and as also illustrated in the schematic diagram of the stator core and frame, which is shown in Figure 10.22.

The r.m.s. velocity and acceleration levels around the stator frame at no load and full load are presented in Figure 10.23 and the key observations are as follows:

(i) The r.m.s. velocity of the 100 Hz component is low at no load and full load, with the maximum being only 0.4 mm/s (0.57 mm/s peak) or 0.022 inch/s peak at the −125° position; this is anticlockwise from the zero reference position, which is taken as 12 o'clock (i.e. the 0 position on Figure 10.23) looking from the NDE of the SCIM shown in Figure 10.21.

(ii) The maximum difference between the magnitudes of the velocities at no load and full load is only 13% at the −100° position. This is expected since the flux density

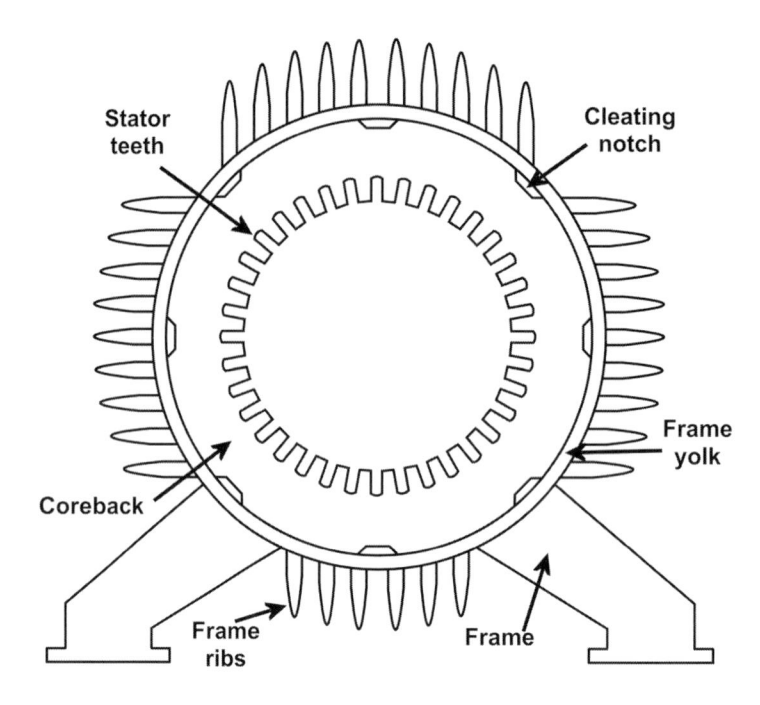

Figure 10.22 Schematic diagram of the construction of the stator core and outer frame.

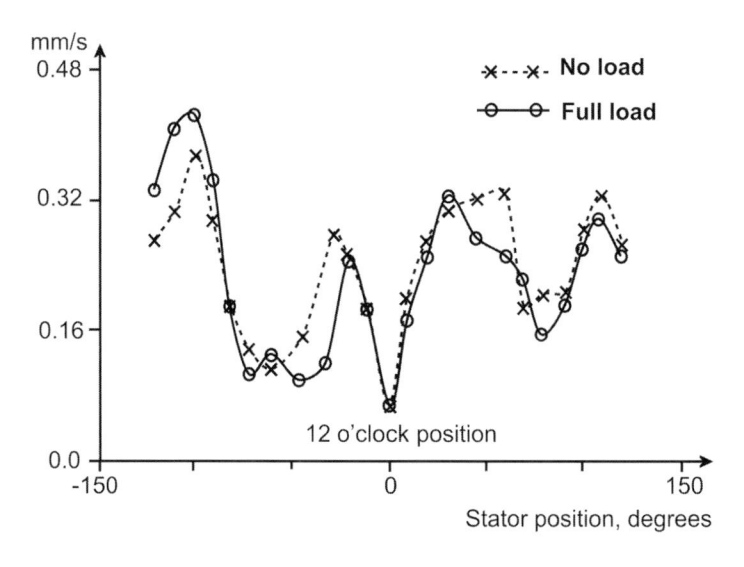

Figure 10.23 Vibration (r.m.s.) levels of the 100 Hz component at no load and full load versus position around the frame of the 4-pole SCIM shown in Figures 10.21 and 10.22.

10.5 Vibration Monitoring to Identify Changes in the Stator Frame Vibration

and flux per pole are virtually constant between no load and full load and are also a function of the applied voltage and frequency.

(iii) The electromagnetic force equation for an induction motor (see references [10.1] to [10.5]) is given by Equation 10.3:

$$E = 4.44 f_1 \Phi N_{ph} k_s k_d, \tag{10.3}$$

where:
E = volts/phase, V
f_1 = supply frequency, Hz
Φ = flux/pole, Webers
N_{ph} = number of turns/phase
k_s = coil span factor of the 3-phase winding
k_d = distribution factor of a group of series connected coils.

At constant voltage and frequency, the flux/pole and therefore the motor's flux density is constant because the number of turns/phase and the coil span and distribution factors are constant.

(iv) It is well known – see references [10.4] and [10.5] – that the electromagnetic force wave at $2f_1$ has a mode wave number equal to twice the number of pole-pairs. The vibration distribution of the 100 Hz ($2f_1$) component around the frame of the motor has nodes (minimum vibration) and anti-nodes (maximum vibration) and this reflects the 4-pole electromagnetic force wave, at 100 Hz.

(v) The results from the tests provide the base-line vibration at 100 Hz around the frame of this motor; this can be used as the reference for vibration measurements on the core as a function of unbalanced voltage.

10.5.3 Measurement of the Velocity and Acceleration of the Twice Supply Frequency Component on the Stator Frame at Full Load as a Function of Unbalanced Voltage Supplies

The operation of induction motors fed from an unbalanced voltage supply was reported in references [10.31] and [10.32], and it is unnecessary to repeat here basic information that has already been published.

An informative paper on the application of monitoring and analysing the $2f_1$ component as a vibration diagnostic tool [10.33] is very worthwhile reading in conjunction with this case history.

Of further interest is a relatively recent, and very informative paper, which was published in 2018, presented results on the effect of motor voltage unbalance on motor vibration via tests and evaluation [10.34].

The results that follow were taken by the author in 1983, and the objective at that time was to demonstrate that the stator core vibration was highly sensitive to changes

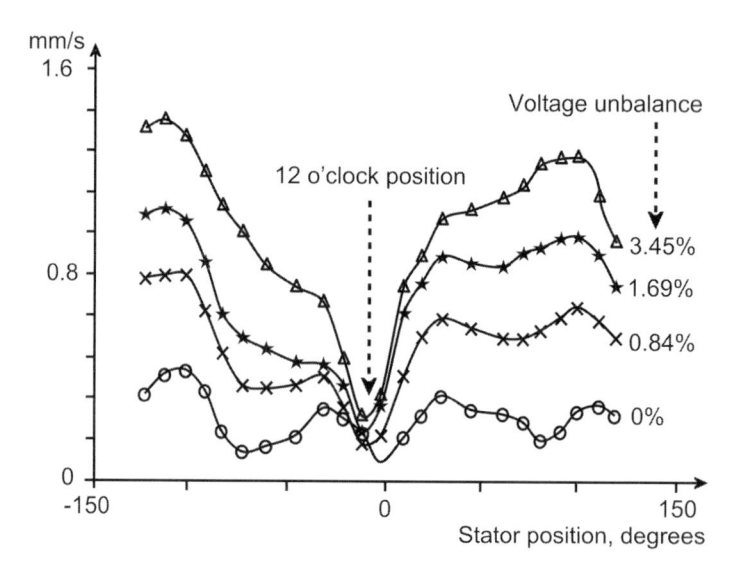

Figure 10.24 Vibration (r.m.s.) levels of the 100 Hz component at full load as a function of voltage unbalance and position around the frame of the 4-pole SCIM.

in voltage unbalance. The supply voltage unbalance as a percentage is normally expressed as:

$$\%\text{Unbalance} = 100 \times \left[\frac{\text{Maximum voltage deviation from the average voltage}}{\text{Average voltage}}\right].$$

10.5.4 Conclusions

The key observations from the velocity plots shown in Figures 10.23 and 10.24 are as follows:

(i) The nodes and anti-nodes are obvious and resemble the expected distribution from the 4-pole electromagnetic force wave profile at 100 Hz, as shown in Figure 10.1a and reference [10.5]. For this 4-pole motor the velocity of the 100 Hz component at the node at 12 o'clock is as expected, a minimum, and there is a negligible change in the velocity at that position as the voltage unbalance increased *in comparison to* the increase in velocity of 100% at the −100° position with only 0.84% voltage unbalance.
(ii) With a voltage unbalance of 3.45% the velocity of the 100 Hz component increased by a factor of three at the −100° position.

The main conclusion is that a small voltage unbalance (e.g. of 0.84%) at full load caused a doubling in the velocity of the 100 Hz component on the stator core and frame, but the optimum position for sensing that velocity increase must be correctly selected.

Appendix 10A Derivation of Twice Supply Frequency Vibration

It has been known for many years that the fundamental electromagnetic force is proportional to the flux density squared, see references [10.1] to [10.5], i.e.:

$$F_{em(\theta,t)} \propto B^2 \qquad (10.4)$$

$B \propto \Phi$,
F_{em} = electromagnetic force (N) – a function of time and space
B = flux density (Tesla) – a function of time and space
Φ = magnetic flux (Webers) – a function of time and space
$B = \frac{\Phi}{A}$
A = cross-sectional area, m².

Let the fundamental flux in the time domain $= \Phi = \Phi_{p1} \sin \omega_1 t$.
The frequency of the fundamental electromagnetic force ($F_{em(\theta,t)} \propto B^2 \propto \Phi^2$), see references [10.6] to [10.9], is therefore given by:

$$\Phi^2 = \Phi_{p1}^2 \sin^2 \omega_1 t = \Phi_{p1}^2 \frac{1}{2}(1 - \cos 2\omega_1 t),$$

where $\omega_1 = 2\pi f_1$.

Therefore, the fundamental flux produces an electromagnetic force at:

twice the supply frequency $= 2f_1$.

10.A.1 Harmonics of Twice Supply Frequency Vibration

The airgap flux waveform will in fact contain odd harmonics (such as the third, fifth, seventh...) of the fundamental flux at the supply frequency and will therefore produce even harmonics of the fundamental twice supply frequency vibration ($2f_1$). As an illustration, let the airgap flux waveform include a third time harmonic:

$$\Phi_{(1,2)t} = \Phi_{p1} \sin \omega_1 t = \Phi_{p3} \sin \omega_3 t.$$

Therefore, the electromagnetic force wave will now contain additional frequency components. Let

$$\begin{aligned}F_{em} \propto \Phi^2 \propto \left(\Phi_{(1,3)t}\right)^2 &= \left(\Phi_{p1} \sin \omega_1 t + \Phi_{p3} \sin \omega_3 t\right)^2 \\ &= \left(\Phi_{p1} \sin \omega_1 t + \Phi_{p3} \sin \omega_3 t\right)\left(\Phi_{p1} \sin \omega_1 t + \Phi_{p3} \sin \omega_3 t\right) \\ &= \Phi_{p1}^2 \sin^2 \omega_1 t + 2\Phi_{p1} \sin \omega_1 t \cdot \Phi_{p3} \sin \omega_3 t + \Phi_{p3}^2 \sin^2 \omega_3 t,\end{aligned} \qquad (10.5)$$

where: $\omega_1 = 2\pi f_1$ and $\omega_3 = 2\pi f_3$.

Expanding Equation 10.5 and using the trigonometric identities

$$\sin^2 \omega_1 t = \frac{1}{2}(1 - \cos 2\omega_1 t)$$
$$\sin^2 \omega_3 t = \frac{1}{2}(1 - \cos 2\omega_3 t)$$
$$2\sin \omega_1 t \cdot \sin \omega_3 t = \cos(\omega_3 t - \omega_1 t) - \cos(\omega_3 t + \omega_1 t)$$

gives additional electromagnetic forces and consequential vibration at the following frequencies:

4f$_1$; 6f$_1$; 8f$_1$...

References

10.1 M. Liwschitz-Garik and C. C. Whipple, *Electric Machinery Vol. II, A C Machines*, Van Nostrand Company, first published Sept. 1946.

10.2 M. G. Say, *Alternating Current Machines*, fourth edition, ELBS and Pitman Publishing, 1976.

10.3 P. L. Alger, *Induction Machines – Their Behaviour and Uses*, Gordon and Breach Science Publications Inc, second edition, published by OPA *Amsterdam, third printing with additions*, 1995.

10.4 S. J. Yang, Low Noise Electric Motors, *Monographs in Electrical and Electronic Engineering*, IEE, Savoy Place, London, 1981.

10.5 P. L. Alger, Magnetic Noise in Poly-phase Induction Motors, *Transactions AIEE*, 73 (Part IIA), 1954, pp. 118–25.

10.6 W. R. Finley, M. M. Hodowanec and W. G. Holter, An Analytical Approach to Solving Motor Vibration Problems, *IEEE Transactions on Industry Applications*, 36 (5), Sept./Oct. 2000, pp. 1467–80.

10.7 M. J. Costello, Understanding the Vibration Forces in Induction Motors, *Proceedings 19th Turbomachinery Symposium*, College Station, TX, Oct. 1989, pp. 179–83.

10.8 W. R. Findlay and R. R. Burke, Troubleshooting Motor Problems, *IEEE Transactions on Industry Applications*, 27, Nov./Dec. 1991, pp. 1204–13.

10.9 R. O. Eis, Electric Motor Vibration – Cause, Prevention and Cure, *IEEE Transactions on Industry Applications*, 1A11 (3), May/June, 1975.

10.10 NEMA MG1: *Motors and Generators*, 2012.

10.11 API Standard 541, *Form-Wound Squirrel Cage Induction Motors – 375 kW (500 H.P.) and Larger*, fifth edition, December 2014.

10.12 British Standard (BS) 60034-14, *Rotating Electrical Machines, Part 14: Mechanical Vibration of Certain Machines with Shaft Heights of 56 mm and Higher – Measurement, Evaluation and Limits of Vibration Severity*, Feb. 2004.

10.13 IEEE 841-2009, *IEEE Standard for Petroleum and Chemical Industry – Premium Efficiency, Severe-Duty, Totally Enclosed Fan Cooled (TEFC) Squirrel-Cage Induction Motors – up to and Including 375 kW (500 H.P.)*, IEEE, New York.

10.14 M. Chisholm *et al.*, Amending API 541 – An Overview of the New Requirements and Improvements of the Fifth Edition, *IEEE Industry Applications Magazine*, 20 (1), 2014, pp. 41–9.

10.15 J. R. Cameron, W. T. Thomson and A. B. Dow, Vibration and Current Monitoring for Detecting Airgap Eccentricity in Large Induction Motors, *IEE Proceedings*, 133 Part B (3), May 1986.

10.16 W. T. Thomson, J. R. Cameron and A. B. Dow, On-Line Diagnostics of Large Induction Motors, NATO ARW (by invitation only), Catholic University of Leuven, Belgium, August 1986, published in the *NATO Api Series*, Pub. Martines Hijhoff, July 1988.

10.17 D. G. Dorrell, W. T. Thomson and S. Roach, Analysis of Airgap Flux, Current and Vibration Signals as a Function of the Combination of Static and Dynamic Airgap Eccentricity in 3-Phase Induction Motors, *IEEE Transactions on Industry Applications*, 33 (1), Jan./Feb. 1997, pp. 24–34.

10.18 W. T. Thomson, A. Barbour, C. Tassoni and F. Filippetti, An Appraisal of the MMF-Permeance Method and Finite Element Models to Study Static Airgap Eccentricity and its Diagnosis in Induction Machines, *Proceedings ICEM'98*, Istanbul, 1998.

10.19 W. T. Thomson, A Review of On-Line Condition Monitoring Techniques for Three-Phase Squirrel-Cage Induction Motors – Past Present and Future, Keynote address at *IEEE Symposium on Diagnostics for Electrical Machines, Power Electronics and Drives*, Gijon, Spain, Sept. 1999, pp. 3–18.

10.20 W. T. Thomson and P. Orpin, Current and Vibration Monitoring for Fault Diagnosis and Root Cause Analysis of Induction Motors, *Proceedings of 31st Turbomachinery Symposium*, Texas, A&M University, USA, Sept. 2002.

10.21 W. T. Thomson and I. Culbert, *Current Signature Analysis for Condition Monitoring of Cage Induction Motors*, Wiley IEEE Press, ISBN: 978-1-11902959-5, 2017.

10.22 G. B. Kliman and J. Stein, Methods of Motor Current Signature Analysis, *Electric Machines and Power Systems*, Hemisphere Publishing, New York, NY, 20, 1982.

10.23 G. B. Kliman, R. A. Koegl, J. Stein, R. D. Endicott and M. W. Madden, Non-Invasive Detection of Broken Rotor Bars in Operating Induction Motors, *IEEE Transactions on Energy Conversion*, 3 (4), Dec. 1988, pp. 873–9.

10.24 W. T. Thomson and R. J. Gilmore, Motor Current Signature Analysis to Detect Faults in Induction Motor Drives – Fundamentals, Data Interpretation and Industrial Case Histories, *Proceedings of 32nd Turbomachinery Symposium*, Texas, A&M University, USA, Sept. 2003.

10.25 M. Fenger, M. Susnik and W. T. Thomson, Development of a Fully Portable Current Signature Analysis Meter to Detect Electrical and Mechanical Faults in Induction Motor Drives, *Iris Rotating Machines Conference*, IRMC 01, Washington DC, June 2001.

10.26 W. T. Thomson and M. Fenger, Current Signature Analysis to Detect Induction Motor Faults, *IEEE Industry Applications Magazine*, 7 (4), 2001, pp. 26–34.

10.27 W. T. Thomson and A. Barbour, The On-Line Prediction of Airgap Eccentricity Levels in Large (MW Range) 3-Phase Induction Motors, *Proc. IEEE, IEMDC Conference*, Seattle, May 1999 (US$1000 best paper award).

10.28 W. T. Thomson and A. Barbour, On-line Current Monitoring and Application of a Finite Element Method to Predict the Level of Airgap Eccentricity in 3-Phase Induction Motors, *IEEE Transactions on Energy Conversion*, 13 (4), Dec. 1998, pp. 347–57 (includes discussion and closure).

10.29 C. Hargis, B. Gaydon and K. Kamish, The Detection of Rotor Defects in Induction Motors, *Proceedings 1st IEE International Conference on Electrical Machines, Design and Application*, London, UK, 1982, pp. 216–20.

10.30 F. Filippetti, G. Franceschini, M. Martelli and C. Tassoni, An Approach to a Knowledge Representation about Induction Machine Diagnostics in Expert Systems, *International Conference on Electrical Machines ICEM'88*, Pisa, Italy, Sept. 1988.

10.31 J. H. Dymond, Operation on Unbalanced Voltage: One Motor's Experience and More, *IEEE Transactions on Industry Applications*, 43 (3) May/June 2007, pp. 829–37.

10.32 J. E. Williams, Operation of 3-Phase Induction Motors on Unbalanced Voltages, *AIEE Transactions*, 73 pt III-A, April 1954, pp. 125–33.

10.33 M. Tsypeakin, Induction Motor Condition Monitoring: Vibration Analysis Technique – a twice line frequency component as a Diagnostic Tool, *IEEE International Conferences on Electrical Machines and Drives*, 2013, pp. 117–24.

10.34 M. Campbell and G. Arce, Effect of Motor Voltage Unbalance; Test and Evaluation, *IEEE Transactions on Industry Applications*, 54 (1), 2018, pp. 905–11.

11 Miscellaneous Industrial Case Histories on Vibration Analysis Applied to Induction Motor Drives

11.1 Industrial Case History – Structural Resonance in a Vertically Mounted 265 kW/355 H.P. SCIM Driving a Fire-Water Pump

11.1.1 Background

Abstract – The fire pump motor is a strategic unit at an LNG plant, and it must always be operationally available with minimum risk of failure. A vibration condition monitoring (CM) sub-contractor had monitored the motor for 14 years from 1992 to 2006. A poorly presented trend plot of the velocity at the NDE of the motor was produced by the sub-contractor; the plot indicates wide variations in the overall r.m.s. velocities, as shown in Figure 11.1. The author was appointed by the owner of the LNG plant to carry out an independent investigation into the results produced by the sub-contractor.

Figure 11.1 shows a random variation of the velocity levels.

Between June and December of 1998 the velocity increased by a factor of seven, from 2 mm/s to 14 mm/s. The end user disputed that result. By March 2006, the velocity had decreased to 8 mm/s r.m.s. but no actions were taken by the end user during that period. Five months later, in August 2006, it had increased to 16 mm/s r.m.s.

The sub-contractor did not attempt to explain to the end user the dramatic variations in velocity levels, and this was a gross omission by the sub-contractor.

The author provided further critical comments on Figure 11.1, to the end user, which were as follows:

- The time spans between the dates when the velocities were actually measured are *unequal on the x-axis* and correspond to the actual velocity points (in mm/s on the *y*-axis!) on the graph. This is *not* the conventional method that should be used to draw a graph, because the reader must work out the dates on which the measurements were taken.
- There were no annotations on the velocity trend plot by the sub-contractor, but the author has added dates when the velocities were widely different.
- It is a vertical motor, and the actual position of the accelerometer at the NDE on the motor was not specified by the sub-contractor who simply stated NDE velocity.

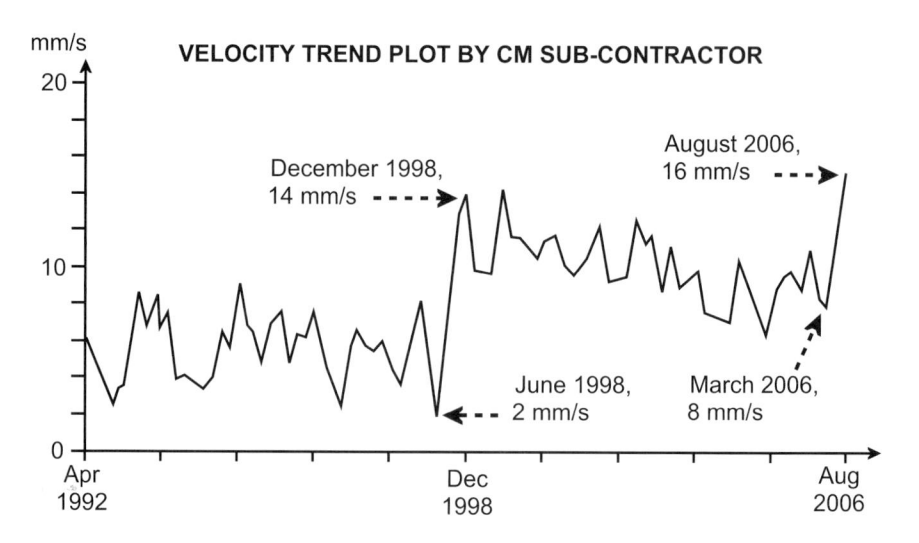

Figure 11.1 Velocity at the NDEH position during a period of 14 years produced by the *vibration CM sub-contractor* (*not* the author) for the oil and gas company.

- A crucial factor is that the operating current and pump flow rate were not recorded by the vibration CM sub-contractor for each velocity measurement. It is well known that operating conditions can change the vibration levels, but this was ignored.
- Figure 11.1 shows a random variation of the velocity levels; for example, in June 1998, the value was 2 mm/s r.m.s. but six months later, in December 1998, it was 14 mm/s r.m.s. but then decreased to 8 mm/s r.m.s. in March 2006. Five months later, in August 2006, it had increased to 16 mm/s r.m.s.

The oil and gas company requested the author to establish the fundamental reasons for the variations presented in Figure 11.1 because the vibration CM sub-contractor had not done so but had simply advised the client to stop the motor on 14th August 2006.

The fire-water pump system (Figure 11.2) operates as follows:

(i) An induction motor (15 kW/20 H.P.) driven pump, which the client refers to as the *jockey pump*, operates continuously to keep the water pressure at 10 bar in the water distribution pipes and at the outlets, but it was not operational due to a major malfunction.

(ii) In its normal mode of operation the main fire-water pump motor (415 V, 265 kW/355 H.P.) automatically comes into operation if the pressure drops to 9 bar when hose outlets are opened. However, because of the faulty jockey pump the main motor was required to continuously drive the fire-water pump to maintain the pressure in the pump system at 10 bar.

(iii) There is also a standby diesel engine fire-water pump which comes into operation if the pressure drops to 8 bar when a high usage of water is required.

Because of the vital necessity for the fire-water pump to be operational at all times, the end user was very concerned that the vibration CM sub-contractor had reported that

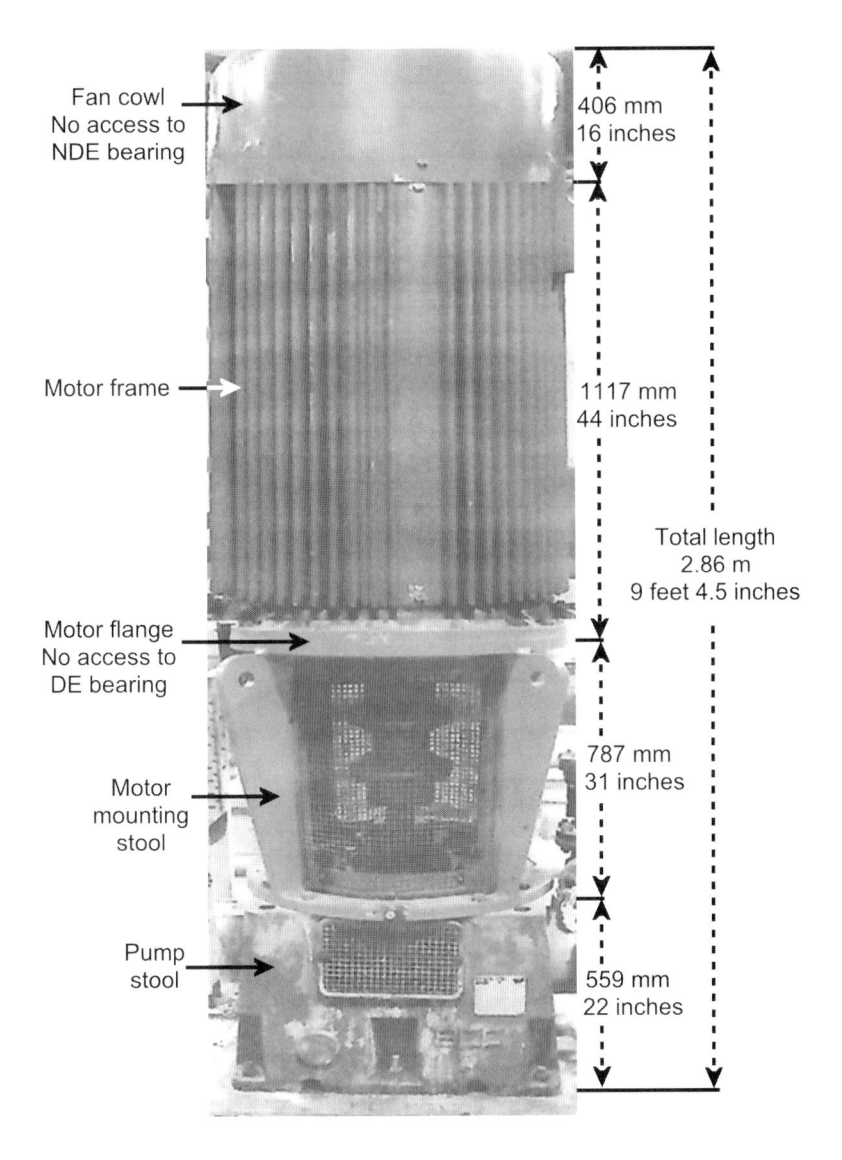

11.1 Industrial Case History – Structural Resonance

Figure 11.2 A 3-phase, SCIM, 415 V, 265 kW/355 H.P. 438 A, 50 Hz, 1480 r.p.m.,1990. Aluminium die cast rotor cage winding.
Note the total vertical length of the drive train is 2.86 metres/9 feet 4.5 inches.

the vibration at the NDE of the main fire-water pump motor was 16 mm/s. This was much too high and was deemed to be in the high-risk category for a potential motor failure.

The author was commissioned to deliver the following objectives:

(1) To determine the root cause of the wide variation in magnitude of the r.m.s. velocities (*as reported by the vibration CM sub-contractor*) in the horizontal direction at the NDE of the motor.

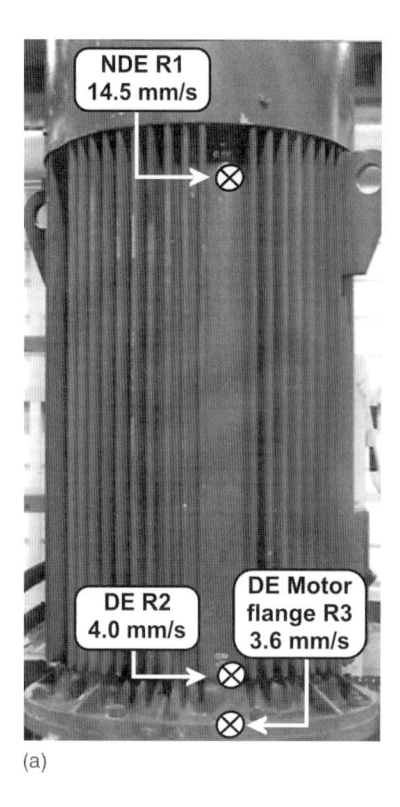

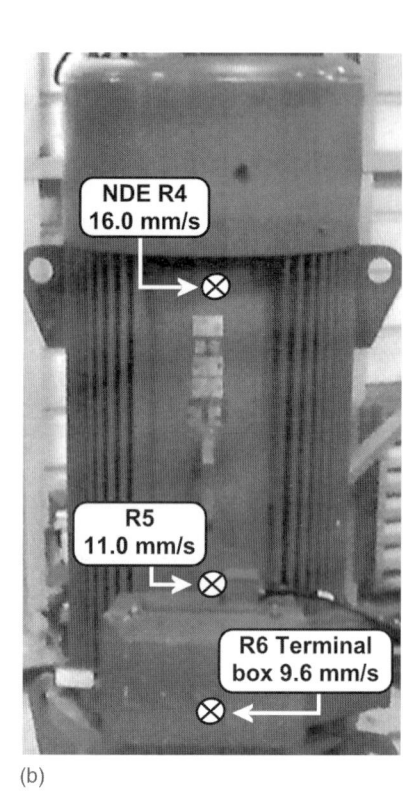

Figure 11.3 Overall r.m.s. velocities during phase (1) – coupled to the pump, operating current was 300 amperes.

(2) To ascertain the fundamental cause of the last velocity, recorded by the sub-contractor in August 2006, being very high at 16 mm/s r.m.s., when five months prior to that date it was 8 mm/s r.m.s. Both readings were taken in the horizontal direction at the NDE of the motor.

11.1.2 Phase One – Vibration Measurements During On-Site Coupled Run

The operating current was 300 amperes compared to the full-load current of 435 amperes. The water flow rate was 150 m³/hour. The DE and NDE bearing housings were not accessible.

The positions of the accelerometers, which were 180° apart on opposite sides of the motor, are shown in Figures 11.3a and b with their overall r.m.s. velocities.

Note that the reference accelerometer position (R1) is at 12 o'clock when viewed by an observer looking downwards from the top of the vertical SCIM at the NDE and facing the opposite side of the outer frame from the side with the terminal box.

The key observations from the overall velocities as presented in Figure 11.3 are as follows:

(i) The highest overall r.m.s. velocity on the motor at this operating current (300 amperes, $I_{f.l.}$ = 438 amperes) and flow rate of 150 m³/hour on 24th August 2006

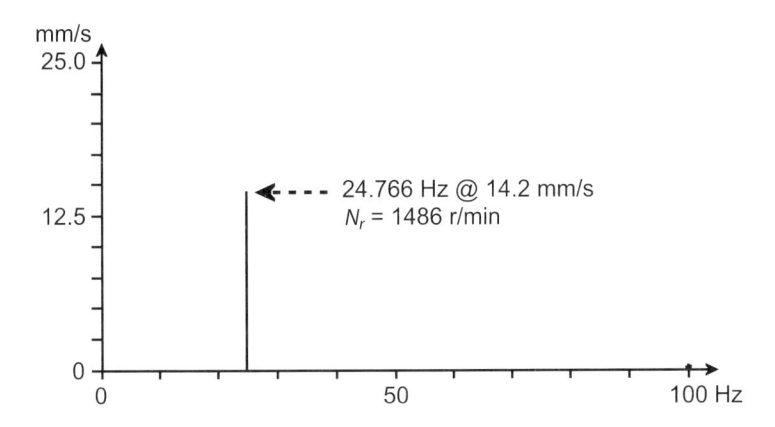

Figure 11.4 Velocity spectrum, position R1 at the NDE, 10–1000 Hz.

(measured by the author), was 16 mm/s r.m.s. at position R4 (Figure 11.3b) at the NDE of the SCIM. This was the same magnitude as reported by the vibration sub-contractor on 14th August 2006.

(ii) The overall velocity at position R1 in Figure 11.3a (180° from R4) was 14.5 mm/s at the NDE.

(iii) The overall velocity at position R2 in Figure 11.3a was 4 mm/s r.m.s. at the DE of the SCIM, which was at a position of 1000 mm/40 inches vertically downwards from R1 (at 14.5 mm/s r.m.s.).

The velocity spectrum for position R1 at the NDE is presented in Figure 11.4, and shows that the *1X* rotational speed frequency component at 24.766 Hz (N_r = 1486 r/min) was at a velocity of 14.2 mm/s (overall velocity – 14.5 mm/s r.m.s.).

This is the fundamental frequency of vibration due to the centrifugal force from the rotor (i.e. $C.F. \propto mw^2r$ (N)) and its magnitude in mm/s is normally due to mechanical imbalance in the rotor. All the vibration spectra from the other positions were also dominated by the *1X* rotational speed frequency component. There was no evidence of any harmonics at *2X, 3X*.

The predicted bearing defect frequencies for the NDE and NDE bearings are presented in Table 11.1. The sample velocity spectrum shown in Figure 11.4 for position R1 at the NDE confirms that there were no bearing defect frequencies.

A rolling element bearing at an advanced stage of degradation can exhibit multiple harmonics of the *1X* component.

11.1.3 Phase Two – Vibration Measurements – On-Site Uncoupled Run

The key observations from the overall r.m.s. velocities recorded during the on-site uncoupled run (see Table 11.2), compared to the coupled run (see Figure 11.3) were as follows:

(a) The velocity at position R4 at the NDE of the motor was 15.6 mm/s compared to 16 mm/s during the coupled run. This is a negligible change in the velocity.

Table 11.1 Bearing defect frequencies

NDE bearing type: 6316 C3 deep groove ball bearing			
Motor speed: 1486 r/min			
Bearing frequencies			
BPFO	BPFI	FTF	2 ×BSF
121.7 Hz	76 Hz	9.6 Hz	102.7 Hz
DE bearing type: 6324 C3 deep groove ball bearing			
Motor speed: 1486 r/min			
Bearing frequencies			
BPFO	BPFI	FTF	2 × BSF
120.6 Hz	77.6 Hz	9.7 Hz	108.6 Hz

Table 11.2 Phase Two – Overall r.m.s. velocities ±10% – span 10–1000 Hz
Date of uncoupled run: 25th August 2006

Operational current on no load = 64 amperes

Reference accelerometer position: 12 o'clock as viewed from a top-down plan view at the NDE of the vertical motor at the side of the motor's frame *without* the terminal box

NDE Radial position 12.00 on motor frame R1	DE Radial position 12.00 on motor frame R2	DE Radial position 12.00 on motor flange R3
13.8 mm/s	4 mm/s	3.5 mm/s

Reference accelerometer position: 12 o'clock as viewed from a top-down plan view at the NDE of the vertical motor at the side of the motor's frame *with* the terminal box.

NDE Radial position 12.00 on motor frame R4	On motor frame Midway down the motor frame under and in line with lifting hook 12.00 R5	Terminal box 12.00 R6
15.6 mm/s	10 mm/s	9 mm/s

(b) The velocity at position R1 at the NDE of the motor was 13.8 mm/s compared to 14.5 mm/s during the coupled run. This is a negligible change in the velocity.

(c) These results confirm that there was an insignificant amount of vibration being transmitted from the pump to the motor during the coupled run.

11.1.3.1 Interim Conclusions and Way Forward

Based on the results and analysis so far, the author predicted that there was a structural resonance of the motor on its mounting.

This is often referred to as a ***motor 'reed' mounting frequency***, whereby the rotational speed frequency of the rotor ($1X$ in Hz) equals, or is close to, a structural natural frequency of the motor on its mounting. The author recommended that the client should hire a structural vibration expert to carry out vibration *bump testing* to determine if a natural frequency existed on the motor that was close to/within its operating speed range.

This was carried out by an independent structural vibration specialist and it was established that there was a natural frequency of the motor's structure at 24.38 Hz with a first-order mode shape, which would produce the maximum velocity at the NDE during resonance.

In summary:

(i) Natural frequency of the motor on its mounting = 24.38 Hz (1463/r/min).
(ii) The nominal operating speed range of the motor is between 1480 r/min (full load) to its no-load speed of 1499 r/min ($1X \cong 25$ Hz). Therefore, the frequency range of the C.F. is between 24.66 Hz and 25 Hz. Note that the nominal full-load speed can be up to 5 r/min less than the nameplate speed [10.21].
(iii) This is certainly within the frequency range that could excite a resonance of the motor on its mounting.

To reduce the vibration at resonance there were two courses of action, namely:

(1) Change the structural natural frequency so that it was, for example, 10 Hz higher than the operating rotational speed frequency range of the motor.
(2) Reduce the magnitude of the C.F. (forcing function) at the $1X$ rotational speed frequency produced by the rotor to as low a magnitude as was practically possible.

The first option was unacceptable to the client because major structural changes to the motor mounting set-up were required.

The second option was selected because the motor, which was manufactured in 1990, had had its rotor balanced to the ISO G6.3 grade. Thus, there was plenty of scope to reduce the mechanical imbalance in the rotor, with the goal being to achieve an ISO balance grade of G0.4.

11.1.4 Phase Three – Motor FAT Vibration Measurements – Rotor Re-balanced to ISO G0.4 Grade – New Bearings Fitted

The motor was sent to a repair shop and the rotor was balanced to G0.4; this was a challenging and lengthy task with this size of rotor – the balance certificate is presented in Appendix 11A. New bearings were fitted. The repair shop could not run this vertical motor because they did not have a suitable vertical mounting set up.

The on-site motor mounting stool was delivered to the repair shop (at the author's insistence) and it was bolted to a solid steel base-plate of 200 mm/8 inches thickness with a surface area of 7.5 m^2/81 $feet^2$.

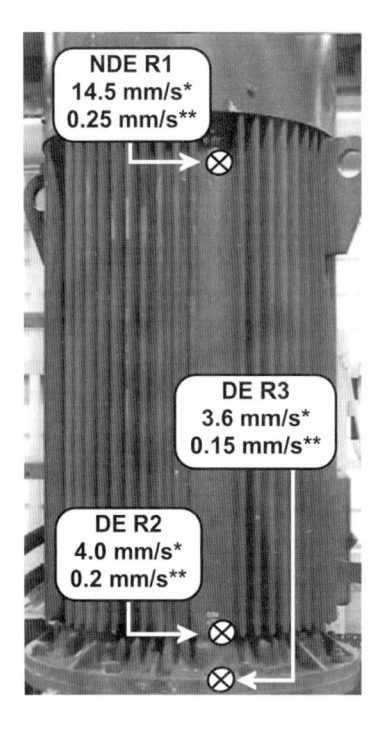

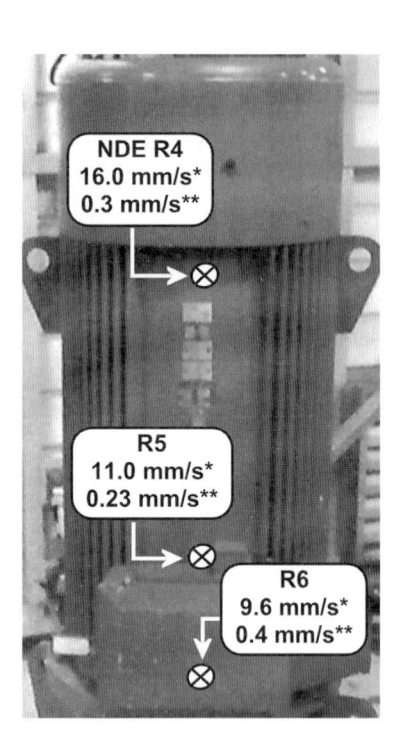

Figure 11.5 Overall r.m.s. velocities from phase (1).
*Coupled to the pump before re-balancing the rotor.
**FAT vibration tests uncoupled at the repair shop after the rotor was re-balanced to ISO grade G0.4 with the motor mounted on its mounting stool, which was bolted to a large, solid steel base-plate.

The repair shop could only carry out a no-load uncoupled run of the motor because they did not have a loading test facility for vertical motors.

The refurbished motor was supplied at rated voltage and frequency and the no-load current was 64 amperes. The overall r.m.s. velocities are presented in Figure 11.5. The key observations from the vibration Factory Acceptance Test (FAT) test are as follows:

(1) The velocities at the NDE at positions R1 and R4 were only 0.25 and 0.3 mm/s r.m.s. respectively due to the very low balance grade of ISO G0.4.
(2) The velocity at position R5 was only 0.23 mm/s r.m.s.
(3) The overall r.m.s. velocities after the refurbishments were all very low with the maximum being on the terminal box (R6) at 0.4 mm/s.

11.1.5 Phase Four – On-Site Vibration Measurements on the Refurbished Motor during a Coupled Run – Operating Current of 300 Amperes

Figure 11.6 presents the accelerometer positions and overall velocity levels. The key differences between the overall velocities during the coupled run before and after re-balancing the rotor to ISO G0.4, and with the motor operating at the same current and flow rate in both cases, were as follows:

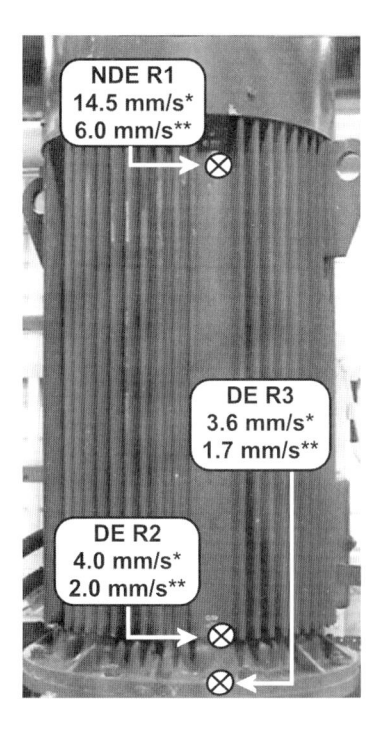

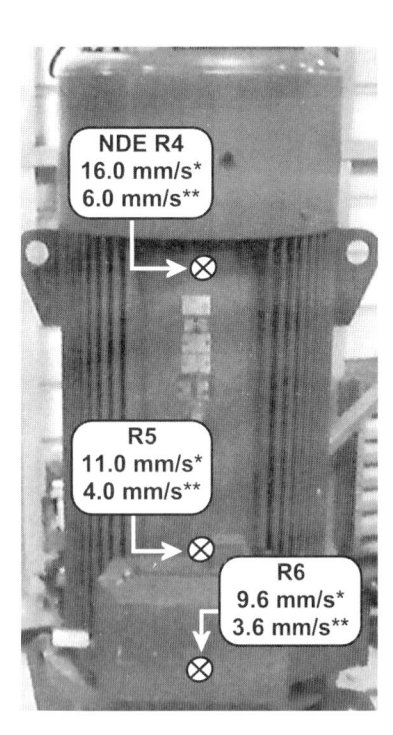

Figure 11.6 Overall r.m.s. velocities.
*Coupled to the pump before the rotor was re-balanced.
**Coupled to the pump after the rotor was re-balanced and new bearings were fitted with the motor operating at 300 amperes.

- NDE R1: before 14.5 mm/s and after 6 mm/s – dropped by a factor of 2.4.
- NDE R4: before 16 mm/s and after 6 mm/s – dropped by a factor of 2.7.
- DE R2: before 4 mm/s and after 1.7 mm/s – dropped by a factor of 2.4.

These results verified that the motor can be operated at loads up to 300 amperes; this was achieved by re-balancing the rotor to ISO G0.4.

11.1.6 Phase Five – On-Site Vibration Measurements on the Refurbished Motor during a Coupled Run – Operating Current of 400 Amperes

The load on the motor was increased to 400 amperes at the maximum flow rate of 175 m³/hour.

The key observations were as follows:

(i) The speed dropped from 1482 r/min at 300 amperes to 1476 r/min at 400 amperes, and therefore the $1X$ rotational speed frequency component was 24.6 Hz.

(ii) At a load current of 400 amperes the rotor speed frequency was 24.6 Hz, which is 1476 r/min, and was obtained from the spectrum analysis presented in

Table 11.3 Overall r.m.s. velocities ±10%, span 10–1000 Hz

Refurbished motor that was fitted with new bearings and the rotor was balanced to the ISO standard of G0.4

On-site date of coupled run: 4th October 2006
Operational current of 400 amperes; delivery of 175 m^3 of water/hour

NDE	DE	DE
Radial position 12.00 on motor frame R1	Radial position 12.00 on motor frame R2	Radial position 12.00 on motor's flange R3
10 mm/s	2.7 mm/s	1.0 mm/s
NDE	**On Motor Frame**	**Terminal box**
Radial position 12.00 on motor frame R4	Midway down the motor frame under and in line with lifting hook 12.00 R5	12.00 R6
10 mm/s	6.6 mm/s	3.0 mm/s

Figure 11.8. This is the excitation forcing frequency due to the centrifugal force (C.F.) produced by the rotor. Thus the frequency (24.6 Hz) of the C.F. is now closer to the first natural frequency (24.38 Hz) of the motor on its mounting. Consequently the velocity at the motor's NDE at 400 amperes will be higher compared to a load current of 300 amperes. This was indeed the case, and Table 11.3 presents the overall r.m.s. velocities for the motor loaded to 400 amperes and the pump delivering a flow rate of 175 m^3/hour.

(iii) The differences between the velocities with a load of 300 amperes and 400 amperes with flow rates of 150 m^3/hour and 175 m^3/hour respectively were as follows:

NDE R1: velocity increased from 6 mm/s to 10 mm/s – a factor increase of 1.67.

NDE R4: velocity increased from 6 mm/s to 10 mm/s – a factor increase of 1.67.

The velocity spectra at position R1 and R4 with operating loads of 300 amperes and 400 amperes respectively are presented in Figures 11.7 and 11.8. The spectra are dominated by the *1X* rotational speed frequency component.

11.1.7 Conclusions

(i) The overall r.m.s. velocity measurements and VSA carried out by the author confirmed that the high velocity of 16 mm/s r.m.s. at the NDE of the motor was due to resonance of the motor on its mounting–pump stool (see Figure 11.2).

(ii) The magnitude of the vibration is a function of load and operating speed of the motor.

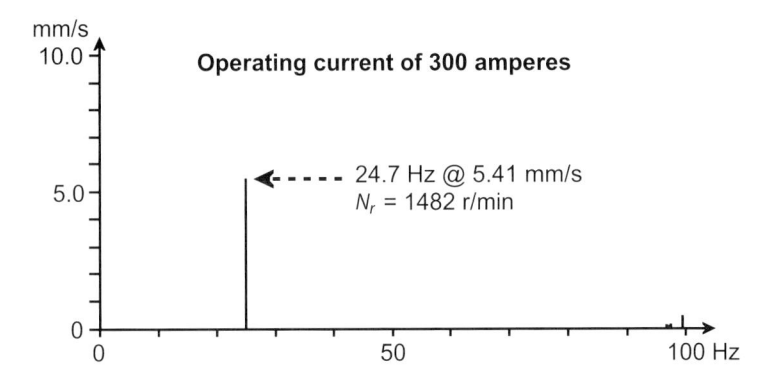

11.1 Industrial Case History – Structural Resonance

Figure 11.7 Velocity spectrum, position R1, operating current of 300 amperes.

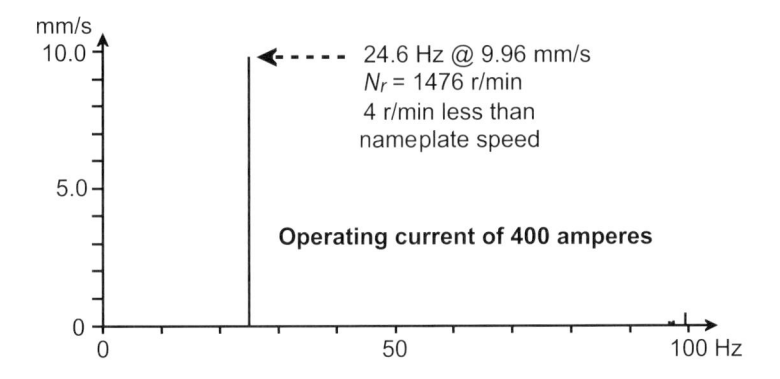

Chart annotation: Operating current of 300 amperes; 24.7 Hz @ 5.41 mm/s; N_r = 1482 r/min.

Figure 11.8 Velocity spectrum, position R4, operating current of 400 amperes.

Chart annotation: 24.6 Hz @ 9.96 mm/s; N_r = 1476 r/min; 4 r/min less than nameplate speed; Operating current of 400 amperes.

(iii) Re-balancing the rotor to ISO G0.4 to reduce the high vibration (NDE: 16 mm/s), which was caused by resonance, was successful because the vibration at the NDE of the motor was reduced from 16 mm/s to 6 mm/s, when the motor was operating at 300 amperes and a flow rate of 150 m³ of water/hour.

(iv) The motor can be operated from 300 amperes down to light load because the vibration drops as the rotational speed frequency moves away from the resonant or reed frequency.

(v) When the motor was operated at 400 amperes with a flow rate of 175 m³/hour, the rotational speed frequency (24.6 Hz) was closer to the resonant or reed frequency (24.38 Hz), and the velocity at the NDE increased from 6 mm/s (at 300 amperes) to 10 mm/s at 400 amperes. The client was advised not to operate the motor above 300 amperes.

(vi) The vibration CM sub-contractor obtained widely different velocities (see Figure 11.1) because vibration measurements were taken when the motor was operating at different operating currents and pump flow rates.

(vii) This case history reinforces the need for vibration CM companies to record the operational load current of an induction motor and the actual mechanical load being delivered at the time when vibration measurements are being taken.

(viii) In many cases, the personnel employed by vibration CM sub-contractors do not record the operational currents of motors or the mechanical load conditions when measuring vibration on electric motors. This observation is based on the author's experience of reviewing hundreds of reports produced by vibration CM companies on behalf of end users.

Appendix 11A Rotor Balance Certificate

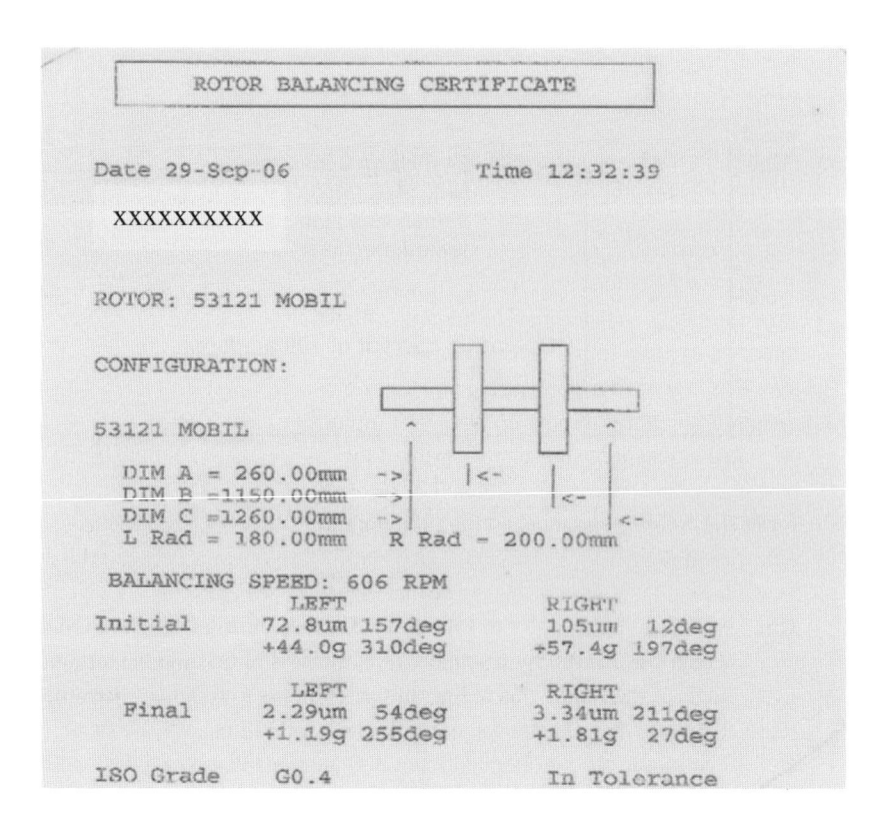

```
            ROTOR BALANCING CERTIFICATE

    Date 29-Sep-06              Time 12:32:39

    XXXXXXXXXX


    ROTOR: 53121 MOBIL

    CONFIGURATION:

    53121 MOBIL

    DIM A = 260.00mm
    DIM B = 1150.00mm
    DIM C = 1260.00mm
    L Rad =  180.00mm     R Rad = 200.00mm

    BALANCING SPEED: 606 RPM
                 LEFT                 RIGHT
    Initial      72.8um 157deg        105um   12deg
                 +44.0g 310deg        +57.4g 197deg

                 LEFT                 RIGHT
    Final        2.29um 54deg         3.34um 211deg
                 +1.19g 255deg        +1.81g  27deg

    ISO Grade    G0.4                 In Tolerance
```

11.2 Industrial Case History – Investigation into the Cause of Loud Acoustic Noise from an Inverter-Fed 3.3 kV, 4500 kW/6032 H.P. Vertically Mounted SCIM Driving a Multi-Phase Pump

11.2.1 Background

Abstract – A major oil company had received reports from personnel on one of its offshore oil and gas production platforms that an unexplained and loud acoustic noise

11.2 Industrial Case History – Investigation into the Cause of Loud Acoustic Noise

Table 11.4 Nameplate data

Type of motor	3-phase SCIM
Voltage	3.3 kV
Power	4500 kW/6032 H.P.
Current (full-load)	2 × 457 amperes; this means two parallel paths in the stator winding, each carrying 457 amperes with an input line current = 914 amperes
Frequency (nominal)	50.33 Hz
Speed (maximum)	1510 r/min (4-pole)
p.f.	0.89
Efficiency	96.8%
Motor full-load rated torque	28,650 Nm
Inertia	265 kg m^2
Method of cooling	IC86W
Mounting arrangement	IP55
DE bearing	6316 M C3 – deep groove ball bearing
NDE bearing	7330 B MP – angular contact ball bearing

(*as perceived by offshore personnel but not measured*) was emanating from an inverter-fed, vertically mounted 3.3 kV, 4500 kW/6032 H.P. SCIM when running at approximately 950 r/min. The motor, which drives a multi-phase pump at different speeds, had previously operated between 1110 r/min and 1510 r/min and the end user confirmed that no acoustic noise problem was ever reported. The process now required the motor to operate from 900 r/min up to 1500 r/min. At the lower speed of 950 r/min, the client was now concerned that the acoustic noise might be indicative of a malfunction that only manifested itself at that speed.

The oil company commissioned the author to determine the root cause of the loud acoustic noise, occurring in the region of an operational speed of 950 r/min. A photograph of the motor is shown in Figure 11.9, and Table 11.4 gives its relevant nameplate data.

It is well known – see references [11.1] to [11.3] – that inverter-fed SCIMs can be the source of unwanted acoustic noise but such noise does not necessarily mean that there is a fault in the motor – see references [11.4] and [11.5].

11.2.2 On-Site Vibration Measurements and Analysis

11.2.2.1 Introduction

The objective was to measure and analyse the vibration at the operational speed of 950 r/min, at which it was reported that an unexplained loud acoustic noise was being produced by the motor. That noise did not exist when the motor was operating within the previous speed range of 1110 r/min to 1500 r/min.

The vibration at 1110 r/min was measured and analysed to provide a comparison with the vibration at 950 r/min. This should verify whether there are unique vibration components that exist only in the vibration spectrum at 950 r/min. If that is the case,

Figure 11.9 Photograph of the 3.3 kV, 4500 kW/6032 H.P. variable-speed, vertically mounted SCIM driving a multi-phase pump on the oil and gas production platform.

these vibration components could be the source that excites the outer frame of the motor, which can then act as an acoustic radiator.

11.2.2.2 Overall Velocities on the Motor

Figure 11.10 shows a schematic layout of the main SCIM as installed on the offshore oil and gas production platform. The cooling fan motors (415 V, 4-pole, 1485 r/min) are fed from a fixed-frequency supply that is independent of the inverter, and they can therefore be discounted as the source of the high acoustic noise at 950 r/min.

Table 11.5 presents the overall r.m.s. velocities on the motor's DE flange and on the NDE bearing assembly at the operational speeds of 950 r/min and 1110 r/min.

The key observations from the overall velocity measurements at 950 r/min and 1110 r/min are as follows:

(i) The overall r.m.s. velocities are all below 1.0 mm/s (0.04 inches/s) r.m.s. at both speeds, and are perfectly acceptable.
(ii) The differences between the overall r.m.s. velocities at the two speeds are insignificant.

However, it is the frequency content of the velocity spectra at the different speeds that is the key factor with respect to specific vibration components exciting the stator core and/or outer box frame to produce acoustic noise. A very brief explanation on the response of the human ear is presented in Appendix 11B. This is necessary because it

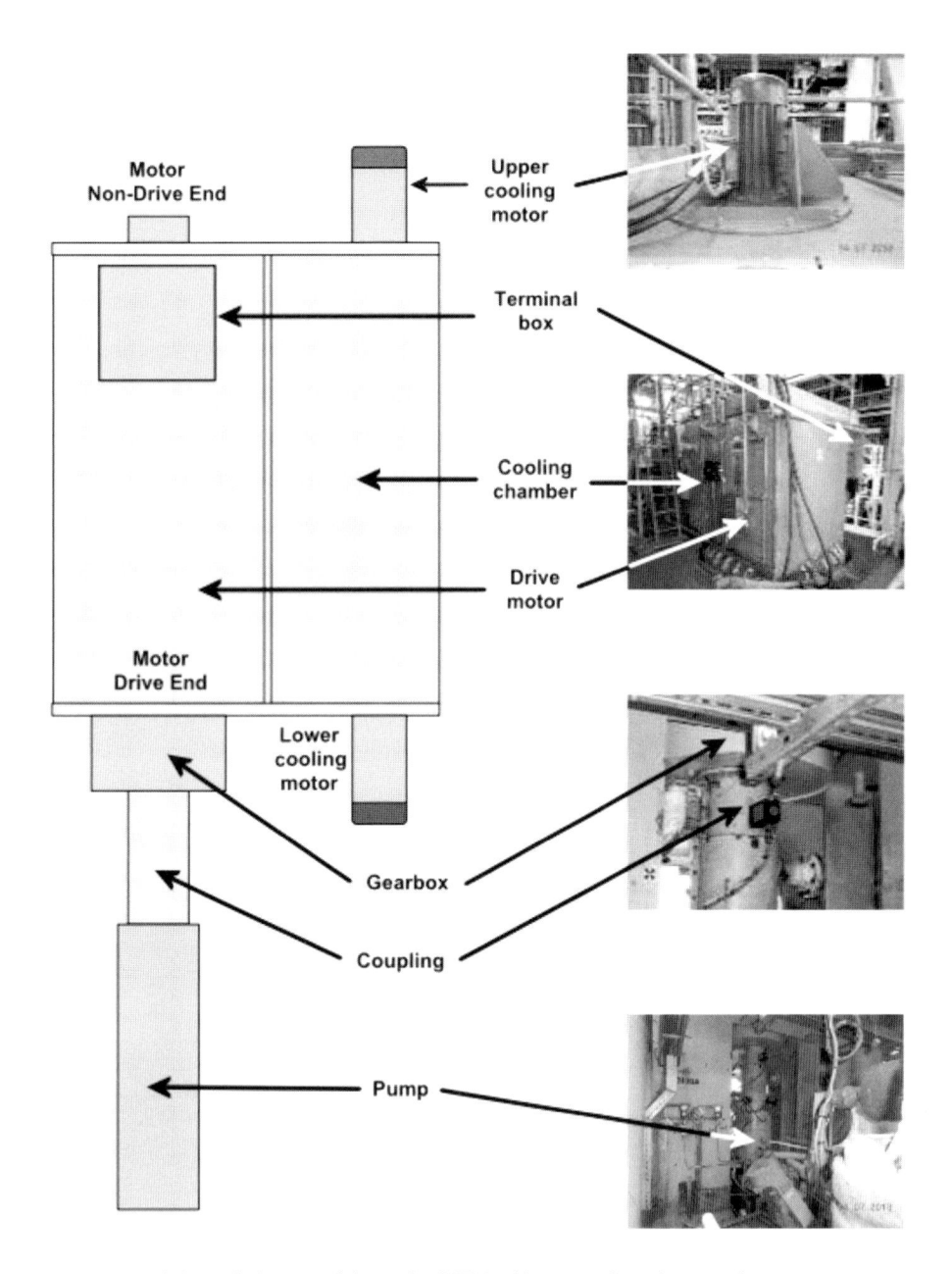

11.2 Industrial Case History – Investigation into the Cause of Loud Acoustic Noise

Figure 11.10 Schematic layout of the main SCIM with supporting photographs.

will be shown that the two major factors that caused the offshore personnel to report a high acoustic noise from the motor were as follows:

(i) The ear is highly responsive to sound pressure levels in the range of 800 Hz to 4 kHz.
(ii) The unique production of a vibration component on the motor at 827 Hz at 950 r/min – see Figure 11.15).

Miscellaneous Industrial Case Histories on Vibration Analysis

Table 11.5 Overall r.m.s. velocities (±10%); frequency span of 5–1000 Hz

Motor speed: 950 r/min Accelerometer positions – on the NDE bearing assembly – see Figure 11.11 (mm/s r.m.s.)			Motor speed: 1110 r/min Accelerometer positions – on the NDE bearing assembly – see Figure 11.11 (mm/s r.m.s.)		
NDE 0	NDE 90	NDE A	NDE 0	NDE 90	NDE A
0.64 mm/s	0.56 mm/s	0.63 mm/s	0.67 mm/s	0.8 mm/s	0.8 mm/s
Motor speed: 950 r/min Accelerometer positions – on the motor's DE flange – see Figure 11.12 (mm/s r.m.s.)			Motor speed: 1110 r/min Accelerometer positions – on the motor's DE flange - see Figure 11.12 (mm/s r.m.s.)		
DE 0	DE 90	DE A	DE 0	DE 90	DE A
0.6 mm/s	0.5 mm/s	0.72 mm/s	0.47 mm/s	0.67 mm/s	0.6 mm/s

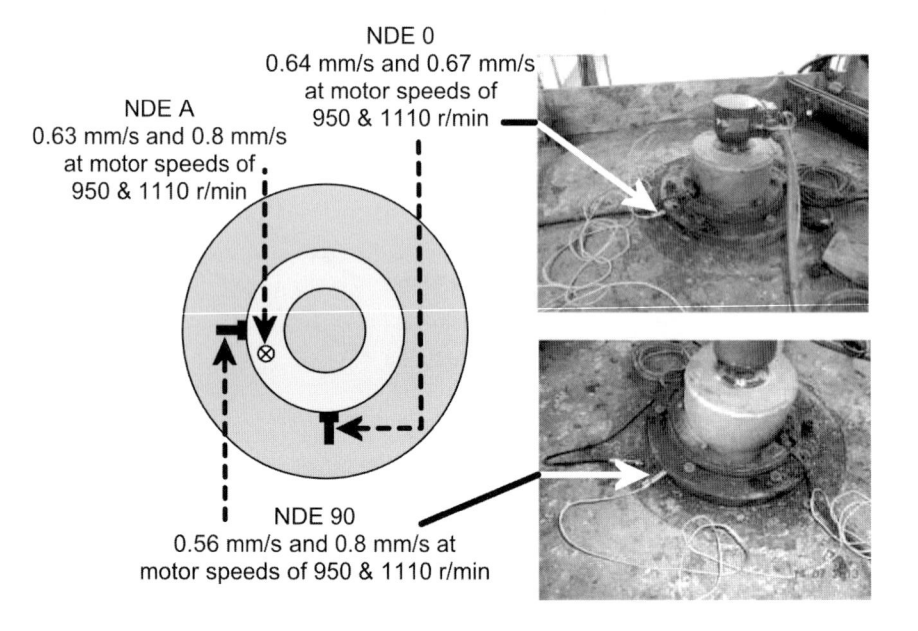

Figure 11.11 Presentation of the overall r.m.s. velocities on the NDE bearing housing at the operational speeds of 950 r/min and 1110 r/min.

11.2.3 Current Spectrum – Magnetic Flux and Electromagnetic Forces

Before carrying out any vibration measurements it was essential to measure the frequency content of the current supplied to the motor from the inverter when the motor was producing the loud (as perceived by offshore engineers) acoustic noise at a speed of 950 r/min.

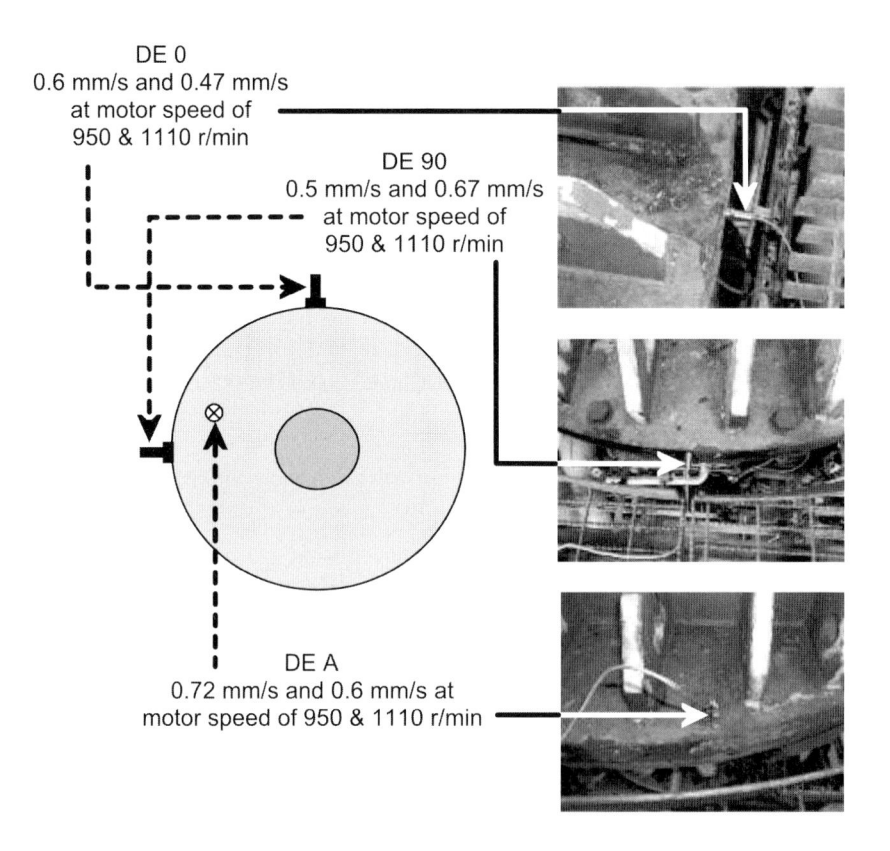

11.2 Industrial Case History – Investigation into the Cause of Loud Acoustic Noise

Figure 11.12 Presentation of the overall r.m.s. velocities on the DE mounting flange at the operational speeds of 950 r/min and 1110 r/min.

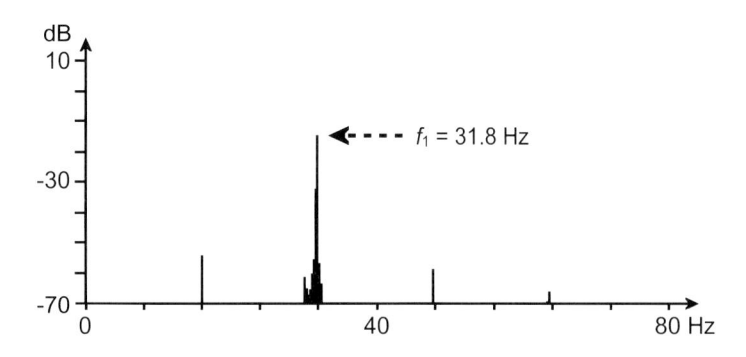

Figure 11.13 Current spectrum – fundamental frequency from the inverter.

Figure 11.13 shows that the fundamental frequency of the current waveform from the inverter with the motor operating at 950 r/min and an input current of 400 amperes ($I_{f.l.}$ = 914 amperes), was f_1 = 31.8 Hz. Recall that the synchronous speed of the rotating magnetic field in this 4-pole (two pole-pairs) SCIM is given by:

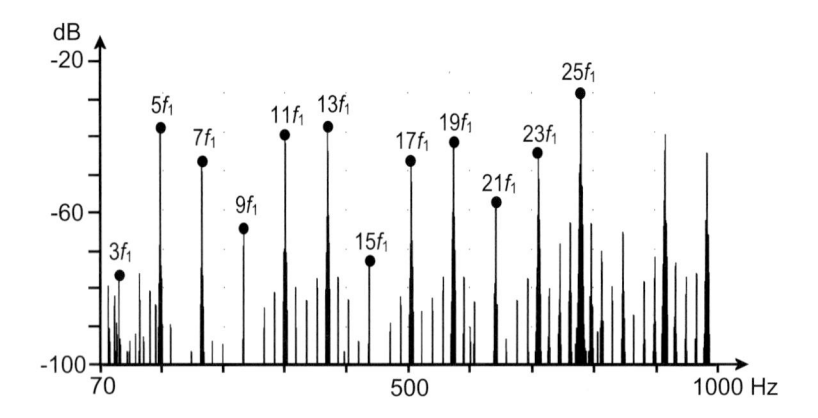

Figure 11.14 Current spectrum to confirm multiple harmonics of the fundamental frequency from the inverter.

$$N_s = (60 \times f_1)/\text{pole-pairs} = (60 \times 31.8)/2 = 954 \text{ r/min}.$$

Figure 11.14 shows the frequency content of the current spectrum up to 1000 Hz and, as expected, the inverter supply current has multiple harmonics of the fundamental frequency, f_1.

From Appendix 10A it was mathematically proven that the fundamental flux produces an electromagnetic force at:

$$\text{twice the supply frequency} = 2f_1.$$

With the motor supplied at a frequency of $f_1 = 31.8$ Hz, the fundamental electromagnetic force will be at a frequency of:

$$2f_1 = 63.6 \text{ Hz}.$$

The corresponding vibration due to this force will be at a frequency = 63.6 Hz.

This is normally referred to as the vibration at *twice the supply frequency* and is the fundamental vibration component due to electromagnetic forces. Figure 11.14 shows the current spectrum produced by the inverter up to 1000 Hz and, as expected, there are numerous odd harmonics.

Each of these current components produces a corresponding flux component and, by applying the previous analysis, this means electromagnetic forces can be produced *at twice the frequency of each of these odd harmonic flux components.*

11.2.4 Vibration Spectrum Analysis

Figures 11.15 and 11.16 show the velocity spectra at the DEO position at 950 r/min and 1110 r/min respectively. The velocity component at 827 Hz is due to the electromagnetic force (F_{em}) at $2(13f_1)$. Its velocity is 0.3 mm/s r.m.s. and its acceleration is 1.6 m/s^2. It is accepted that high frequencies should be measured in acceleration.

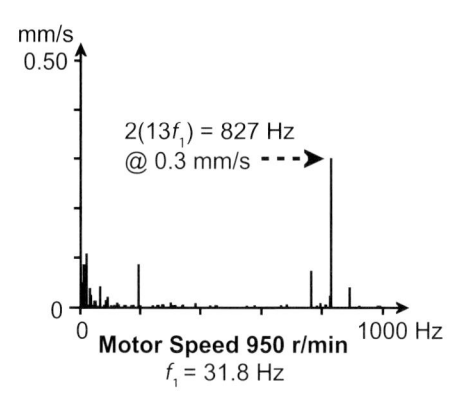

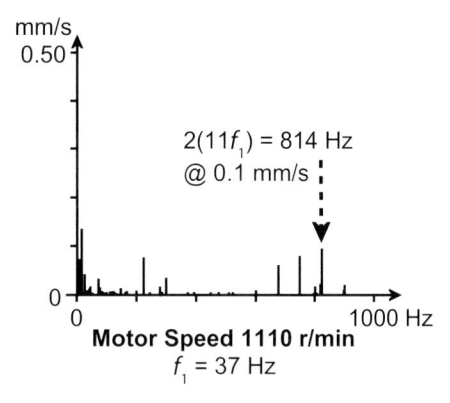

Figure 11.15 Vibration spectrum: DE0 overall velocity level 0.6 mm/s.

Figure 11.16 Vibration spectrum: DE0 overall velocity level 0.47 mm/s.

11.2.4.1 Conclusions

(i) The overall r.m.s. velocities measured at the DE and NDE are perfectly normal.

(ii) Electromagnetic rotating force waves act directly on the stator core and outer frame structure, and are also transmitted to the bearings and flanges of the motor.

(iii) When operated at 950 r/min the motor's vibration spectrum contains a unique component at 827 Hz that is due to an electromagnetic force at that frequency and is a by-product of the complex supply current waveform from the inverter.

(iv) When operated at 1110 r/min, no significant vibration components from electromagnetic forces were evident in the velocity spectrum, and it can be concluded that the stator core or frame were not responsive (negligible vibration, no vibration at $2(13f_1)$) to the electromagnetic forces at that higher speed of operation, hence no acoustic noise problem was previously reported.

(v) It is concluded from the analysis of current and vibration spectra that the acoustic noise from the motor at a speed of 950 r/min is due to an electromagnetic force (proportional to flux^2) at $2\ (13f_1)$, which causes vibration of the stator core and outer frame of the motor at that frequency.

  (a) This noise was of concern to the end user because the motor had only recently been operated at this speed, and had not been heard at the higher operating speeds, which were previously used.

(vi) The fundamental inverter supply frequency was $f_1 = 31.8$ Hz, with a corresponding rotor speed of 950 r/min, thus the thirteenth current harmonic and corresponding flux component in the airgap flux waveform is $13f_1 = 413.5$ Hz. An electromagnetic force, F_{em} (due to flux^2), occurs at $2(13f_1) = 827$ Hz, and this is clearly seen in the vibration spectrum from the DE.

(vii) Acoustic pressure waves at a frequency of 827 Hz are perceived to be much louder than those in lower frequency ranges for the same acoustic sound pressure level (SPL) in N/m^2 or in dB – see Figure 11.A and associated explanation.

(viii) The motor can be operated continuously in the range 900 r/min to 1100 r/min and at higher speeds up to a maximum of 1510 r/min, as specified on the motor's nameplate, because the vibration and acoustic noise at 827 Hz were **not due to a fault** but were induced due to the inverter supply.

Appendix 11B Human Perception of Acoustic Noise

(i) The physical and neural mechanisms of hearing are extremely complex. The human ear has a logarithmic response and can detect sounds spanning six decades of intensity and in fact is a very non-linear transducer.

(ii) In terms of frequency, the audible range extends over three decades, from 20 Hz to 20 kHz for a teenager, but the upper limit of frequency declines as humans age and, for example, by the age of 70 it has typically dropped to 12 kHz.

(iii) For a given constant sound pressure level (SPL), the *perceived* loudness of a sound is not constant over the audible range because it is frequency dependent, with sounds of equal SPL appearing to be much louder to the human ear in the 800 Hz to 4 kHz range than the sound at lower frequencies, as shown in Figure 11.A.

(iv) The variability of loudness perception with frequency is formalized in 'equal loudness contours' as shown in Figure 11.A. These contours were developed from original plots known as 'Fletcher–Munson curves', named after early researchers in the field.

(v) As an example of loudness perception variability, the contours show that a sound at a given sound pressure level is perceived as twenty-five times louder at 1 kHz than at 100 Hz.

11.3 Industrial Case History – VSA of a Repaired 1000 kW/1340 H.P. SRIM Diagnosed a High Twice Supply Frequency Vibration and Cracks in the Concrete Mounting Plinth

11.3.1 Background and Objectives

Abstract – The author was contacted by a cement factory to carry out a vibration investigation into a repaired 1000 kW/1340 H.P. SRIM during a FAT followed by vibration measurements and VSA during on-site re-commissioning. The motor was originally removed from service because of an unacceptably high velocity of the $2f_1$ vibration component (the magnitude was not provided to the author). A photograph of the SRIM during re-commissioning is shown in Figure 11.17.

11.3 Industrial Case History – VSA of a Repaired 1000 kW/1340 H.P. SRIM

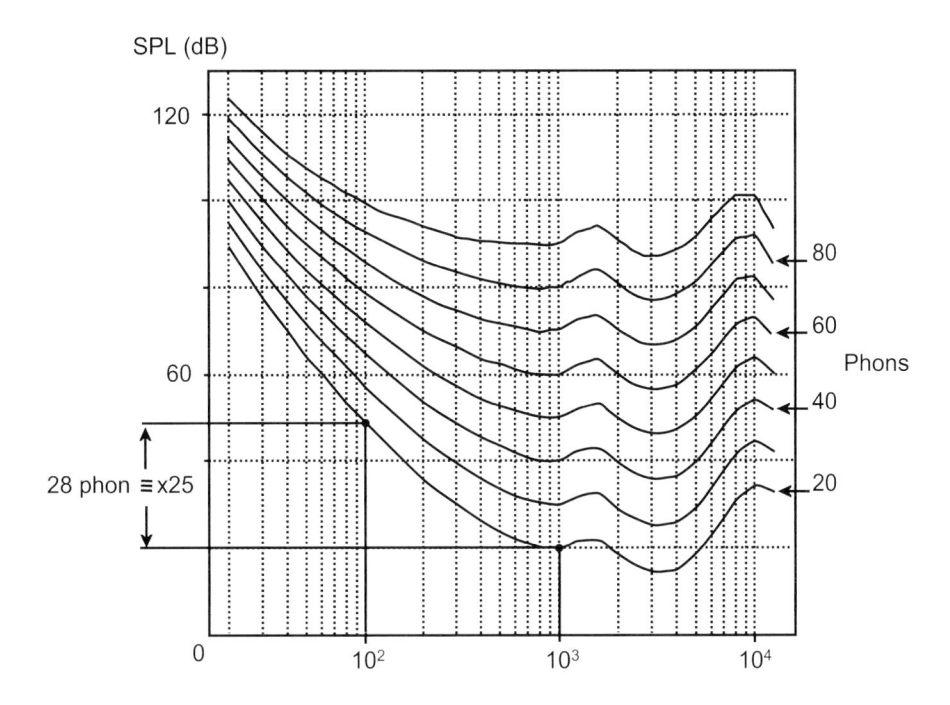

Figure 11.A SPL (sound pressure levels) and loudness level contours in Phons versus frequency.

Figure 11.17 SRIM during re-commissioning in the cement factory.

The motor repair shop found cracks at the fixing positions of the stator frame to the main frame, but no photos were provided to the author. The radial airgap was measured via feeler gauges, and it was confirmed that the airgap eccentricity was 28% at the NDE. This is a very high level of airgap eccentricity when compared to 5% airgap eccentricity, which OEMs aim to achieve in a brand-new motor. The cement factory required the motor to be back in service as soon as practically possible, and therefore the cracks from the stator frame to main frame were temporarily repaired by welding. Also, the airgaps were set to within 10% of the radial airgap length at the NDE and DE.

The author's objectives were as follows:

(i) To carry out vibration measurements and analysis on the repaired SRIM during a FAT programme.
(ii) To determine, based on the FAT results, whether the motor was fit for purpose with respect to vibration levels.
(iii) To carry out VSA during no-load and coupled runs of the SRIM after re-commissioning.
(iv) To advise the client if the motor could continue to run and recommend any future actions.

11.3.2 Vibration Measurements and Analysis – Uncoupled Run at Repair Shop

Motor nameplate data were as follows:
- 3-phase SRIM (slip-ring induction motor); Year of manufacture: 1988
- 3.3 kV, 1000 kW/1340 H.P., 217 A, 50 Hz, 987 r/min (6-pole)
- Star connected, insulation class: F
- Frame size: DW 450; enclosure: IP55, CACA
- DE bearing: NU6332E M C3 (E = extra capacity, M = brass cage)
- NDE bearing: 6332 M C3
- Rotor amperes: 400 A, open circuit rotor volts: 1530 V.

The following data were obtained from the OEM:

- Nominal airgap length: 2 mm ($\pm 10\%$) or 80 mils/thou ($\pm 10\%$)
- Number of rotor slots: 54
- Number of stator slots: 72.

The motor's cooler had been removed and full access to the stator core back (see Figure 11.18) and the DE and NDE bearing pedestals were accessible – see Figures 11.19 and 11.20, which also show the positions of the accelerometers.

The motor was not bolted to the large base-plate because it was running uncoupled and the only torque developed is to overcome windage and friction, therefore the reaction torque at the feet is negligible. It was not bolted down so that a comparison could be made between the tests in the motor shop on no load with a no-load

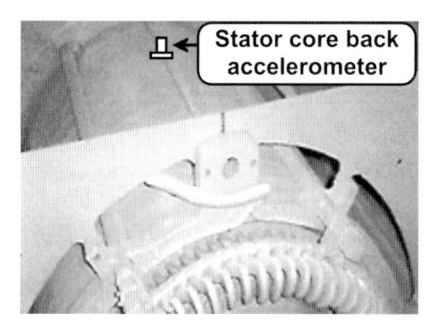

11.3 Industrial Case History – VSA of a Repaired 1000 kW/1340 H.P. SRIM

Figure 11.18 Stator core assembly.

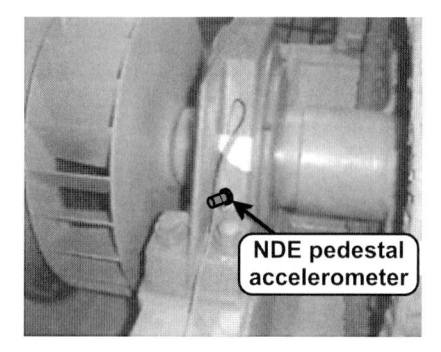

Figure 11.19 NDE bearing housing.

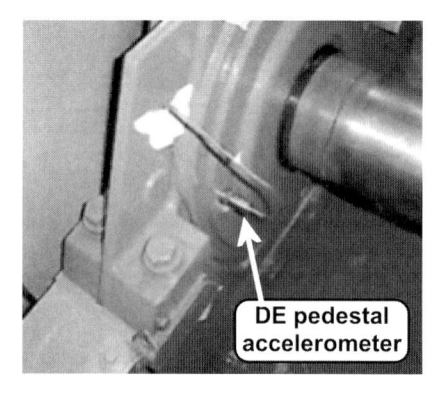

Figure 11.20 DE bearing housing.

uncoupled run when bolted down to the large concrete plinth at the cement factory, as shown in Figure 11.17.

The motor repair shop could not supply the motor at 3.3 kV due to supply limitation, but the client accepted the 3 kV, which was available at a supply frequency of nominally 50 Hz. Thus, it was operating below its rated flux density, therefore the

Table 11.6 Overall r.m.s. velocities (±10%), span 5–1000 Hz and the velocities of the $2f_1$ component on the bearing housings

DE overall r.m.s. velocities			
DE bearing housing	DEV: 1.2 mm/s	DEH: 1.4 mm/s	DEA:1.9 mm/s
DE velocity r.m.s. $2f_1$ component			
DE bearing housing	DEV: 0.9 mm/s	DEH: 1.2 mm/s	DEA:1.7 mm/s
NDE overall r.m.s. velocities			
NDE bearing housing	NDEV:2.0 m/s	NDEH: 0.6 mm/s	NDEA:1.0 mm/s
NDE velocity r.m.s. $2f_1$ component			
NDE bearing housing	NDEV:1.8 mm/s	NDEH: 0.4 mm/s	NDEA: 0.8 mm/s

velocity of the $2f_1$ vibration component was slightly reduced in magnitude compared to its value at 3.3 kV.

The no-load currents for the red, yellow and blue phases were 49 amperes, 48 amperes and 54 amperes; this unbalance can be expressed – see references [11.6] to [11.8] – as:

$$\% \text{ Unbalance} = 100 \times \left[\frac{\text{Maximum current deviation from the average current}}{\text{Average current}} \right]$$
$$= 7.3\%.$$

11.3.2.1 Overall R.M.S. Velocities and Magnitude of the $2f_1$ Component on the Bearing Housings – Uncoupled Run in Repair Shop

Table 11.6 presents the overall r.m.s velocities in the vertical, horizontal and axial positions on the bearing housings. The vibration standard, BS 60034-14-2004 [11.9], was used as the criterion for assessing the velocity and the maximum allowable for this motor is 2.3 mm/s r.m.s. The velocities of the $2f_1$ component are also presented in the same table.

A sample of the velocity spectra is presented in Figures 11.21 and 11.22 for the NDE and DE bearing housing in the vertical direction.

The key observations of the vibration measurements on the bearing housings during the uncoupled run in the repair shop are as follows:

(i) The overall r.m.s. values on both bearing housings were normal and below the acceptable limit of 2.3 mm/s r.m.s.
(ii) Figures 11.21 and 11.22 show that the $2f_1$ component dominates the spectra, for example, at the NDE in the vertical direction the $2f_1$ component is 90% of the overall value. This is unusual for a 6-pole induction motor.

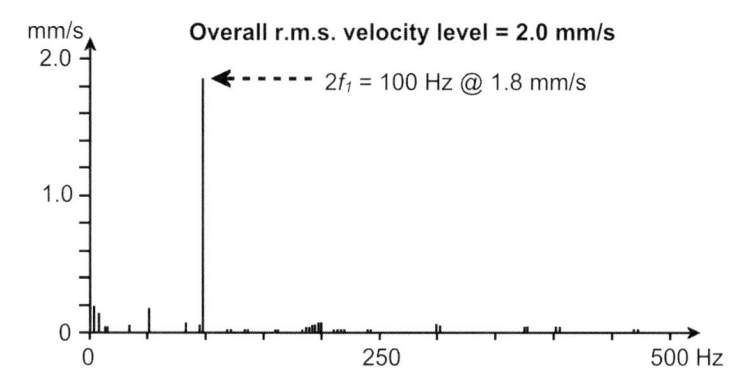

Figure 11.21 NDE vertical bearing housing, velocity spectrum, uncoupled in repair shop.

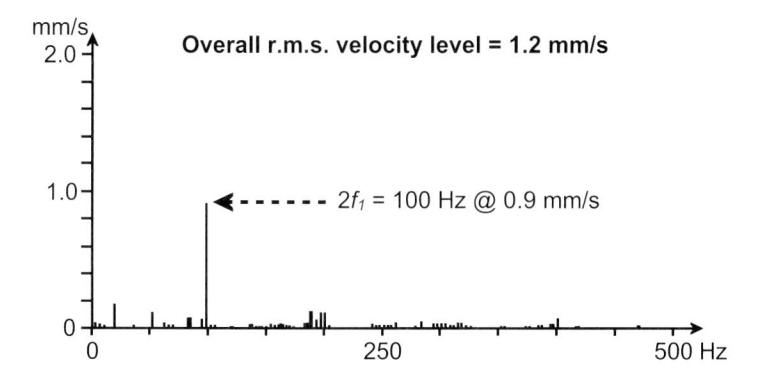

Figure 11.22 DE vertical bearing housing, velocity spectrum, uncoupled in repair shop.

The stator core back overall r.m.s. velocity at position X in Figure 11.18 was 4.2 mm/s and the *2f₁* component was 4 mm/s, as shown in Figure 11.23.

11.3.3 Vibration Measurements and Analysis – Uncoupled and Coupled Runs at the Cement Factory

It was not possible to measure the vibration on the NDE bearing housing due to the restricted access imposed by the totally enclosed slip ring assembly. Figure 11.24 illustrates the positions of the accelerometers during the uncoupled and coupled runs at the cement factory, and Table 11.7 presents the overall r.m.s. velocities.

11.3.3.1 Key Observation from the On-Site Uncoupled and Coupled Runs

The velocity at position 5 was 4 mm/s r.m.s., and it was observed that there were cracks in the concrete mounting plinth directly below the motor's fixing bolts, as

Table 11.7 Overall r.m.s. velocities (±10%), span 5–1000 Hz

On-site uncoupled run
No-load currents: $I_R = 62$ A, $I_Y = 58$ A, $I_B = 58$ A

Position 1	Position 2	Position 3	Position 4	Position 5	Position 6	Position 7
0.5 m/s	1.9 m/s	0.7 m/s	1.0 m/s	4 mm/s	0.9 m/s	0.3 m/s

On-site coupled run
No-load currents: $I_R = 130$ A, $I_Y = 135$ A, $I_B = 130$ A

Position 1	Position 2	Position 3	Position 4	Position 5	Position 6	Position 7
1.0 m/s	1.9 m/s	1.2 m/s	0.7 m/s	4 mm/s	0.9 m/s	0.5 m/s

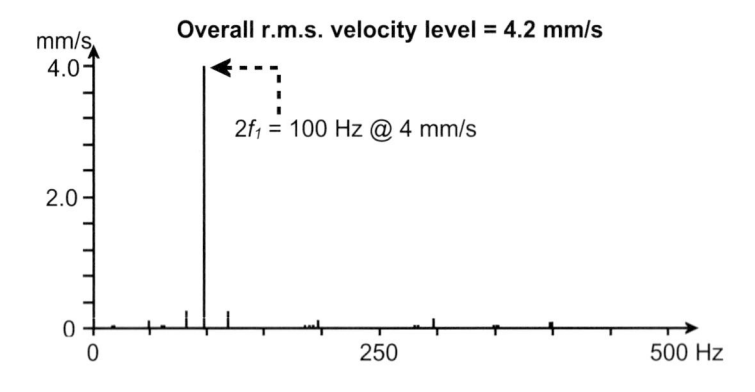

Figure 11.23 Stator core back, velocity spectrum, uncoupled run in repair shop. Note that the repair shop could not load test this 1000 kW/1340 H.P. SRIM.

shown in Figures 11.25 and 11.26. This was the first time the author had been on-site at the cement factory.

One of the holding-down bolts was slackened at position 1 and the acoustic noise dramatically increased. The overall velocity at position 5 increased from 4 mm/s to 8 mm/s r.m.s., which was not normal.

This increase was dominated by the $2f_1$ component as shown when Figures 11.27 and 11.28 are compared; this reveals that the $2f_1$ component increased from 3.8 mm/s to 6.2 mm/s r.m.s., an increase of 63%.

11.3.3.2 Conclusions

(i) The results presented in Section 11.3.3 verified that the velocity at $2f_1$ is highly sensitive to the tightness of the motor's fixing-down bolts.

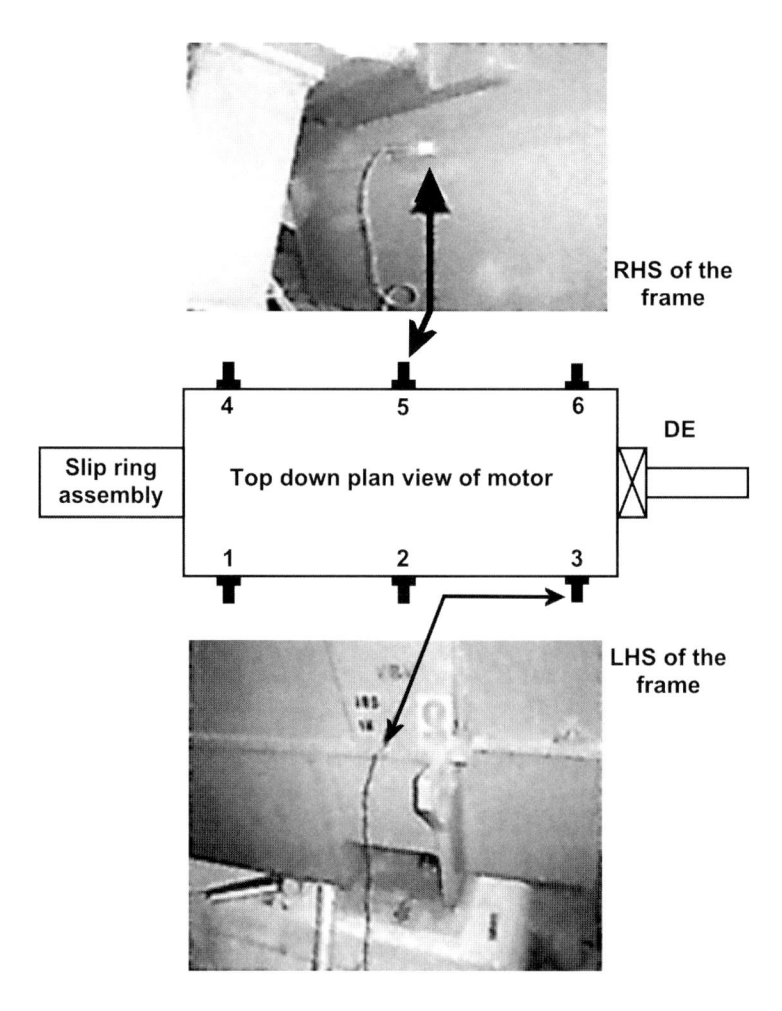

11.3 Industrial Case History – VSA of a Repaired 1000 kW/1340 H.P. SRIM

Figure 11.24 Positions of accelerometers.

Figure 11.25 Cracked concrete mounting plinth.

Figure 11.26 Cracked concrete mounting plinth.

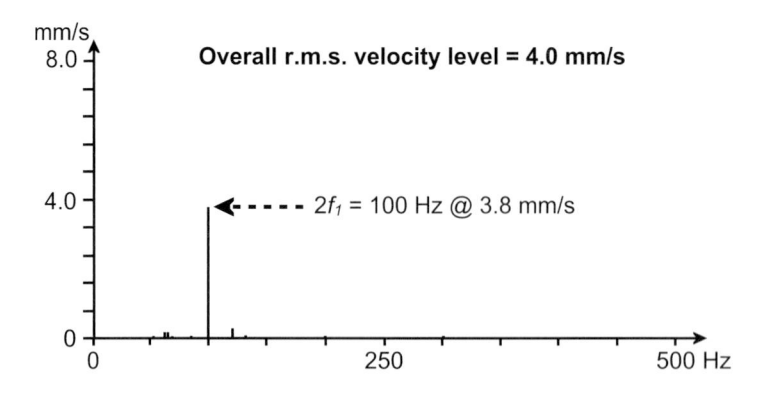

Figure 11.27 Velocity spectrum, position 5; all holding-down bolts were tight.

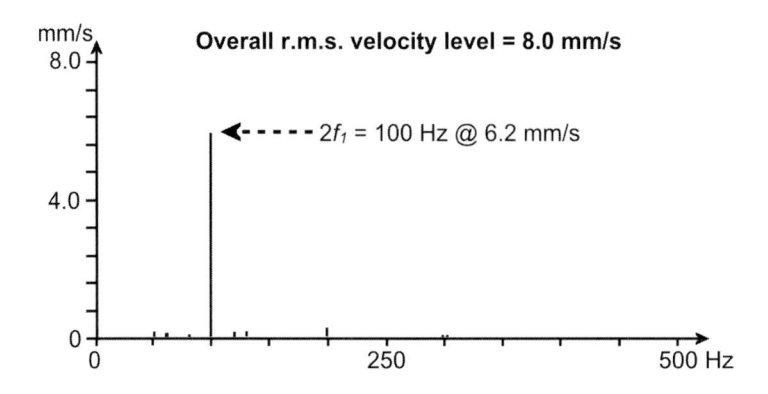

Overall r.m.s. velocity level = 4.0 mm/s

$2f_1$ = 100 Hz @ 3.8 mm/s

Figure 11.28 Velocity spectrum, position 5; bolt at position 1 was slackened.

Overall r.m.s. velocity level = 8.0 mm/s

$2f_1$ = 100 Hz @ 6.2 mm/s

(ii) Cracks were identified in the concrete plinth directly under the holding-down bolts.
(iii) Cracks were also found by the repair shop between the fixing positions of the stator core assembly to the main base of the motor.
(iv) Items (i) and (ii) indicate that there are abnormal levels of force and vibration being transmitted to these positions within this motor.
(v) The client should remove the motor at the next major outage to establish the extent and depth of the cracks under the fixing-down bolts. Carry out repairs as required.

References

11.1 M. Liwschitz-Garik and C. C. Whipple, *Electric Machinery Vol. II, A-C Machines*, Van Nostrand Company, first published Sept. 1946.

11.2 M. G. Say, *Alternating Current Machines*, fourth edition, ELBS and Pitman Publishing, 1976.

11.3 P. L. Alger, *Induction Machines – Their Behaviour and Uses*, Gordon and Breach Science Publications Inc., second edition, published by OPA Amsterdam, third printing with additions, 1995.

11.4 S. J. Yang, Low Noise Electric Motors, *Monographs in Electrical and Electronic Engineering*, IEE, Savoy Place, London, 1981.

11.5 P. L. Alger Magnetic Noise in Poly-phase Induction Motors, *Transactions AIEE*, 73 (Part IIA), 1954, pp. 118–25.

11.6 J. H. Dymond, Operation on Unbalanced Voltage: One Motor's Experience and More, *IEEE Transactions on Industry Applications*, 43 (3), May/June, 2007, pp. 829–37.

11.7 E. Williams, Operation of 3-Phase Induction Motors on Unbalanced Voltages, *AIEE Transactions*, 73, pt III-A, April 1954, pp. 125–33.

11.8 M. Campbell and G. Arce, Effect of Motor Voltage Unbalance; Test and Evaluation, *IEEE Transactions on Industry Applications*, 54 (1), 2018, pp. 905–11.

11.9 British Standard (BS) 60034-14, *Rotating Electrical Machines, Part 14 Mechanical Vibration of Certain Machines with Shaft Heights of 56 mm and Higher – Measurement, Evaluation and Limits of Vibration Severity*, Feb. 2004.

12 Overview of Key Features of Vibration Monitoring of SCIMs

Abstract – An appraisal is presented on VSA to diagnose faults in rolling element bearings. A discussion is provided on the key outcome that end users hope to achieve from vibration monitoring, which is the prognosis of remaining operational life of a SCIM after a fault is diagnosed.

It will also be reiterated, via more photographic evidence, that access to mount temporary accelerometers directly on the bearing housings of rolling element bearings used in induction motors can be very difficult.

Guidelines are given for grease management of rolling element bearings because of the predominance of failures being caused by incorrect greasing practices.

12.1 Appraisal on VSA to Diagnose Faults in Rolling Element Bearings Used in SCIMs

Rolling element bearing faults, which constitute the largest percentage of failures in SCIMs – see references [12.1] to [12.6] – can be diagnosed by measuring the combination of the overall r.m.s. velocities and velocities of the bearing defect frequencies. To achieve this, the trending of overall velocities and velocities of the bearing defect frequencies is essential to reliably diagnose bearing faults.

Trending is the taking of repeatable measurements under the same conditions at regular intervals so that change, particularly a sudden major increase in vibration levels, in any of these values can be readily and accurately seen.

Key points to be applied when trending vibration measurements are as follows:

(i) When no permanent accelerometers are fitted, it is vital that temporary accelerometers are put in exactly the same positions every time measurements are taken.

(ii) Ideally, the motor should be operating at the same loading (input current) each time vibration measurements are taken.

(iii) If possible the time interval between the taking of sets of measurements should be constant.

(iv) Trending only the overall velocity levels very often does not indicate a high rate of rise of the velocities of bearing defect frequencies because the *1X* component dominates the overall velocity.

(v) If the overall velocity level doubles, this can be used as a practical guide to indicate that a problem exists in a SCIM; and likewise for the velocity level of a bearing defect frequency, this is a good indicator that the fault has increased in severity. This is verified in a number of the case histories (e.g. see the case histories in Sections 5.1 and 7.1).

(vi) As the percentage velocity levels of any of the bearing defect frequencies increases with respect to the overall velocity levels, this can also be used as an indicator of bearing faults. This is verified in the case histories (e.g. see the case history in Section 6.4).

Note that a similar **vibration monitoring and VSA strategy** can be applied to the diagnosis of shaft misalignment (see the case histories in Chapter 1), faults in sleeve bearings (see the case histories in Chapter 9) and vibration problems caused by abnormal electromagnetic forces (see the case histories in Chapter 10).

One-off measurements of the overall velocities on the bearing housings and vibration spectrum analysis (VSA) with no historical records are an inefficient application of vibration monitoring.

12.2 Predictions and Prognosis of Remaining Run Life

When VSA and the vibration analyst predict that a specific fault exists in a SCIM, the end users often require answers to the following:

(i) How severe is the fault?
(ii) How long can the motor operate with this type of fault?
(iii) How quickly will the severity of the fault increase?
(iv) How long will the motor run with this fault before it fails?
(v) What is the risk of a rolling element bearing fault causing a bearing collapse and a consequential rotor to stator rub?
(vi) Should the motor be immediately switched off and removed from service to eliminate a catastrophic motor failure and a risk to the health and safety of personnel and other items of plant?

The author's opinions with respect to VSA and the vibration analyst being able to provide **definitive and quantifiable** answers to items (i) to (vi) are as follows:

(i) The application of trending vibration measurements when analysed by a knowledgeable and professional vibration analyst can provide an answer to item (i), and sometimes an answer to item (vi).
(ii) It is not technically or practically possible to provide a definitive and quantifiable prognosis for items (ii) to (v) because of the many variables that can subsequently affect the rate at which the fault severity increases in, for example, a rolling element bearing with continued use of the motor.

What is normally achievable by a vibration analyst is to indicate that either an inner or outer race fault, a rolling element fault or a faulty cage exists.

However, the analyst can only provide a generalised categorisation of the fault severity with recommendations, such as:

- Very early stages of a bearing fault: check the greasing records.
- Low level of fault severity: repeat vibration measurements at least once/month and check the greasing records.
- Medium level of fault severity: repeat vibration measurements at least every two weeks.
- Severe level of fault severity: repeat vibration measurements on a daily basis.
- Very severe level of fault severity: remove the motor from service.

It is again emphasised that trending of the velocities of the bearing defect frequencies is essential after the initial estimate of fault severity via VSA to indicate subsequent degradation of a rolling element bearing.

12.2.1 Variables that Affect the Remaining Run Life of Faulty Rolling Element Bearings in SCIMs

Vibration spectrum analysis (VSA) or any other vibration analysis technique, such as *envelope analysis*, cannot quantify the actual severity of the fault, which can only be reliably done by removing the bearing for a forensic inspection.

The following are typical examples of variables that prevent the prognosis of the remaining run life of faulty rolling element bearings:

(i) Under or over greasing the bearing after the fault is diagnosed.
(ii) Abnormal mechanical load dynamics, which transmit high forces to the bearing.
(iii) Cyclic and transient load dynamics from the driven load.
(iv) Forces due to airgap eccentricity in a SCIM, in particular rotating forces wave due to an increase in dynamic airgap eccentricity.

12.3 Difficulties of Access to Measure Vibration Directly on the Bearing Housings of Rolling Element Bearings in Induction Motors

To diagnose faults in rolling element bearings by using VSA it is strongly recommended that the vibration is measured directly on the bearing housings. However, induction motors with rolling element bearings and power ratings up to for example 1000 H.P./746 kW (Figure 12.1) are normally cooled by an external fan at the NDE, as shown in the case histories and further illustrated in the photographs in Figures 12.1 to 12.4. While the motor is running, access to the NDE bearing is impossible.

As shown in the case histories, these motors do not normally have permanently mounted accelerometers on the bearing housings. For example, the end users carry out

12.3 Difficulties of Access to Measure Vibration 273

Figure 12.1 A 6.6 kV 746 kW/1000 H.P. SCIM with rolling element bearings.

Figure 12.2 A 415 V 185 kW/250 H.P. SCIM with rolling element bearings.

Figure 12.3 A 415 V 132 kW/175 H.P. SCIM with rolling element bearings.

Figure 12.4 A 3.3 kV 400 kW/540 H.P. SCIM with rolling element bearings.

vibration measurements on a monthly sequence while the motors are running, using temporary accelerometers mounted via strong magnets or fixing studs. However, access to the NDE bearing housings is often not possible. This is a major constraint, because the vibration can only be measured on the outer frame close to the fan cowl at the NDE, or alternatively on the bolts that secure the fan cowl to the end frame, which houses the bearing.

The latter (i.e. the bolt) does provide a direct transmission path for the vibration on the bearing housing to the bolt that is bolted into the outer periphery of the NDE end frame as shown in Figures 12.5a and b.

The mechanical stiffness and frequency response of the end frame in the radial direction between the bearing housing and the outer periphery of the end frame, where the vibration can be measured, will normally attenuate the vibration produced by the bearing. Consequently, the very early stages of bearing faults are difficult to detect by outer frame vibration measurements, but the case histories verified that as the bearing faults become more serious the bearing defect frequencies can be detected via VSA.

The measurement of vibration on the DE bearing housing requires the use of a small accelerometer because the surface area is often limited, as shown in Figure 12.6. The coupling guard can also prohibit access to the DE bearing housing, as shown in Figure 12.7.

With respect to the use of envelope analysis or shock pulse techniques to detect the inception of bearing defects the special accelerometers should **be mounted directly on the bearing housings**. If not a false diagnosis of a bearing fault can occur (see the case history in Section 8.2).

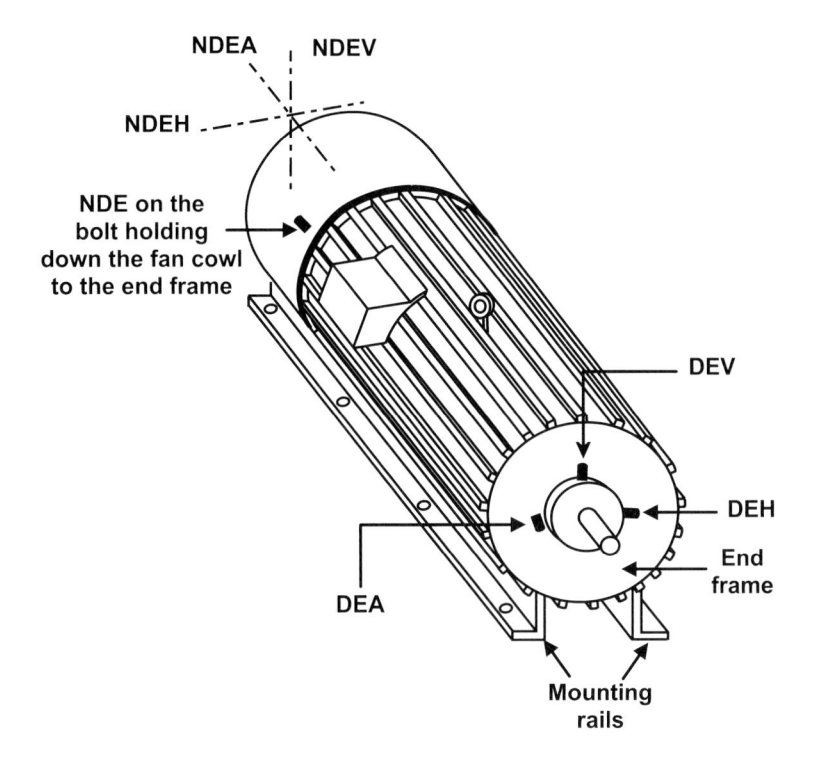

12.3 Difficulties of Access to Measure Vibration

Figure 12.5a Illustration of accelerometer positions.

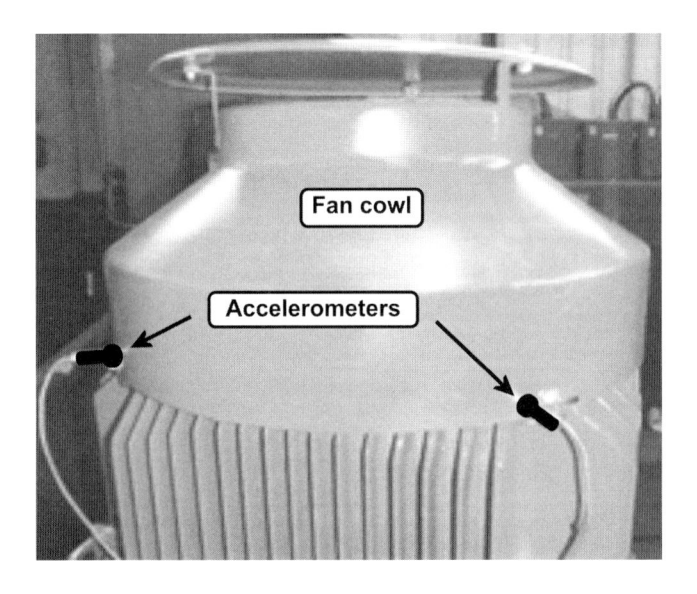

Figure 12.5b Position of accelerometers on the bolts holding down the fan cowl to the NDE end frame.

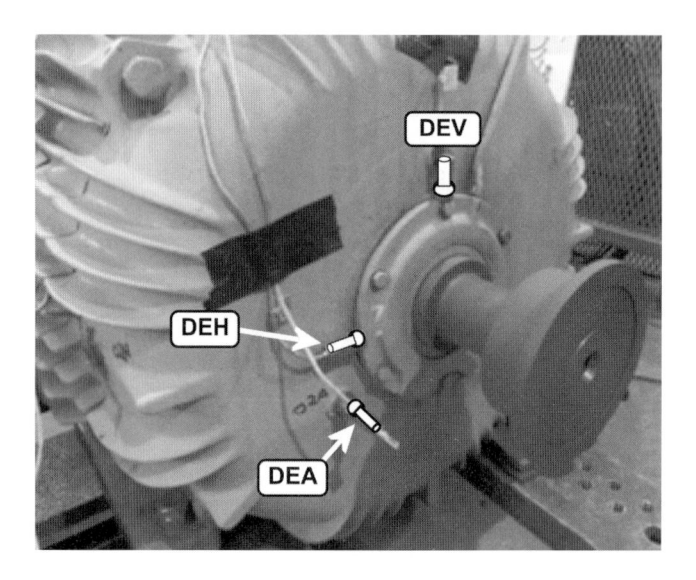

276   Overview of Key Features of Vibration Monitoring of SCIMs

Figure 12.6 Photograph showing the small surface area that can exist on the DE bearing housing.

Figure 12.7 Photograph showing a coupling guard which prohibits access to the DE bearing housing.

It is strongly recommended to *not apply* envelope analysis or SPM to diagnose bearing faults by measuring the vibration on the outer frame.

12.4 Guidelines for Successful Grease Management of Rolling Element Bearings in Induction Motors

In many cases, as shown in Figure 3.1 and in the case histories, the major cause of rolling element bearing failures is due to greasing issues such as:

- over greasing
- under greasing
- mixing of greases
- incompetent re-greasing of rolling element bearings
- lack of a stringent planned maintenance procedure for greasing of bearings.

Approximately 50% of faults in induction motors are caused by bearing failures, of which a significant proportion are the result of greasing problems in rolling element bearings, as demonstrated in the case histories. Many of these could be eliminated by personnel with proper knowledge who have been trained on the correct greasing procedures. The following points are the author's guidelines for successful greasing of rolling element bearings in induction motors:

(i) The starting point is the OEM's operation and maintenance manual for the motor. It should state the type of grease, the re-greasing interval based on actual run hours of the motor, and the volume of grease to be inserted.
(ii) It is normal for manual grease guns to be used. One pump of the grease gun will release one shot of grease and the weight of that one shot for that grease gun should be measured and clearly marked on the grease gun.
(iii) The type of grease to be used in each grease gun should be clearly specified on the gun, and different greases should never be mixed in a grease gun or applied to a bearing.
(iv) Re-greasing of rolling element bearings should be done when the motor is running.
(v) Grease relief apertures may exist at the 6 o'clock position in the bearing housing in horizontal motors and they should be checked when greasing the bearing.
(vi) It is essential that end users have proper training schemes for personnel who are assigned to grease bearings in electrical machines. It is the author's experience that not enough serious attention is given to training by end users and they wrongly assume that anyone can properly grease bearings – that is NOT the case.

Normally there are no automatic run counters fitted to motor drives, and therefore there is a lack of accurate records on the number of run hours. End users often re-grease the bearings every three months irrespective of run hours, but this does not follow the OEM's guidelines.

Accurate records should be kept of run hours; this should be the responsibility of operations and maintenance personnel, and re-greasing carried out as per the operation and maintenance manual.

It is often the case that, with vertically mounted induction motors (see Figure 12.4), the end user has retrofitted extension grease pipes so that the bearings can be greased by maintenance personnel standing at ground level so that temporary scaffolding is unnecessary. The author has seen cases where the extension pipes have several right-angled bends (the worst case was six), which is bad practice for successful re-greasing and should be avoided. If it is not possible to avoid this then the extension pipes should be cleaned out at least every six months to remove the possibility of blockages.

12.5 Incorrect Bearings Fitted by Motor Repair Shop

When new bearings are fitted to induction motors, the end user should insist that the repair shop fits identical bearings in accordance with the information on the motor's nameplate.

It can be the case that certain repair shops do not keep diligent records and photographs of the removed faulty bearings, or of the actual new bearings that were fitted. The author requires a repair shop to photograph the bearing numbers on the new bearings that they have fitted. The reasons for this are as follows:

(i) The OEM of the motor selected the bearings (as stamped on the nameplate) to suit the motor's design, operational speed and dynamic loading on the bearing.

(ii) Two major manufacturers, who supply bearings worldwide, have replaced certain off-the-shelf, stock items of steel cage bearings with extra capacity (E rated), brass cage bearings. For example, they have replaced an N324 C3 with an N324 E M C3, and other cylindrical roller bearings are now E rated as standard off-the-shelf cylindrical roller bearings. It may seem that an E rated cylindrical roller element bearing is better than a non-E rated bearing, but the crucial point is that an E rated bearing is unnecessary and will be overrated in a motor that was designed for a non-E rated bearing. Excessive skidding can occur (e.g. see case history in Section 7.2) because the dynamic loading on the bearing is insufficient to effectively rotate the rolling elements.

(iii) An N324 E MC3 has 13 rollers and an N324 C3 has 12 rollers, and the vibration analyst will not know that an E rated bearing has been fitted, unless the repair shop has updated the nameplate, and that is very unlikely indeed. The prediction of the bearing defect frequencies will be based on the nameplate information, and the vibration analysis no longer becomes valid to detect bearing defects in an E rated bearing. This is based on the author's knowledge of what can happen in practice.

References

12.1 A. H. Bonnett, Root Cause Methodology for Induction Motors; A Step-by-Step Guide to Examining Failure, *IEEE Industry Applications Magazine*, 18 (6), 2012, pp. 50–62.

References

12.2 http://new.abb.com/docs/librariesprovider53/about-downloads/motors_ebook.pdf?sfvrsn=4.

12.3 IEEE Committee Report, Report of Large Motor Reliability Survey of Industrial and Commercial Installations, *IEEE Transactions on Industry Applications*, 1A-21 (4), Parts I and II, July/Aug. 1985.

12.4 IEEE Committee Report, Report of Large Motor Reliability Survey of Industrial and Commercial Installations, *IEEE Transactions on Industry Applications*, 1A-23 (1), Part III, Jan./Feb. 1987.

12.5 O. V. Thorson and M. Dalva, A Survey of Faults on Induction Motors in Offshore Oil Industry, Petrochemical Industry, Gas Terminals and Oil Refineries, *IEEE Transactions on Industry Applications*, 31 (5), Sept./Oct. 1995.

12.6 https://www.efficientplantmag.com/2012/03/large-electric-motor-reliability-what-did-the-studies-really-say/ *Large Electric Motor Reliability: What Did The Studies Really Say?*, EP Editorial Staff March 23, 2012; *Efficient Plant Magazine* formerly *Maintenance Technology*: https://www.efficientplantmag.com/.

Index

abnormal misalignment, 9, 23
acoustic noise, 2, 88, 112, 135, 137, 140, 144, 252–4, 256, 259, 266
alignment, 9, 19, 25, 36, 126, 182
aluminium die cast, 31, 222, 226–7
American Petroleum Institute, 189, 210, 212
angular contact bearing, 55, 57, 76, 90, 95, 103
angular misalignment, 8
API, 189, 210, 212, 214, 219
attenuation, 116, 171, 173, 213
auxiliary sea water lift pump, 120
axial clearance, 49, 182
axial load, 52, 54, 58, 90, 103, 179, 182
axial misalignment, 8

Babbitt, 175, 179–80, 182–3, 187, 198–9
ball diameter, 81
ball pass frequency inner race, 115
ball pass frequency outer race, 115
ball spin frequency, 79, 115
Bearing Analysis Handbook, 6
bearing clearances, 4, 49, 176
bearing defect frequency, 83, 116, 271
bearing degradation, 75, 78, 161
bearing dynamic capacity, 65
bearing housing vibration, 185, 191, 214, 216
bearing insulation, 162
bearing life, 63–5
bearing load zone, 49
bearing shells, 175, 182
boiler forced draft fan, 115
Bow Thruster, 109–10, 114
brass cage, 12, 49, 54, 57, 83, 90, 95, 103, 138, 158–9, 262, 278
British Standard, 15, 156, 186
broken rotor bars, 2, 202, 206, 221–2, 226–7, 231
broken shaft, 75, 159, 161, 166

cement factory, 260, 263, 265–6
centrifugal force, 10, 35, 133, 136, 146, 164, 192, 245, 250
coil span factor, 235

condition monitoring, 1, 6–7, 11, 27, 31–2, 76, 79, 92, 160–1, 221, 226, 231
contact angle, 55, 58, 110
corrosion, 9, 36, 68, 152, 167
coupling guard, 20, 79, 81, 115, 153, 274
cylindrical roller element bearing, 47, 53, 55, 78, 90, 103, 127, 152, 158, 278

damaged ball, 115
decibel, 86
deep grove ball bearing, 52

eddy current displacement probe, 183
electromagnetic force, 2, 6, 17, 19, 208, 210–11, 216, 221, 225, 233, 235–7, 258, 271
electrostatic discharge, 71–2
envelope analysis, 76, 78–9, 152, 154, 159, 274
equivalent bearing load, 65
excessive load, 66, 182
extra capacity bearing, 54, 82, 133, 136, 138, 154, 159

Factory Acceptance Test, 28, 68, 90, 163, 183, 185, 202, 209, 211–12, 262
false brinelling, 66, 120, 127, 132, 140, 163, 166–7
Fast Fourier Transform, 15
fatigue failure, 65, 67, 79, 133
fluid film bearing, 175
flux density, 208, 233, 235, 237, 263
4-point contact ball bearing, 55
FPSO, 109
frequency conversion, 193
frequency spectra, 11, 84, 86, 107
frequency spectrum, 87, 103, 107
fretting, 68, 167
fundamental train frequency, 81

gas recirculating fan, 19
grease lubrication, 50, 95
grease management, 270, 277

ingress protection, 198
inner race, 80

Index

inner race defects, 119
inverter, 72, 193, 253–4, 256, 258–9

journal bearing, 4, 175–6, 192, 211

Kramer, 193

magnetic force, 208, 211, 233
MCSA, 222, 226–7, 231
mechanical imbalance, 10, 15, 25, 31, 106, 192, 206, 245, 247
mechanical impedance, 6, 225
mechanical stiffness, 90, 103, 116, 171, 173, 274
mechanical unbalance, 118, 146, 151, 164
misalignment, 7–8, 10, 25, 31, 53, 59, 72, 124, 182
Motor Current Signature Analysis, 202, 221, 226–7, 231

NEMA, 16, 28, 172, 185, 210, 213

offshore installation manager, 27
oil drainage, 179
oil-fed pressure bearing, 177
oil feed, 179–80
oil film wedge, 176
oil scoop ring, 181
oil viscosity, 180
OIM, 27, 31

parallel misalignment, 7
pitch diameter, 81
pitting, 65, 72, 163

QJ, 57–8

radial clearance, 12, 49
RCFA, 141, 160
resonance, 12, 83, 192, 197, 250
ring lubricated bearing, 181
roller bearing, 37, 82, 117, 133
rolling element bearing, 3–4, 6–7, 47, 49–50, 52, 61, 63, 75, 78, 122, 154, 174, 270–2, 277
rotational speed frequency, 11, 13, 29, 81, 83, 92, 110, 113, 124, 163, 197, 214, 245, 247, 249, 251
rotor slot vibration, 211, 215, 222, 225, 227, 231
rotor slotting, 210, 221, 231
RTD, 204

shaft alignment, 7
shaft currents, 68
shaft displacement, 4, 75, 185–6, 189, 192, 197–8, 200–2, 204
shaft displacement probe, 189
shaft misalignment, 7–8, 10–11, 17, 19, 32, 35–6, 39, 41, 106, 271
shielded ball bearing, 141
skidding, 68, 80, 108, 133, 135–7, 158, 278
sleeve bearing, 3, 20, 27, 75, 161, 174, 177–8, 180, 182–3, 185–6, 190, 271
slip energy recovery, 193
slip frequency, 193, 227, 231
Slip Ring Induction Motor, 192–3
slipping, 67
soap, 52, 95
soft foot, 9, 11, 19, 32, 36–7, 39–42, 208
spalling, 65, 163
spherical roller bearing, 59
spike energy, 79
SRIM, 195, 198, 201, 260, 262
SRP, 165, 167
stator core, 2, 114, 142, 208, 210
stator core back, 208, 210–14, 216, 219–20, 222, 227, 262, 265
stator frame vibration, 230, 232
stator winding temperature, 204
structural resonance, 241, 247
sulphate removal pump, 160, 166
supply voltage unbalance, 208, 236
SWIP, 192

tapered roller bearing, 58
TEFC, 78
Tesla, Nikola, 1
thickener, 52
time domain, 13, 213, 216, 237
true brinelling, 67
twice supply frequency, 17, 210, 218, 232, 237

vibration spectrum analysis, 1, 7, 10, 15, 37, 105, 129, 232, 258, 271

white metal, 175, 179–80, 182–3, 187, 199

zoom spectrum, 131, 136–7, 157